Saber Ben Abdessalem · Mohamed Hamdaoui ·
Ayda Baffoun · Adel Elamri
Editors

# Proceedings of the Second International Conference of Innovative Textiles and Developed Materials-ITDM'2; 05-06 May 2023; Tunisia

## Research Innovations in Smart, Technical and Ecological Textiles

INNOVATIVE TEXTILES AND DEVELOPED MATERIALS

Springer

*Editors*
Saber Ben Abdessalem
National Engineering School of Monastir
Monastir, Tunisia

Mohamed Hamdaoui
National Engineering School of Monastir
Monastir, Tunisia

Ayda Baffoun
National Engineering School of Monastir
Monastir, Tunisia

Adel Elamri
National Engineering School of Monastir
Monastir, Tunisia

ISBN 978-981-99-7949-3 ISBN 978-981-99-7950-9 (eBook)
https://doi.org/10.1007/978-981-99-7950-9



This Springer imprint is published by the registered company Springer Nature Singapore Pte Ltd.
The registered company address is: 152 Beach Road, #21-01/04 Gateway East, Singapore 189721, Singapore

Paper in this product is recyclable.

Proceedings of the Second International Conference of Innovative Textiles and Developed Materials-ITDM'2; 05-06 May 2023; Tunisia

# Organization

## Conference Chairman

Prof. Saber Ben Abdessalem, MPTex, ENIM, University of Monastir, Tunisia

## Conference Co-chairman

Prof. Mohamed Hamdaoui, MPTex, ENIM, University of Monastir, Tunisia

## Steering Committee

Assoc. Prof. Ayda Baffoun, MPTex, ENIM, University of Monastir, Tunisia
Assoc. Prof. Adel Ghith, MPTex, ENIM, University of Monastir, Tunisia
Asst. Prof. Adel Elamri, MPTex, ENIM, University of Monastir, Tunisia
Asst. Prof. Imed Feki, MPTex, ENIM, University of Monastir, Tunisia
Asst. Prof. Naoufel Bhouri, MPTex, ENIM, University of Monastir, Tunisia
Dr. Khaled Hammami, MPTex, ENIM, University of Monastir, Tunisia
Dr. Houda Barhoumi, MPTex, ENIM, University of Monastir, Tunisia
Dr. Nesrine Bhouri, MPTex, ENIM, University of Monastir, Tunisia
Dr. Arwa Elaissi, MPTex, ENIM, University of Monastir, Tunisia
Ph.D. Student Nada Hajri, MPTex, ENIM, University of Monastir, Tunisia
M.S. Amel Ben Abdallah, MPTex, ENIM, University of Monastir, Tunisia

## Scientific Committee

Prof. Paul Kickens, Textile Department, Ghent University, Belgium
Prof. Xianyi Zeng, GEMTEX, ENSAIT, University of Lille, France
Prof. Abdelaziz Lallam, LPMT, ENSISA, University of Haute Alsace, France
Prof. Boris Mahltig, University of Applied Sciences—Hochschule Niederrhein, Germany
Prof. Omar Anis Harzallah, LPMT, ENSISA, University of Haute Alsace, France
Prof. Adnene Labed, LGM, University of Biskra, Algeria
Prof. Ivelin Rahnev, E. Miroglio EAD, Bulgaria
Prof. Moez Feki, LAMMDA, University of Sousse, Tunisia
Prof. Abdelmajid Jemni, LESTE, ENIM, University of Monastir, Tunisia
Prof. Faten Fayala, LESTE, ENIM, University of Monastir, Tunisia
Prof. Saber Ben Abdessalem, MPTex, ENIM, University of Monastir, Tunisia
Prof. Mondher Zidi, LGM, ENIM, University of Monastir, Tunisia
Prof. Anis Sakly, LASEE, ENIM, University of Monastir, Tunisia
Prof. Faouzi Askri, LESTE, ENIM, University of Monastir, Tunisia
Prof. Salah Mezlini, LGM, ENIM, University of Monastir, Tunisia
Prof. Zouhaier Affi, LGM, ENIM, University of Monastir, Tunisia
Prof. Neji Ladhari, LGTEX, ISET Ksar Hellal, University of Monastir, Tunisia
Prof. Hatem Majdoub, LIMA, FSM, University of Monastir, Tunisia
Prof. Raoui Mounir Maaroufi, LGBVB, ISBM, University of Monastir, Tunisia
Prof. Mustapha Majdoub, LIMA, FSM, University of Monastir, Tunisia
Prof. Mohamed Hamdaoui, UR MPTex, ENIM, University of Monastir, Tunisia
Prof. Hassen Baoueb, UR MSO, IPEIM, University of Monastir, Tunisia
Prof. Amel Babay, LGTEX, ISET Ksar Hellal, University of Monastir, Tunisia
Prof. Khelifa Slimi, LESTE, ENIM, University of Monastir, Tunisia
Prof. Chedli Boudokhane, Faculty of Pharmacy, University of Monastir, Tunisia
Prof. Slah Msahli, LGTEX, ISET Ksar Hellal, University of Monastir, Tunisia
Prof. Mohamed Ben Hassen, LGTEX, ISET Ksar Hellal, University of Monastir, Tunisia
Prof. Noamen Guermazi, LGME, ENIS, University of Sfax, Tunisia
Dr. Aminoddin Haji, Textile Department, Yazd University, Iran
Dr. Walid Hassen, LMST, ENIM, University of Monastir, Tunisia
Dr. Adel Ghith, UR MPTex, ENIM, University of Monastir, Tunisia
Dr. Ayda Baffoun, UR MPTex, ENIM, University of Monastir, Tunisia
Dr. Riadh Zouari, LGTEX, ISET Ksar Hellal, University of Monastir, Tunisia
Dr. Fayçal Hamdaoui, LASEE, ENIM, University of Monastir, Tunisia

## Editors

Prof. Saber Ben Abdessalem, MPTex, ENIM, University of Monastir, Tunisia
Prof. Mohamed Hamdaoui, MPTex, ENIM, University of Monastir, Tunisia
Dr. Ayda Baffoun, MPTex, ENIM, University of Monastir, Tunisia
Dr. Adel Elamri, MPTex, ENIM, University of Monastir, Tunisia

## Organizers

## Partners

## Sponsors

# Preface

We were very delighted to introduce the conference proceedings book of the Second International Conference on Innovative Textiles and Developed Materials (ITDM'2) which had been held at the Royal Thalassa Hotel, Monastir, Tunisia from 5 to 6 Mai, 2023. We were grateful to all distinguished participants for their contribution to the conference. There were participants from Europe, Asia and Africa to present and discuss their works provided in areas of four main topics: Ecological textiles dyeing and finishing, innovative materials and polymers, smart and technical textiles, and 4.0 industry and textile processes.

It was worth nothing to have an opportunity to welcome our invited keynote speakers, Prof. Aref Meddeb (from ENISo, University of Sousse, Tunisia), Prof. Mohamed Hassen V. Baouab (from IPEIM, University of Monastir, Tunisia) and Prof. Raoui Mounir Maaroufi (ISBM, University of Monastir, Tunisia). Then, 58 accepted communications were presented and discussed. After a rigorous peer review process, only 30 full papers were selected and requested for publication in this volume.

The editors hope that this book will provide the reader a broad overview of the latest research works and findings in innovative textiles and developed materials, and that it will be a valuable reference source for further research.

We are optimistic and full of hope about getting the proceedings of the ITDM'2 Conference covered by Springer in due course and excellent quality.

The editors would like to express their sincere appreciations and thanks to all the authors for their contributions to this publication. We would like to express our gratitude and appreciation for all of the reviewers and scientific committee members for their constructive comments on the papers. We would also like to extend our thanks to the members of the steering team for their hard work.

We highly believed that this conference was an engrossing and captivating discussion for every participant.

Monastir, Tunisia

Prof. Saber Ben Abdessalem
Conference Chairman

Prof. Mohamed Hamdaoui
Conference Co-chairman

Assoc. Prof. Ayda Baffoun
President of Steering Committee

Asst. Prof. Adel Elamri
Corresponding Editor

# Contents

**Ecological Textiles Dyeing & Finishing**

**Multiple Reuse of Residual Dyebaths for Cotton Dyeing with Reactive Dyestuffs: A Promising Technique for Environmental Sustainability** .......... 3
Rania Moussa, Marwa Souissi, Sabrine Chaouch, and Ali Moussa

**Effect of Polyester Structure on Drying Behavior** .......... 15
Mohamed Ghaith Chakroun, Sofien Benltoufa, Adel Ghith, and Faten Fayala

**A Comparative Study of Environmental Sustainability Initiatives of Two Major Jeanswear Brands** .......... 25
Mouna Haj Nasr, Hassen Hedfi, Ayda Baffoun, and Mohamed Hamdaoui

**Investigation and Optimization of a New Bleaching Process for Sulfur-Dyed Denim Fabric** .......... 37
Sarra Said, Imed Feki, Mohamed Hamdaoui, Sabri Halaoua, and Walid Sahraoui

**Investigation of an Eco-Friendly Acid Dyeing Process of Cotton Fabric** .......... 51
Adel Elamri, Imed Feki, Mohamed Amine Touati, Mohamed Hamdaoui, and Omar Harzallah

**Dyeing of the Polyester Fabric with the Fluorescein Molecule** .......... 65
Fredj Saad, Ayda Baffoun, and Mohamed Hamdaoui

**Study of a New Dyeing Process Based on the Capillary Rise Technique** .......... 73
Sarra Said, Imed Feki, Mohamed Hamdaoui, and Walid Sahraoui

**Innovative Materials & Polymers**

**Ballistic Impact Response of Multi-Layers Armors Against Piercing Projectile: Finite Element Modeling** ... 87
Hassouna Amira and Salah Mezlini

**Effect of the Cultivar on the Development Hemp Stem and Textile Fiber in the East of France** ... 97
Aurélie Decker, Adrien Tritter, Vivien Sarazin, Jean-Yves Drean, and Omar Anis Harzallah

**Energetic Valorization of Cutting Textile Waste by Transforming It into Thermal Insulation Products** ... 107
Nour Sghaier and Nourhene Dibej

**Functionalization of a Textile Surface by N2 Jet Technology to Improve Its Roughness for Medical Applications** ... 119
Maleke ZidI, Foued Khoffi, Yosri Khalsi, Abdel Tazibt, Frédéric Heim, and Slah Msahli

**The Living Design: The Soul of Regenerative Design and the Key to an Ecological Transition** ... 125
Mariem Fakhfakh and Amine Hadj Taieb

**Using Fabric Warping in Architecture: A Contemporary Conception for Technical Use** ... 137
Amine Hadj Taieb and Wouroud Turki

**Use of Date Palm Waste to Produce Bio-Composite Material** ... 147
Ferhat Maroua, Labed Adnane, and Djemai Hocine

**Application of Biopolymers in Medical Textiles: Myriad of Opportunities** ... 153
Adel Elamri, Khmais Zdiri, and Mohamed Hamdaoui

**Smart & Technical Textiles**

**Study of the Weight Loss of Abrasive Non-wovens from Industrial Waste** ... 173
Elaissi Arwa and Ghith Adel

**Numerical Investigations of the Tribological Behavior of G1151 Fabric Through Frictional Sensitivity Study During Drawing Using a Hybrid Elastic Discrete Model** ... 183
Walid Najjar, Mondher Nasri, Philippe Dal Santo, Xavier Legrand, and Damien Soulat

**Effect of the Twist and Composite Process on the Tensile Behavior of Carbon Multifilament Rovings. An Experimental Investigation Toward the Identification of Constituents' Behaviors** ................. 193
Mohamed Medhat Salem, Haifa Mensi, Chayma Soltani, Adel Ghith, and Xavier Legrand

**Investigation of the Properties of Developed Abrasives After Abrasion Test** ................................................ 201
Elaissi Arwa and Ghith Adel

**Contribution to the Improvement of Certain Comfort Characteristics of Baby Diapers** .................................... 211
Wala Kallala, Adel Ghith, and Faten Fayala

**DMAIC Approach to Optimize the Production Capacities of a Tunisian Clothing Manufacture: A Case Study** .................. 229
Nesrine Bhouri, Naoufel Bhouri, and Dorra Oualha

**The Impact of Acid Blue Dye Concentration on Color Strength and the Chromaticity Coordinates of Raw and Tannery Wool: A Comparative Study** ........................................... 237
Olfa Abdellaoui, Harizi Taoufik, Feriel Bouatay, and Msahli Slah

**Smart Textile Industry: Threats and Opportunities** .................. 245
Aref Meddeb

**4.0 Industry & Textile Processes**

**Fabric Texture Reconstruction Via Multilayer Dictionary Learning** .... 267
Bo Xing, Qingqing Shao, Xianyi Zeng, and Jun Wang

**Application of the Desirability Function for Optimizing the Use of Calcium Carbonate Used in the Production Process of PVC-Coated Textile** ............................................ 275
Stambouli Mouna, Chaouech Walid, Gargoubi Sondes, Zouari Riadh, and Msahli Slah

**Ergonomic Study on Ironing Stations in Tunisian Clothing Companies** ........................................................ 289
Slim Boussaadoun, Nejib Sejri, and Mohamed Hamdaoui

**An Algorithm at Incoming Quality Control of Raw Woven Fabrics—Cottonade Jacquard** ................................... 305
Tashka Ivanova Koleva and Ivelin Rahnev

**Application of Multi-objective Ant Colony Algorithm to Resolve Textile Color Matching Problem** ................................... 313
Sabrine Chaouch, Ali Moussa, and Neji Ladhari

**Deep Learning for DENIM Defects Detection** ........................ 325
Khaled Hammami, Imed Feki, and Saber Ben Abdessalem

**Modeling of Textile Materials and Structures: Some Numerical and Experimental Aspects** ........................................ 335
Hassen Hedfi and Hédi BelHadjSalah

# Ecological Textiles Dyeing & Finishing

# Multiple Reuse of Residual Dyebaths for Cotton Dyeing with Reactive Dyestuffs: A Promising Technique for Environmental Sustainability

**Rania Moussa, Marwa Souissi, Sabrine Chaouch, and Ali Moussa**

**Abstract** This research investigates the feasibility of reusing residual dyebaths for cotton dyeing with reactive dyes. For this, three reactive dyestuffs, namely C.I. Reactive Yellow 145, C.I. Reactive Red 238, and C.I. Reactive Blue 235, were used, separately and in mixtures, for dyeing cotton fabrics. After each dyeing, exhausted dyebaths were analyzed, reconstituted, and reused several times for various shades and mixtures. To evaluate the obtained results, two criteria were evaluated and analyzed such as dyebath exhaustion and color conformity of dyed samples. Analysis of the results showed that, in spite of the multiple reuses of dyebaths, the color of samples dyed with the reused dyebaths are comparable to those obtained from the initial dyebath with acceptable values of dyebath exhaustion and color conformity. As a result, this dyebath reuse method has then the potential to significantly reduce the use of water, dyes, and auxiliary quantities while ensuring that the performance of the dyeing process is maintained.

**Keywords** Dyeing · Dyebath reuse · Color · Reactive dyes

R. Moussa (✉) · S. Chaouch · A. Moussa
Textile Engineering Laboratory (LGTex), University of Monastir, Ksar-Hellal, Tunisia
e-mail: raniamoussa219@gmail.com

R. Moussa · M. Souissi · A. Moussa
National Engineering School of Monastir (ENIM), University of Monastir, Monastir, Tunisia

M. Souissi
Laboratory of «Chimie de l'Environnement et des Procédés Propres» (LCE2P), University of Monastir, Monastir, Tunisia

S. Chaouch
Higher Institute of Technological Studies (ISET) of Ksar-Hellal, Ksar-Hellal, Tunisia



S. B. Abdessalem et al. (eds.), *Proceedings of the Second International Conference of Innovative Textiles and Developed Materials-ITDM'2; 05-06 May 2023; Tunisia*,
https://doi.org/10.1007/978-981-99-7950-9_1

# 1 Introduction

The textile industry is one of the most significant sources of pollution in the world. The finishing process, in particular, is notorious for its significant environmental impact, including high water and energy consumptions, chemical wastes, etc. To avoid these problems, several methods have been developed such as wastewater treatment, green chemistry, and process optimization [1–3]. However, these methods often require high costs and complex implementation, making them challenging for many textile companies to adopt.

One approach that could significantly reduce the environmental impact of textile dyeing is the direct reuse of residual dyebaths. This method consists of reusing the exhausted dyebaths obtained from previous dyeings through the addition of the necessary amounts of water, dyes, and auxiliaries until the initial dyebath is reconstituted [4–6]. This reuse allows to reduce the amount of water, dyes, and chemicals needed for each dyeing process. Consequently, this method offers several advantages, including reduced water consumption, lower chemical waste, and lower operating costs. In the literature, existing works have shown that the reconstitution and the reuse of residual dyebaths can lead to similar colors and fastness as the initial dyeing process [7, 8].

The aim of this work is to assess the viability of reusing residual dyebaths for cotton dyeing with reactive dyes. Three different reactive dyestuffs were used, separately and in mixtures, for dyeing cotton fabrics. Exhausted dyebaths will be analyzed, reconstituted, and reused multiple times to produce various shades and mixtures. After each reuse, the dyebath exhaustion and the color conformity of dyed samples are evaluated and discussed compared to the results obtained using initial dye baths.

# 2 Material and Methods

## *2.1 Material*

Pre-bleached woven 100% cotton samples were used for dyeing. Their main characteristics are as follows: 2/2 twill weave pattern, fabric weight of 206.25 g/m$^2$, fabric thickness of 0.59 mm, and warp and weft densities of 25 and 19 threads/cm, respectively. For dyeing, three cold bifunctional reactive dyestuffs were used and studied in this work: C.I. Reactive Yellow 145 ($\lambda_{max}$ = 433 nm), C.I. Reactive Red 238 ($\lambda_{max}$ = 543 nm), and C.I. Reactive Blue 235 ($\lambda_{max}$ = 610 nm). These dyes stuffs were provided by Huntsman Company (Switzerland). Other chemical auxiliaries were added in dye baths including CHTT MRS as a wetting agent, NaCl as an electrolyte, and NaOH as an alkali.

## 2.2 *Dyebath Reuse Procedure*

**Dyeing**. Dyeing recipe and process used in this study are presented in Table 1 and Fig. 1, respectively. All dyeings were carried out using an "Ahiba Nuance Top Speed II-B" machine (Datacolor Company, USA). The liquor-to-fiber ratio was maintained at 10:1. After each dyeing, a finishing treatment is conducted in order to eliminate unfixed dyes from the sample. This finishing treatment contains the following steps: hot rinsing at 50 °C for 10 min, neutralization at 50 °C for 10 min using acetic acid to adjust the pH to 6, soaping at 95 °C for 10 min using 2 g/L of detergent and 2 g/L of sodium carbonate, second hot rinsing at 50 °C for 10 min, and finally cold rinsing for 10 min. All these post-treatments were carried out with a liquor-to-fiber ratio of 40:1. After each dyeing, residual dyebaths were analyzed in order to deduce the required quantities of dyes, auxiliaries, and water to add in order to reconstitute the initial dyebath.

**Readjustment of the Dye Concentration**. To restore the exhaustion dyebaths, it is necessary to know their composition, in particular their dye concentrations. In this study, the residual dye concentration ($C_{res}$) of the exhausted dyebath was determined using a spectrophotometer HACH-DR-3900 (HACH, USA) at the wavelength $\lambda_{max}$

**Table 1** Dyeing recipe

| Levels | Products | Values |
|---|---|---|
| A | Reactive dye | X%* |
| | Wetting agent (CHTT MRS) | 3 mL/L |
| B | Electrolyte (NaCl) | Y%** |
| C | Caustic soda (NaOH) | Z%** |

* X%: color shade
** Y% and Z% are the concentration of electrolyte and caustic soda, determined according to the manufacturer's instructions

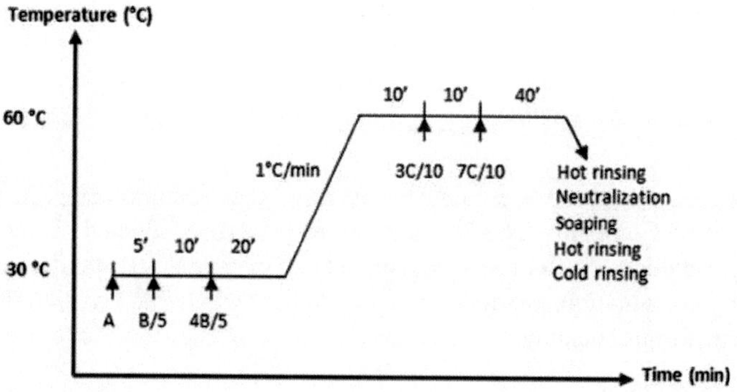


**Fig. 1** Dyeing process

corresponding to the maximum absorbance of each studied dye. The added concentration of the dye ($C_{ad}$) required to reconstitute the original dyebath was determined as follows:

$$C_{ad} = C_0 - C_{res} \quad (1)$$

where $C_0$ (mg.L$^{-1}$) is the initial dye concentration specified in the recipe.

**Readjustment of the Bath Volume**. After each dyeing process, the decrease in water volume due to fiber retention and evaporation was compensated by adding approximately 10 mL of fresh water, which was necessary to restore the initial dyebath volume of 50 mL.

**Readjustment of Auxiliary Quantities**. After readjusting the pH of the residual dye bath to its initial value, the residual dye bath was also reconstituted to the desired quantities of auxiliaries. In this study, we have used three auxiliaries: CHTT MRS as a wetting agent, NaCl as an electrolyte, and NaOH as an alkali. The added amount of each auxiliary was calculated based on the desired shade of dye.

## 2.3 *Determination of Dyebath Exhaustion*

After each dyeing, residual dyebath was cooled at ambient temperature and analyzed with a spectrophotometer HACH-DR-3900 at the $\lambda_{max}$ for each dye. The dyebath exhaustion is determined using Eq. 2:

$$E(\%) = \frac{C_0 - C_{res}}{C_0} \times 100 \quad (2)$$

where $C_0$ (mg.L$^{-1}$) is the initial dye concentration and $C_{res}$ (mg.L$^{-1}$) is the residual concentration of the dye after dyeing.

## 2.4 *Color Measurement*

Colors of dyed samples were evaluated using a spectrophotometer Spectraflash 500 (Datacolor Company, USA). Measurements were conducted under specific measuring conditions, including an illuminant of D65 and a 10° standard observer. To perform the evaluation, textile fabrics were folded twice (four layers of thickness) to achieve maximum opacity.

# 3 Results and Discussion

## *3.1 Dyeing with a Single Reactive Dye*

**Analysis of Exhaustions of Various Reused Dyebaths.** Figure 2 illustrates the dyebath exhaustions obtained for the studied dyes during the reuse cycles of the dyebath. It can be observed that the exhaustions of the dyebaths remained relatively high even after five multiple reuses, with mean values ranging from 58.98% to 93.85%. Among the studied dyes, the C.I. Reactive Yellow 145 dye presented the highest initial exhaustion percentage of 93.85%, followed by C.I. Reactive Blue 235 and C.I. Reactive Red 238 dyes with initial exhaustion values of 90.73% and 77.68%, respectively. During the five reuse cycles, the evolution of the dyebath exhaustion is different according to the dye; it remained relatively stable for the blue dye with a mean exhaustion value of 92.16%, exhibited a small decrease from reuse 1 (90.12%) to reuse 5 (79.5%) for the yellow dye, and a high decrease from reuse 1 (70.6%) to reuse 5 (58.98%) for the red dye with the lowest overall mean exhaustion percentage of 67.56%.

In addition, it should be noted that there are some differences in exhaustion values across the dyebath reuse cycles, indicating some degree of variability, but they remain relatively small with coefficients of variation below the threshold of 5% (for blue and yellow dyes). These results suggest that while the reuse of residual dyebath could potentially be useful in reducing the amount of dye required for cotton dyeing, the effectiveness of this technique could vary depending on the dye type and its affinity to the cotton fiber. The dyebath reuse technique economizes about 10% of the used

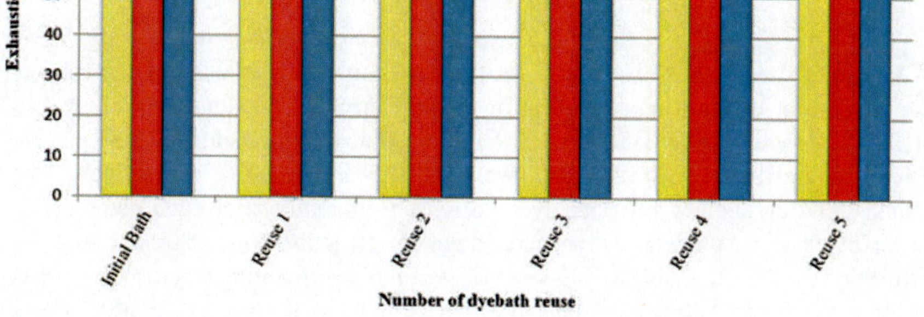


**Fig. 2** Dye exhaustions $E(\%)$ for the initial and the 5 reused dye baths

quantity of dye. It was also observed that the volume of water added to the residual baths between reuses did not exceed 10 mL versus an initial volume of 50 mL; so, a water saves of 80% compared to conventional dyeing. Moreover, this reuse method allows to avoid the rejection of residual dyebaths and their treatment.

**Analysis of the Color Conformity** Color measurements were conducted on the dyed samples during the five reuse cycles. For each reuse, corresponding values of CIELAB coordinates are the mean values of two different samples dyed separately. Table 2 shows the colorimetric coordinates and the color differences (in D65/10°) between samples dyed with the initial dyebath and those dyed with reused baths (case of 0.5% of dye) for the three studied dyes. As can be observed, the majority of the $\Delta E_{cmc(2:1)}$ values are lower than 1.5 with average values of 1.054, 0.926, and 1.536 for yellow, red, and blue dyes, respectively. These mean values can be considered acceptable for yellow and red dyes but high for blue dye.

## 3.2 *Dyeing with a Binary Mixture of Reactive Dyes*

In this section, we examined the dyeing of cotton fabrics with binary mixtures of reactive dyes, in particular mixtures containing both yellow and blue dyes (C.I. Reactive Yellow 145 and C.I. Reactive Blue 235). Yellow and blue dyes were chosen because they have distant values of $\lambda_{max}$ and obviously compatible, making it possible to use them in mixture without any adverse effects on the color or quality of dyed samples. The principle of reconstruction of residual dyebaths used in this case was the same as that for single dyes, where the dye mixture was added to the dyebath and allowed to react with the cotton fabric. Figure 3 shows an example of the absorbance spectra measured in the case of the mixture containing 0.25% of yellow dye and 0.25% of blue dye. Absorbance values measured at $\lambda_{max}$ of each dye were used to deduce the exhaustion values of dye baths for these dyes.

So, the residual bath from the dyeing process was also reused in subsequent dyeing cycles to economize the quantities of dye and water used. Overall, the use of binary mixtures of reactive dyes in combination with the reuse of residual baths presents a promising approach to achieving sustainable and cost-effective cotton dyeing processes.

**Analysis of the Dyebath Exhaustions.** Figure 4 shows the exhaustion percentages for a series of dyebath reuse cycles for the mixture "0.25% Reactive Yellow 145 + 0.25% Reactive Blue 235 dyes." The initial exhaustion percentages for Reactive Blue 235 and Reactive Yellow 145 were 93.03% and 91.21%, respectively. The exhaustion percentages for both dyes decreased gradually after each reuse cycle. After five reuses, the exhaustion percentage for Reactive Blue 235 was 89.42%, while that for Reactive Yellow 145 was 80.94%. It is worth noting that the exhaustion percentages for both dyes exhibited some variation from one reuse to another. These results suggest that the reuse of residual dyebath containing mixtures of dyes could potentially be useful in reducing the amount of dyes required for cotton dyeing, but

**Table 2** CIELAB coordinates and color differences (in D65/10°) between samples dyed with the initial dyebath and those dyed with reused baths (case of 0.5% of dye)

| Dye | Dyebaths | $L^*$ | $a^*$ | $b^*$ | $C^*$ | $h$ | $\Delta L^*$ | $\Delta a^*$ | $\Delta b^*$ | $\Delta C^*$ | $\Delta H^*$ | $\Delta E_{CMC(2:1)}$ |
|---|---|---|---|---|---|---|---|---|---|---|---|---|
| Yellow | Initial | 77.48 | 24.34 | 58.29 | 63.17 | 67.34 | … | … | … | … | … | … |
| | Reuse 1 | 77.40 | 24.70 | 62.80 | 67.48 | 68.54 | −0.08 | 0.36 | 4.51 | 4.31 | 1.19 | 1.847 |
| | Reuse 2 | 76.09 | 25.60 | 62.47 | 67.52 | 67.72 | −1.39 | 1.27 | 4.18 | 4.35 | 0.38 | 1.647 |
| | Reuse 3 | 77.30 | 23.96 | 60.75 | 65.31 | 68.47 | −0.18 | −0.38 | 2.46 | 2.14 | 1.13 | 1.237 |
| | Reuse 4 | 77.60 | 24.09 | 58.57 | 63.33 | 67.65 | 0.12 | −0.25 | 0.28 | 0.16 | 0.31 | 0.270 |
| | Reuse 5 | 77.61 | 23.94 | 58.17 | 62.90 | 67.63 | 0.13 | −0.40 | −0.13 | −0.27 | 0.29 | 0.269 |
| Red | Initial | 63.64 | 41.03 | −10.64 | 42.38 | 345.46 | … | … | … | … | … | … |
| | Reuse 1 | 63.60 | 44.30 | −11.73 | 45.83 | 345.16 | −0.04 | 3.27 | −1.09 | 3.45 | −0.30 | 1.458 |
| | Reuse 2 | 63.81 | 43.62 | −11.65 | 45.16 | 345.00 | 0.16 | 2.59 | −1.01 | 2.78 | −0.46 | 1.189 |
| | Reuse 3 | 63.49 | 41.79 | −12.84 | 43.72 | 342.93 | −0.16 | 0.76 | −2.20 | 1.34 | −2.54 | 1.224 |
| | Reuse 4 | 63.25 | 41.13 | −11.68 | 42.76 | 344.15 | −0.40 | 0.10 | −1.04 | 0.38 | −1.31 | 0.598 |
| | Reuse 5 | 63.77 | 40.67 | −10.57 | 42.02 | 345.43 | 0.13 | −0.36 | 0.07 | −0.36 | −0.04 | 0.160 |
| Blue | Initial | 61.95 | −4.39 | −22.53 | 22.95 | 258.96 | … | … | … | … | … | … |
| | Reuse 1 | 59.32 | −4.25 | −25.43 | 25.78 | 260.52 | −2.64 | 0.14 | −2.90 | 2.83 | 1.56 | 2.001 |
| | Reuse 2 | 59.17 | −4.34 | −25.32 | 25.69 | 260.29 | −2.79 | 0.06 | −2.79 | 2.74 | 1.33 | 1.976 |
| | Reuse 3 | 60.99 | −4.42 | −24.18 | 24.58 | 259.65 | −0.96 | −0.02 | −1.65 | 1.63 | 0.69 | 1.026 |
| | Reuse 4 | 60.63 | −4.40 | −23.66 | 24.07 | 259.47 | −1.32 | −0.01 | −1.13 | 1.11 | 0.51 | 0.849 |
| | Reuse 5 | 59.62 | −4.39 | −25.24 | 25.62 | 260.12 | −2.34 | 0.01 | −2.71 | 2.67 | 1.16 | 1.830 |

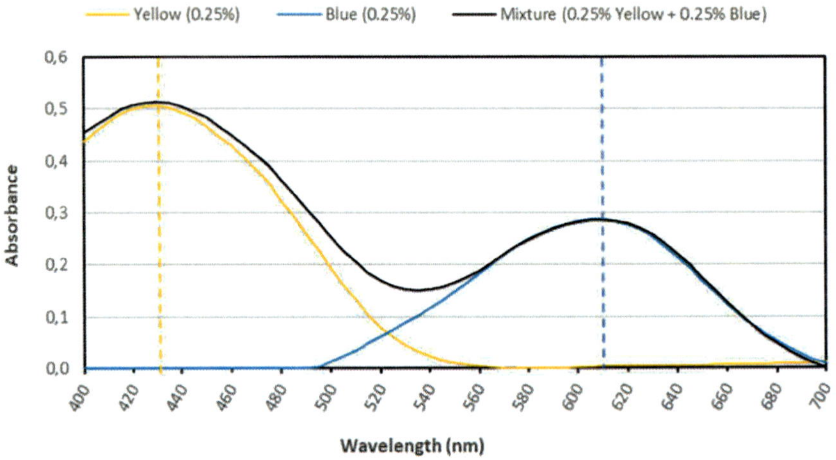


**Fig. 3** Absorbance spectra of the studied mixture (solutions with 500 mg/L of dyes)

the effectiveness of this technique may vary depending on the dye type and its affinity to the cotton fiber.

**Analysis of the Color Conformity**. Same as the previous section, a series of color measurements were carried out on the samples dyed using the studied mixture throughout the five cycles of reuse. Table 3 shows the CIELAB coordinates and the color differences between samples dyed with the initial dyebath and those dyed

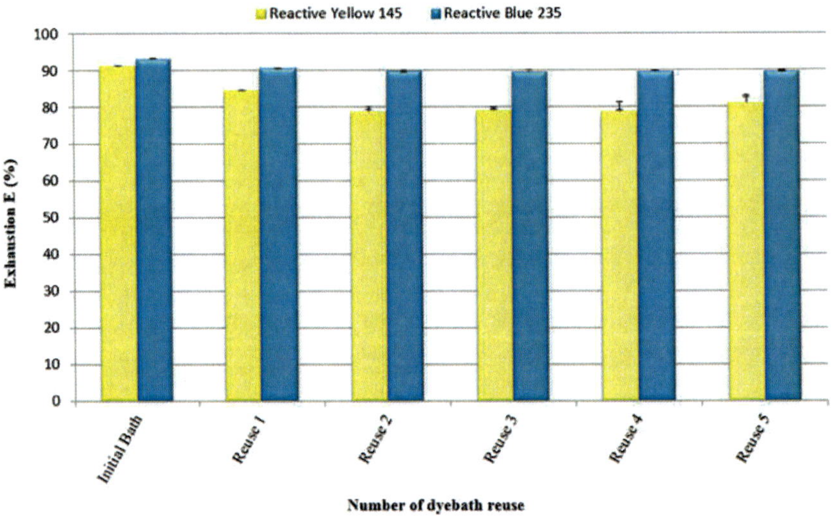


**Fig. 4** Dye exhaustion E (%) for the initial and the 5 reused dyebaths (case of the mixture "0.25% Reactive Yellow 145 + 0.25% Reactive Blue 235)

with reused baths (case of the mixture “0.25% Reactive Yellow 145 + 0.25% Reactive Blue 235). As can be observed, the $\Delta E_{CMC(2:1)}$ values obtained after the different reuses were relatively high, ranging from 1.6 to 3.287. These high values of color differences could be attributed to some errors in measurements, in particular when measuring absorbance values at $\lambda_{max}$ of each dye. Further studies are then necessary to improve the accuracy and the conformity of the dyed samples and to better understand the factors that influence the results obtained with mixtures.

## 4 Conclusions

The objective of this study was to evaluate the potential of reusing residual dyebath for the economical and sustainable use of reactive dyes for cotton dyeing. Three reactive dyes such as C.I. Reactive Blue 235, C.I. Reactive Red 238, and C.I. Reactive Yellow 145, as well as a binary mixture of C.I. Reactive Blue 235 and C.I. Reactive Yellow 145, were studied. The exhaustion percentages of the dyebaths were evaluated over five reuse cycles to determine the effectiveness of the reuse technique. The results showed that the exhaustion percentages remained relatively high for all dyes even after multiple reuses, with mean values ranging from 58.98% to 93.89%. The study also found that the volume of water added to the residual baths between reuses did not exceed 10 mL vs an initial volume of 50 mL, resulting in a water save of 80% compared to conventional dyeing. Overall, these findings suggest that the reuse of residual dyebath could potentially be useful in reducing the amount of dyes required for cotton dyeing, as well as for saving water and avoiding the costs of treating and disposing of residual dyebath. Further research is necessary to evaluate the uniformity and the color fastness of the dyed samples, which are crucial aspects to determine the overall quality of the final product. Other areas for investigation could include the performance of other shades and mixtures, as well as the economic feasibility of this method on a larger scale. Additionally, efforts could be made to encourage more textile companies to adopt this approach. By continuing to develop and refine sustainable solutions like the reuse of residual dyebaths, the textile industry can work toward a more environmentally responsible future.

**Table 3** CIELAB coordinates and color differences (in D65/10°) between samples dyed with the initial dye bath and those dyed with reused baths (case of Mixture 0.25% Yellow + 0.25% Blue)

| Dyebaths | $L^*$ | $a^*$ | $b^*$ | $C^*$ | $h$ | $\Delta L^*$ | $\Delta a^*$ | $\Delta b^*$ | $\Delta C^*$ | $\Delta H^*$ | $\Delta E_{CMC(2:1)}$ |
|---|---|---|---|---|---|---|---|---|---|---|---|
| Initial | 64.25 | −4.48 | 21.06 | 21.53 | 102 | … | … | … | … | … | … |
| Reuse 1 | 61.50 | −5.52 | 20.10 | 20.85 | 105.35 | −2.75 | −1.04 | −0.96 | −0.68 | 3.35 | 1.881 |
| Reuse 2 | 62.36 | −5.80 | 20.36 | 21.17 | 105.90 | −1.89 | −1.32 | −0.70 | −0.36 | 3.90 | 1.889 |
| Reuse 3 | 62.38 | −5.57 | 20.54 | 21.28 | 105.18 | −1.87 | −1.09 | −0.52 | −0.25 | 3.18 | 1.60 |
| Reuse 4 | 63.49 | −5.56 | 18.55 | 19.37 | 106.67 | −0.76 | −1.08 | −2.51 | −2.16 | 4.67 | 2.358 |
| Reuse 5 | 63.60 | −6.36 | 18.29 | 19.37 | 109.18 | −0.65 | −1.88 | −2.77 | −2.16 | 7.18 | 3.287 |

## References

1. Environmental Protection Agency [EPA]: Best management practices for pollution prevention in textile industry. Washington, DC: Publication United States Environmental Protection Agency (EPA/625/R-96/004) (1996).
2. Riou, S.: Énergie et environnement [Energy and environment]. L'industrie Textile, 1308, 65–68 (1999).
3. Simonet, G.: Guide des techniques de l'ennoblissement textile. Chap. 5 L'eau dans l'ennoblissement [Guide of textile finishing techniques. Chapter 5: Water in finishing]. (pp. 97–124). Paris: SPIET (1982).
4. Bergenthal, J.F., Tawa, A.J.: Investigation of textile dye bath reconstitution and reuse. Volume 2. Operations manual. Research Triangle Park, NC: US Environmental Protection Agency (EPA-600/2 84 100b) (1984).
5. Cook, F.L. Direct dye bath reuse: The future is now. Textile World, 133, 144–147 (1983).
6. White, B., Tincher, W.C., Clark, J.L.: Demonstration of automated dye bath reuse in a production facility. Textile Chemists and Colorists, 30(12), 25–30 (1998).
7. Moussa, A., El-Ghali, A., Ellouzi, S. and Sakli, F.: Color and fastness study of wool dyeing in multiple reuse dye baths using acid and reactive dyestuffs in laboratory scale. Journal of the Textile Institute.;104(3):260–269 (2013).
8. Moussa, A., El-Ghali, A., Ellouzi, S. and Sakli, F.: Multiple reuse of exhausted acid dyebaths for wool dyeing: colorimetric properties, leveling agent effect, and material savings. AATCC Journal of Research.;1(4):1–21 (2014).

# Effect of Polyester Structure on Drying Behavior

**Mohamed Ghaith Chakroun, Sofien Benltoufa, Adel Ghith, and Faten Fayala**

**Abstract** The drying capacity of fabrics is an important feature for sportswear. Furthermore, the environmental impact of these products is crucial for the sustainability of the humankind. Combining the eco-friendly feature with clothing comfort is the golden solution. Claims have been made on the comfort level of COCONA® new biodegradable polyester yarn. This study is dedicated for investigating these claims. Six knitted fabrics were made in this work, three were made with COCONA® polyester yarn and three others were made with its biodegradable version. A comparison between the drying time, as a clothing comfort property, of these two groups of fabrics was done on finished and domestically washed fabrics. Results indicate that after finishing and washing, knits made with the biodegradable COCONA® yarn have a superior capacity for drying than knits made with the conventional COCONA® polyester yarn. After the domestic washing cycles, drying times of biodegradable fabrics show a slight improvement compared to the regular ones.

**Keywords** Clothing comfort · Drying behavior · Sustainability · Biodegradable polyester

## 1 Introduction

Global warming and climate change are the most environmental hazards that attracted the world's attention regarding the integrity of the overall ecosystem. The textile industry is one of the highest polluted manufacturing sectors. The pesticides used in farming and chemicals used in textile processing harm the environment and can

M. G. Chakroun (✉) · S. Benltoufa · F. Fayala
LESTE (Laboratoire d'étude des Systèmes Thermiques et Énergétiques), National Engineering School of Monastir, University of Monastir, Monastir, Tunisia
e-mail: medghaithchakroun@gmail.com

A. Ghith
Textile Materials and Processes Research Unit, National Engineering School of Monastir, University of Monastir, Monastir, Tunisia


S. B. Abdessalem et al. (eds.), *Proceedings of the Second International Conference of Innovative Textiles and Developed Materials-ITDM'2; 05-06 May 2023; Tunisia*,
https://doi.org/10.1007/978-981-99-7950-9_2

contribute to the air and water contamination. The overuse of natural resources in the entire supply chain of a textile product depletes or disturbs the ecological balance. After use and old clothes are thrown away to landfill sites and they take up precious space in these sites [1]. Due to polymers' slow rate of degradation, the majority of discarded clothing stays in the ecosystem for years [2, 3]. The major focus of the apparel industry was the fashionable design and advanced functionalities of some type of clothes. The textile industry is attempting to provide ecologically friendly and sustainable goods that need to be supported by the general public as consumer awareness of this environmental threat grows.

The United Nations World Commission defines sustainability as "the continuous development that satisfies the present without compromising the capacity of future generations to meet their own needs" [1, 4]. Reduced negative effects on the quality of the air, water, and other natural resources are a key component of sustainable development [1]. Brands are using sustainability labels on their products to persuade customers in a competitive marketing environment.

The sportswear manufacturing field has a major part in the ecosystem menace, considering its great use of synthetic fibers like polyester and polyamide. Man-made fibers are dominant over natural ones in sportswear products owing to their low cost and inexpensive treatment in one hand and on the other hand, to their superior properties such as wicking capacity, high drying rate, and resistance to microorganisms [1, 2]. With some finishing treatments, synthetic fibers provide better clothing comfort than natural fibers. Unfortunately, most of them are generated from petrochemical raw materials which are non-biodegradable and responsible for the emission of $CO_2$ during their product life cycle. Therefore, the great challenge for sportswear manufacturers is to provide an eco-friendly product with a satisfying level of comfort to the wearer.

The amount of comfort provided by a piece of clothing while engaging in physical activity is an important factor in performance, well-being, and health of the user. Physiological comfort refers to the thermal and wet state of the clothing–human body system. The effectiveness of heat and moisture exchange processes between the human body and the environment via the layer(s) of clothing is connected to these states. A body temperature of 37 °C and an average skin temperature of 33 °C without sweat are considered physiologically comfortable for a human body [5–7]. The layer(s) of the clothing must maintain the body and skin temperatures in their suitable states and ensure the thermal balance of the body with its environment. The physiological clothing comfort of sportswear can be evaluated through subjective wear trials or by assessing some properties like the water vapor and air permeability, the wetting and wicking capacity, and the drying behavior.

Investments are made in making innovative eco-friendly fibers with sustainable practices that provide a high level of comfort. Eco-friendly fibers can be considered recyclable or biodegradable. Recyclability refers to the recovery of useful materials from waste through chemical or mechanical decomposition [4]. Nowadays, the most prominent fibers for sportswear are polyester, followed by polyamide. Polyester fiber recycling offers a more environmentally friendly production method by reducing landfill volume and plastic waste. A comparison done by Hatamlou et al. [8] between the moisture transfer behavior of knitted fabrics made from recycled and regular polyester fibers showed that the recycled polyester fabric had better absorbency, wettability, drying speed, and capillary rise than regular polyester fabric. The moisture management tester evaluation gave both fabrics a good grade.

Biodegradable materials are defined as their ability to naturally break down into simpler materials. Substances ingested by organisms (bacteria, fungi) must be harmful to decomposers and must decompose within a relatively short period of time on human timescales [3, 4]. Despite being synthetic, PLA fibers are made from raw materials that are both natural and regenerative. It is made by separating and fermenting sugars from annually renewable crops like maize, sugar beet, or wheat starch. As a result, it is a biodegradable and bioabsorbable polymer. During its lifetime, Earth's carbon dioxide levels might decrease [2, 9, 10].

The heat and moisture transmission characteristics of cotton/PLA mixed knitted textiles were studied by Guruprasad et al. [9]. They discovered that adding PLA fibers to cotton fibers improves the comfort of knitted garments, with 65/35 being the ideal mix ratio for materials with enhanced moisture transport capacity.

Demanding the advocacy of their product, "go green" manufacturers promote the excellent functionalities and the sustainability offered. The intention of this study is investigating the claims of 37.5® technology producers, and they state that enhancing the intrinsic biodegradable characteristic of initially non-biodegradable yarn to develop the biodegradation rate does not affect the clothing comfort overall performance of their product. To accomplish our purpose, a comparison between the drying time, as a wear comfort property, of a non-biodegradable polyester yarn with its biodegradable model was achieved.

## 2 Materials and Methods

The polyester yarn COCONA® is made of polyester filaments that has energetic particles permanently incorporated into it. These particles actively respond to changes in body temperature by capturing and releasing water vapor. Those particles use the human body's infrared heat to hasten the evaporation of liquid sweat and the motion of its vapor form.

In July 2020, COCONA® delivered a biodegradable version to all 37.5® technology polyamide and polyester filament and staple fibers yarns. According to the industry standard for biodegradation, ASTM D-5511 [11]; According to independent testing, this innovative fiber dramatically dissolves within a shorter time than non-biodegradable fiber, rather than taking centuries to decades as advertised. This results in less waste building up in landfills. If dumped in a landfill that mimics the conditions found in the standardized test, 37.5® technology biodegradable fibers disintegrate by 54% in 341 days, decreasing to methane and carbon dioxide, and ultimately to a biomass in about 3.35 years.

In this study, three knitted structures were designed and produced on a single jersey machine **J1**, **J2,** and **J3**. The three fabrics were made firstly by the regular

**Table 1** Details of the employed fabrics

| Pattern | Code | Yarn fraction |
|---|---|---|
| Jersey crepe | J1B | 50% Polyester<br>50% COCONA® biodegradable |
| Jersey crepe | J2B | |
| Jersey mesh | J3B | |
| Jersey crepe | J1R | 50% Polyester<br>50% COCONA® |
| Jersey crepe | J2R | |
| Jersey mesh | J3R | |

COCONA® polyester yarn, the fabrics are coded as **J1R, J2R,** and **J3R**. Then, the yarn was replaced with its biodegradable version and three other fabrics were made, coded as **J1B, J2B,** and **J3B**. The manufacturing process and machinery settings were the same for fabrics made with the same structure. The pattern and yarn fraction of these fabrics are shown in Table 1.

Washing and bleaching were done simultaneously on a small dyeing machine. Finishing and heat setting were carried out on a stenter machine successively. Furthermore, the durability of the fabrics was tested by undergoing five cycles of domestic washing and drying to the finished fabrics. The structural characteristics experiences and the drying time evaluation of the six fabrics were performed after the finishing and the domestic washing phases.

## 2.1 *Yarn's Structural Characteristics*

The yarn's porosity was considered as the percentage of air volume from the entire yarn volume following the Eq. (1) [12, 13]:

$$\text{Porosity}(\%) = 100\left(1 - \left(d^2.N/D^2\right)\right) \quad (1)$$

Here, *d* and *D* are, respectively, the fibers and yarns average diameters. The diameters were measured using a digital Leica microscope. The linear density in dtex and the average number of filaments per cross-section (*N*) were provided by the yarn suppliers. The characteristics of the employed yarns are shown in Table 2.

**Table 2** Characteristics of the yarns

| Yarn | Number of filaments (filament) | Linear density (dtex) | *d* (μm) | *D* (μm) | Porosity (%) |
|---|---|---|---|---|---|
| Polyester | 72 | 83 | 12.17 | 190 | 70.46 |
| COCONA® | 48 | 83 | 13.93 | 200 | 76.71 |
| COCONA® biodegradable | 48 | 83 | 13.27 | 260 | 87.50 |

**Table 3** Structural properties of the finished and washed **J1R** and **J1B**

| Phase | Thickness (mm) | | Mass per unit area (g/m$^2$) | | Course density (Courses/cm) | | Wale density (Wales/cm) | | Stitch density (Stitches/cm$^2$) | | Loop length (cm) | |
|---|---|---|---|---|---|---|---|---|---|---|---|---|
| | *J1R* | *J1B* | *J1R* | *J1B* | *J1R* | *J1B* | *J1R* | *J1B* | *J1R* | *J1B* | *J1R* | *J1B* |
| Finished | 0.61 | 0.46 | 76.6 | 65.56 | 15.8 | 15.6 | 15.1 | 15 | 238.6 | 233.8 | 26.98 | 25.89 |
| CV (%) | 2.19 | 1.96 | 1.95 | 1.59 | 2.83 | 3.51 | 1.48 | 4.71 | | | 1.35 | 3.06 |
| Washed | 0.67 | 0.51 | 80.4 | 69.54 | 16.2 | 15.4 | 15.5 | 15.6 | 251 | 240.2 | 26.93 | 25.2 |
| CV (%) | 1.69 | 2.16 | 2.12 | 1.37 | 2.76 | 3.56 | 3.23 | 3.51 | | | 1.32 | 2.32 |

**Table 4** Structural properties of the finished and washed **J2R** and **J2B**

| Phase | Thickness (mm) | | Mass per unit area (g/m$^2$) | | Course density (Courses/cm) | | Wale density (Wales/cm) | | Stitch density (Stitches/cm$^2$) | | Loop length (cm) | |
|---|---|---|---|---|---|---|---|---|---|---|---|---|
| | *J2R* | *J2B* | *J2R* | *J2B* | *J2R* | *J2B* | *J2R* | *J2B* | *J2R* | *J2B* | *J2R* | *J2B* |
| Finished | 0.48 | 0.45 | 108.56 | 100.98 | 24 | 25.2 | 17.5 | 17.6 | 420 | 443.6 | 18 | 18.6 |
| CV (%) | 1.88 | 1.97 | 2.88 | 0.69 | 4.17 | 3.32 | 2.86 | 3.11 | | | 1.65 | 1.49 |
| Washed | 0.5 | 0.48 | 111.66 | 106.14 | 23 | 24.6 | 18 | 18.4 | 414 | 452.8 | 18.3 | 18.9 |
| CV (%) | 0.9 | 1.13 | 0.74 | 1.37 | 0 | 3.64 | 0 | 2.98 | | | 2.45 | 1.61 |

## 2.2 *Fabric's Structural Characteristics*

The standard atmospheric conditions for testing were considered as, 65% ± 2% relative humidity and 20 °C ± 2 °C room temperature. The fabric's samples were conditioned in the same atmosphere for 24 h before testing. The fabric thickness was considered as the distance between the parallel presser foot of the FAST1 and the reference plate, the presser foot exerts a pressure equal to 0.196 kPa with an area of 10 cm$^2$. The measure of the mass per unit area was done following the standard method EN 12127 [14]. The number of courses, wales and stitches per centimeter were measured according to the standard EN 14971:2006 method A [15]. A 100 stitches loop length was calculated in accordance with the standard EN 14970:2006 [16]. The structural properties of J1, J2, and J3 fabrics made with the two versions of yarn are presented, respectively, in Tables 3, 4, and 5.

## 2.3 *The Drying Behavior*

The drying behavior of fabrics is the clothing comfort property examined in this study. The drying behavior of fabrics was portrayed by estimating the drying time. At first, the mass of a circular 100 cm$^2$ specimen was recorded. Then, the distilled

**Table 5** Structural properties of the finished and washed **J3R** and **J3B**

| Phase | Thickness (mm) | | Mass per unit area (g/m²) | | Course density (Courses/cm) | | Wale density (Wales/cm) | | Stitch density (Stitches/cm²) | | Loop length (cm) | |
|---|---|---|---|---|---|---|---|---|---|---|---|---|
| | *J3R* | *J3B* | *J3R* | *J3B* | *J3R* | *J3B* | *J3R* | *J3B* | *J3R* | *J3B* | *J3R* | *J3B* |
| Finished | 0.49 | 0.53 | 96.76 | 98.38 | 24.2 | 23.6 | 16.7 | 18 | 404.3 | 424.8 | 18.75 | 18.51 |
| CV (%) | 2.24 | 1.03 | 0.83 | 0.99 | 3.46 | 2.32 | 2.68 | 0 | | | 0.72 | 0.90 |
| Washed | 0.53 | 0.56 | 100.22 | 104.36 | 26 | 25.2 | 16.6 | 17.6 | 431.6 | 443.6 | 18.45 | 18.84 |
| CV (%) | 1.33 | 2.97 | 0.37 | 2.01 | 2.72 | 4.35 | 3.30 | 3.11 | | | 1.09 | 1.14 |

water was used to drop a 20 ± 2 mg of water on the digital scale. The drop must be placed in the center of the specimen when it is placed backwards. The chronometer was launched at this point. The mass was recorded every two minutes until it returned to the initial sample's mass. According to this method, the drying time is defined as the time recorded when the mass has returned to the fabric initial mass.

## 3 Results and Discussions

The drying times of garments created with non-biodegradable and biodegradable COCONA® polyester yarn after finishing and following domestic washing cycles will be compared in this section.

The fast-drying capacity of knitted fabrics intended for sportswear is crucial for providing a high level of comfort. The new biodegradable yarn must thus conserve or improve the drying behavior by shortening the time required for water to evaporate in order to guarantee a greater comfort performance. Figures 1 and 2 show, respectively, how long it takes for knits to dry after being finished and after being cleaned.

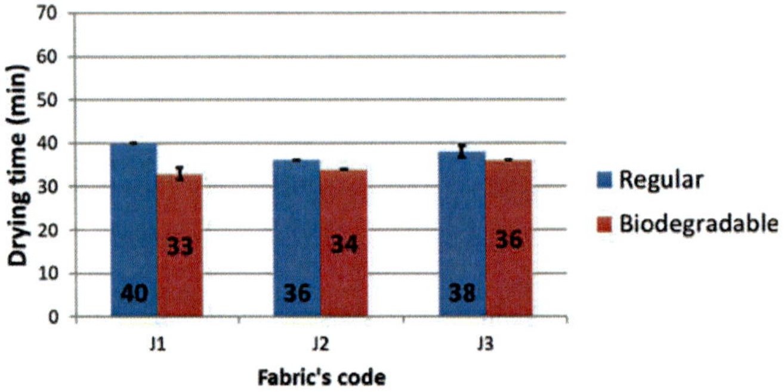


**Fig. 1** The drying time of finished fabrics

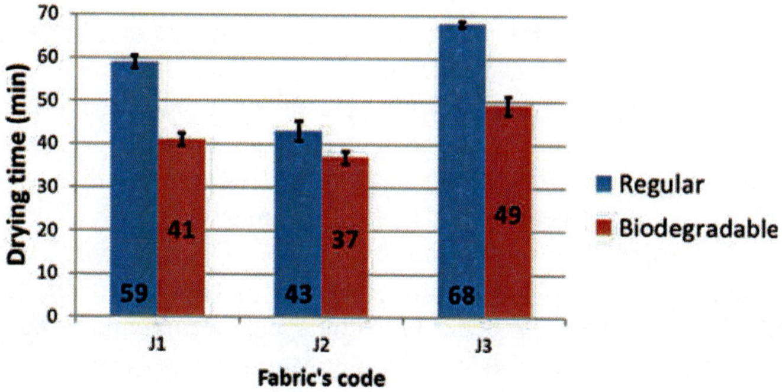


**Fig. 2** The drying time of washed fabrics

Knits produced with the non-biodegradable COCONA® yarn require longer drying periods at the finished and washed phases than knits made with the biodegradable version. In the case of finished fabrics, the difference between J1R and J1B is greater than the other knits, 7 min between J1R and J1B, 2 min between J2R and J2B, and the same between J3R and J3B. This can be explained by the loose bonding structure of the J1 knits. The distance between threads is large in structures with high loop length value (Structure J1), and there are fewer linking points between yarns than in structures with low loop length value (Structures J2 and J3), so, the liquid moisture management of a single yarn can significantly govern the overall drying capacity of the fabric. The biodegradable COCONA® yarn is more porous than its regular version; hence, its absorption capacity and wicking behavior is more efficient, and ultimately, the drying time of knits made by this yarn is lower. This interpretation is confirmed by the results found by Kim and Kim [17]. They examined the wear comfort properties of knits made with different types of yarns, having different porosity values. They found that if there is more void in the structure of the yarn, then the transmission of water vapor by diffusion will be smoother. And they added that with this fast transmission, the time needed for drying will be reduced. So, following the results and the previous interpretations, we can conclude that the biodegradable COCONA® polyester yarn affects the drying time of the loose structures in a more significant way than the tight structures.

After the domestic washing cycles, an alteration in the drying capacity of fabrics was observed, which is supported by the findings of Duru and Candan [18], who stated that after washing, the drying time of knitted fabrics decline, which is mainly provoked by the changes in the size and volume of the micro- and macro-pores as well as their tortuosity. Moreover, they add in [19] that repeated washing cycles affect the characteristics of fibers and fabrics at various extents, which can generate a change in the liquid transfer capabilities of knits.

When comparing the drying times before and after the domestic washing cycles, a glaring difference exists between values of knits made with the regular COCONA® yarn, 19 min in the **J1** structure, 7 min in the **J2** structure, and 30 min in the **J3** structure. Instead, a smaller difference is observed in the fabrics made with the biodegradable COCONA® yarn, 7 min in the **J1** structure, 3 min in the **J2** structure, and 13 min in the **J3** structure. These results indicate that the biodegradable COCONA® yarn resists to domestic washing cycles better than the regular yarn and that there have been no severe alterations in its pores structure.

Additionally, the repeated household washing cycles can deteriorate or eliminate the moisture management agent applied in the finishing process, which could hinder the absorption and spreading of moisture by the fabrics. Consequently, the drying time is increased. Accordingly, the employment of the COCONA® biodegradable yarn in knitted fabrics enables a better finishing agent fixation and provides to the final consumer a comfortable product even after use.

## 4 Conclusion

In this study, a comparison between the drying time of two types of polyester functional yarns was done. The purpose was to investigate the effect of biodegradable chemical additives on the clothing comfort performance of knitted fabrics.

According to the results, all fabrics made with biodegradable COCONA® yarn dried faster than fabrics made with regular COCONA® yarn both after finishing and after domestic washing cycles. However, after domestic washing cycles, drying times of fabrics made with biodegradable COCONA® yarn increased slightly.

**Acknowledgements** This project is carried out under the **MOBIDOC** scheme, funded by The Ministry of Higher Education and Scientific Research through the **PromEssE** project and managed by the **ANPR** (Agence Nationale de la Promotion de la Recherche Scientifique).

The **VTL** group provided the resources and technologies required for this study, for which the authors are grateful. We also like to express our gratitude to the working group and managers for sharing their expertise with us during this study.

## References

1. Subramanian Senthilkannan Muthu: Sustainable Fibres and Textiles. 1st edn. Woodhead Publishing, Cambridge (2017).
2. R. S. Blackburn: Biodegradable and sustainable fibres. 1st edn. Woodhead Publishing Limited in association with The Textile Institute, Cambridge (2005).
3. R. Nayak: Sustainable Technologies for Fashion and Textiles. 1st edn. Woodhead Publishing, Cambridge (2020).
4. M. I. Tobler-Rohr: Handbook of sustainable textile production. 1st edn. Woodhead Publishing Limited in association with The Textile Institute, Cambridge (2011).
5. A. Das, R. Alagirusamy: Science in clothing comfort. 1st edn. Woodhead Publishing India, New Delhi (2011).
6. R. Sinclair: Textiles and Fashion: Materials, Design and Technology. 1st edn. Woodhead Publishing Limited in association with The Textile Institute, Cambridge (2015).
7. K. Parsons: Human thermal environments: The effects of hot, moderate, and cold environments on human health, comfort, and performance. 3rd edn. CRC Press, Boca Raton (2002).

8. M. Hatamlou, A. T. Özgüney, N. Özdil, G. S. Mengüç: Performance of recycled PET and conventional PES fibers in case of water transport properties. Industria Textila 71(6),538–544 (2020).
9. R. Guruprasad, M. v. Vivekanandan, A. Arputharaj, S. Saxena, S. K. Chattopadhyay: Development of cotton-rich/polylactic acid fiber blend knitted fabrics for sports textiles. Journal of Industrial Textiles 45(3), 405–415 (2015).
10. B. Ozgen: New biodegradable fibres, yarn properties and their applications in textiles: A review. Industria Textila 63(1), 3–7 (2012).
11. ASTM D-5511-18 Standard Test Method for Determining Anaerobic Biodegradation of Plastic Materials Under High-Solids Anaerobic-Digestion Conditions.
12. E. A. Elzaher Eltahan, M. Sultan, A.B. Mito: Determination of loop length, tightness factor and porosity of single jersey knitted fabric. Alexandria Engineering Journal 55(2), 851–856 (2016).
13. B. Özgen, S. Altaş: The investigation of thermal comfort, moisture management and handle properties of knitted fabrics made of various fibres. Tekstil ve Konfeksiyon 24(3), 272–278 (2014).
14. EN 12127:1997 Textiles - Fabrics - Determination of mass per unit area using small samples.
15. EN 14971:2006 Textiles. Knitted fabrics. Determination of number of stitches per unit of length and unit area.
16. EN 14970:2006 Textiles. Knitted fabrics. Determination of stitch length and yarn linear density in weft knitted fabrics.
17. A. K. Hyun, J. K. Seung: Physical Properties and Wear Comfort of Bio-Fiber-Embedded Yarns and their Knitted Fabrics According to Yarn Structures. Autex Research Journal 19(3), 1–9 (2019).
18. S. Cimilli Duru, C. Candan: Wicking and drying behaviors of knitted fabrics produced with different poliamide yarns. Tekstil ve Konfeksiyon 26(3), 280–286 (2016).
19. S. C. Duru, C. Candan: Effect of repeated laundering on wicking and drying properties of fabrics of seamless garments. Textile Research Journal 83(6), 591–605 (2013).

# A Comparative Study of Environmental Sustainability Initiatives of Two Major Jeanswear Brands

**Mouna Haj Nasr, Hassen Hedfi, Ayda Baffoun, and Mohamed Hamdaoui**

**Abstract** The fashion industry's supply chain is extremely extended, with many different suppliers, distributors and retailers frequently involved. Hence, there are several barriers when attempting to monitor and improve sustainability progress within the whole supply chain. Numerous fashion brands acknowledge the importance of preserving the environment for the future of the fashion industry, as well as the rising environmental awareness of consumers. To address this, brands are proactively working to integrate environmental sustainability practices into their manufacturing operations. This paper highlights the environmental damage caused by the jeans industry and examines the sustainability initiatives of Levi's and Kontoor, two leading jeans brands. Through an analysis of their published sustainability reports and official websites, we managed to provide insights into their innovative approaches and environmental strategies implemented. Both brands are actively working toward reducing their environmental footprint and promoting sustainability within the industry. They have set challenging goals to reduce their carbon emissions, conserve water, eliminate harmful chemicals from their supply chain and lower their waste generation. They also commit to sourcing their materials responsibly and improving their circularity.

**Keywords** Fashion industry · Environment · Sustainability · Jeanswear

M. Haj Nasr (✉) · A. Baffoun · M. Hamdaoui
Textile Materials and Processes Research Unit MPTex, National School of Engineers of Monastir, University of Monastir, 5019 Monastir, Tunisia
e-mail: hadjnasrmouna@gmail.com

H. Hedfi
Mechanical Engineering Laboratory LGM, National School of Engineers of Monastir, University of Monastir, 5019 Monastir, Tunisia

 
S. B. Abdessalem et al. (eds.), *Proceedings of the Second International Conference of Innovative Textiles and Developed Materials-ITDM'2; 05-06 May 2023; Tunisia*,
https://doi.org/10.1007/978-981-99-7950-9_3

# 1 Introduction

Denim has become a staple in the global apparel market and is most often used for jeans [1], one of the most recognizable and iconic products that appeal to consumers of all ages. Denim fabric production is an international business, with most of it taking place in countries like China, its main global exporter. India, Bangladesh, Pakistan, Turkey, and the United States are also key denim-producing countries [2]. In 2022, the denim fabric market accounted for 4.3 billion meters of fabric [3], and was valued at $27.1 billion. This market is likely to grow to over $35 billion by 2027 [1].

Similarly, the global Jeanswear market is forecasted to grow from 100 billion US dollars in 2018 to around 128 billion US dollars by 2023 [4]. A 2012 report stated that roughly 1.8 billion pairs of denim pants were sold annually [2], but in 2022, the number had risen to about 3.1 billion pairs. For the ensuing years, it was anticipated to increase steadily, surpassing 3.46 billion units by 2026 [5].

For decades, a pair of jeans has been the backbone of countless fashion collections. Yet, it remains a part of the fashion industry's "take, make, waste" approach [6]. The process of making jeans is complicated since it encompasses many steps such as raw materials extraction, fabric production, garment manufacturing, finishing and transportation [2].

Each of these steps has a substantial environmental impact in terms of resources usage (such as energy, water and chemicals), waste generation and greenhouse gas (GHG) emissions. Moreover, the specific design and construction of jeans make it difficult to recycle and repurpose them [6].

The fact is that jeans production leads to ecosystem pollution and health risks is worrying. Research indicates that the jeans market accounts for about 1% of the world's total greenhouse gas emissions. To be precise, each pair of jeans produced results in the emission of 33.4 kg of carbon dioxide equivalent (CO2e), utilizes 3781 L of water, and requires 12 square meters of land [7].

This paper aims to outline and compare some of the key environmental sustainability initiatives implemented by two jeans labels—Levi Strauss & Co. and Kontoor Brands—in order to reduce their ecological footprint.

The rest of the paper is structured as follows: initially, the methodology of the study is outlined. Then, a comparative analysis of the environmental strategies and future targets of both brands is conducted. The paper concludes with a presentation of the findings.

## 2 Methodology

### 2.1 Selection of Brands

Levi Strauss & Co. is a highly recognizable label in the denim industry [1] and is considered as one of the top competitors in the jeans market alongside Wrangler and Lee, which are both owned by Kontoor Brands [8].

Still, Levi's outperformed Kontoor Brands in terms of net revenue. In fiscal year 2021, the brand reported a net revenue of 5.8 billion [9], while Kontoor Brands reported 2.5 billion during the same period. This is an indication that Levi's has a greater global presence when compared to Kontoor Brands. However, it is crucial to keep in mind that Kontoor Brands is a younger company, as it was spun off from VF Corporation in 2019 [1], meaning it may still be in a phase of growth.

### 2.2 Collection of Data

The selected labels, Levi Strauss & Co., and Kontoor Brands, are both US-based companies that produce Jeanswear and are regarded as major competitors. They can therefore be considered a representative sample and compared accordingly.

To conduct this research, we obtained the necessary data from their official websites and their latest publicly available sustainability reports. Specifically, we reviewed the sustainability reports of Levi's for the year 2021 and Kontoor Brands for the year 2020.

### 2.3 Analysis Method

The environmental sustainability strategy of both brands revolves around two major pillars: Planet and Product. As a result, we conducted a thorough comparative assessment by focusing on their efforts to reduce greenhouse gas (GHG) emissions, improve chemical management, increase waste diversion and minimize resource use (e.g., energy, water). In addition, we examined their procurement of sustainable materials and their commitment to circularity throughout the product life cycle.

This type of analysis is essential to understand how these brands are striving to reduce their environmental impact and can help influence decision making and encourage sustainable practices.

# 3 Environmental Sustainability Initiatives

Renowned denim apparel manufacturers Levi's and Kontoor Brands have a significant environmental impact. Although their initiatives, strategies, and long-term objectives may vary significantly, both brands share the common goal of reducing their environmental impact and promoting sustainability across their supply chain.

## *3.1 GHG Emission and Energy Use*

In order to prevent the most severe outcomes of climate change, it is vital that we intensify our endeavors in the coming years to limit the global temperature rise to 1.5 °C [10].

Textile brands are expressing concern about the impact of climate change on their production facilities. This is because, high temperatures can lower operator productivity and increase absenteeism rates. In addition, severe weather events (such as floods, storms and wildfires) can affect logistics and worker health, delay the supply of raw materials and finished products and endanger production sites [11]. Both brands are aware of these consequences, thus, they have set future goals to reduce their carbon footprint and implemented measures to achieve them.

Levi's has established science-based goals to address its absolute Scope 1 and Scope 2 greenhouse gases emissions by 90% and to expand the adoption of renewable electricity by 100% (across its company-operated facilities). Also, it aims to reduce its absolute emissions of Scope 3 by 40% (across its supply chain). All these targets are set to be met in 2025 (from the 2016 baseline) as part of Levi's plan to become climate neutral by 2050 [9].

In 2021, Kontoor Brands presented its ambitious greenhouse gas emission reduction targets, pending approval by the Science Based Targets (SBTi) initiative. It has also set a target of using 100% renewable energy, across its owned and operated facilities, by 2025 [11].

Both Levi's and Kontoor Brands have implemented some key initiatives to reduce their GHG emissions. For instance, both brands are shifting to renewable energy through generating onsite solar energy or purchasing Renewable Energy Certificates (RECs). Furthermore, in order to improve their energy efficiency, both brands have taken some measures across their facilities, such as installing LED lighting and motion-sensing light. They are also upgrading HVAC systems, to reduce their energy consumption and GHG emissions [9, 11].

On top of this, Kontoor Brands is known for supporting its suppliers by providing training to improve their energy efficiency and implementing programs to track and report their emissions and energy use, which is also the case with Levi's, both of which use the Higg Facility Environmental Module (Higg FEM) assessment tool [9, 11].

## 3.2 Water Use

The denim sector remains a water-intensive industry, as evidenced by several studies [12]. Cotton farming, denim fabric production (which involves bleaching, dyeing and finishing) and garment washing, a crucial step in obtaining the desired look and softness, require substantial amounts of water [11]. Furthermore, the laundering habits of consumers are seen as a major threat to the availability of freshwater resources [7].

Both Levi's and Kontoor Brands have committed to reduce their water use within their production processes. Even though their initiatives vary, each brand has implemented a thorough strategy to mitigate its water footprint.

On the one hand, Levi's has been actively promoting and implementing Water < less® finishing techniques and applying water Recycle & Reuse Standard at its manufacturing facilities since 2011. As a result, the company has saved a total of 11.5 billion liters of water between 2011 and 2021. Moreover, the brand has pledged to decrease its water consumption by 50% in areas with limited water resources (starting from the baseline of 2018) and plans to support its main suppliers in becoming certified as Water < Less® facilities by 2025.

Levi's is proactively promoting sustainable practices within the denim industry by openly sharing its water-saving techniques such as the Water Action Strategy, Water < Less® Technique Manual, and Recycle & Reuse Standard and Guidelines, with the aim of inspiring the entire industry to join water conservation efforts [9].

On the other hand, Kontoor Brands has set itself the goal of halving its water consumption by 2030 through the use inventive techniques and by ensuring that all water used in their production processes is adequately purified before being released into the environment. Over the period of 2008 to 2020, the company was able to recycle 8 billion liters of water, which has paved the way for achieving their target of conserving ten billion liters of water by 2025 [11].

In 2019, Kontoor Brands launched an inventive approach known as Indigood™ Foam-Dyeing Technology, which entails using foam instead of water to dye denim. This technique yields a reduction of up to 90% in water usage [13]. It also reduces energy consumption and waste generation by over 60% and eliminates the need for chemical baths during the dyeing process [11].

It is necessary to maintain continuous monitoring and supervision to pinpoint areas where water usage can be reduced and establish strategies to decrease its footprint. To this end, Kontoor Brands and Levi's have adopted Jeanologia's EIM software to track the quantity of water used during finishing processes [9, 11].

## 3.3 Chemicals Use and Wastewater Discharge

According to the researchers, the production of one kilogram of clothing requires between 1.5 and 6.9 kg of chemicals. This indicates that the amount of chemicals used during the manufacturing process is greater than the weight of the final product [14]. To put it in perspective, one pair of Levi's® 501® medium stonewash jeans weighs around 340 g [15].

The wet processing of denim involves the use of chemicals such as synthetic indigo dyes, bleach, wetting and sequestering agents and sulfuric acid, which can be detrimental to the environment. Mishandling these substances can lead to water contamination, while disposing them improperly can add to the pollution of soil and air [14].

Considering this, Kontoor Brands and Levi's have taken noteworthy steps toward producing more eco-friendly products. This involves minimizing the use of harmful chemicals during production, with the ultimate goal of decreasing their chemical footprint and wastewater output.

In 2012, Levi's set a target for its major wet finish producers to reach zero discharge of hazardous chemicals (ZDHC), which was achieved by 2020 [16]. Following this success, the brand has set a challenging goal for all fabric mills and key garment wet finishing producers to use 100% certified Screened Chemistry by 2026 [9].

In 2022, Levi's embraced the Restricted Substances List (RSL) of "Apparel and Footwear International RSL Management" (AFIRM) and mandated that its suppliers comply with it, aiming to encourage consistency in the use of safer chemicals throughout the industry.

In addition, Levi's uses tools like "CleanChain" and the "ZDHC Gateway" to ensure that its first- and second-tier suppliers are using chemicals responsibly and managing their wastewater properly. It also mandates its suppliers to submit monthly reports demonstrating their compliance with the ZDHC's "Manufacturing Restricted Substance List" (MRSL) and to conduct tests of their effluent treatment plants bi-annually [16].

Kontoor Brands, like Levi's, has established a goal of utilizing 100% preferred chemicals by 2023 and is implementing various measures to achieve it. For instance, the brand requires its manufacturers and material suppliers to comply with its Restricted Substances List (RSL).

Kontoor Brands also adopted an innovative chemical management program known as CHEM-IQ℠ that aims to eliminate the use of harmful chemicals from its supply chain. This program includes a compilation of over 430 hazardous substances referred to as the "Substances to Avoid List (STA)." The brand has also partnered with its suppliers to build wastewater treatment facilities and require them to comply with its wastewater regulations. Moreover, it requires all its suppliers to undergo bi-annual wastewater testing by a certified third party [11].

On top of this, Levi's and Kontoor are exploring eco-friendly alternatives such as laser technology to make faded denim rather than using potentially harmful potassium permanganate sprays [9, 11]. In addition, both brands avoid using water-repellent substances (e.g., poly- and perfluorinated substances (PFCs)) [11, 16].

Meanwhile, Levi's is going one step further by collaborating with Stony Creek Colors to test the use of pre-reduced plant-based indigo as an alternative to synthetic indigo [9].

### 3.4 Waste Management

Modern fashion consumption practices create a significant amount of textile waste, which often ends up being burned, landfilled, or shipped to developing countries [11]. Consequently, Levi's and Kontoor Brands have committed to reducing waste in their supply chains and operations as part of their efforts to address these issues.

Both brands have pledged to remove single-use plastics [9, 11], with Levi's setting a target to accomplish this by 2030. As an alternative, they are exploring other types of packaging that are environmentally friendly (e.g., fully reusable, recyclable or home compostable plastics) [9].

Both companies have also committed to lowering their product packaging waste by actively promoting the use of recycled materials and reducing the size of their packaging [9, 11].

In 2021, Levi's took further action by reducing the width of its care labels by 25% and the size of some of its hangtags on Levi's® pants by 62% [9]. Meanwhile, Kontoor Brands has been using 3D design for the past decade, which has proven effective in minimizing the need for physical samples during the development process [11].

On top of this, Levi's and Kontoor Brands have set ambitious targets to reduce their diversion rates in the near future [9, 11]. In fact, Levi's plans to achieve zero waste to landfill at its own facilities and a 50% diversion rate across its key suppliers by 2030. To reach these goals, Levi's has implemented a program to segregate and recycle denim scraps [9]. Meanwhile, Kontoor Brands has established a "closed-loop" approach to recycle or repurpose most of its waste internally. As a result of these efforts, Kontoor Brands has successfully diverted 79% of its waste from landfills [11].

In an effort to further reduce waste, both labels have eliminated the use of pumice [9, 11]. In fact, Levi's opted for a cost-effective alternative by using synthetic stones that can be reused up to 3,000 times and eventually recycled and replaced [9]. Kontoor Brands, on the other hand, swapped pumice with an enzyme [11].

The implementation of these ambitious measures demonstrates a strong commitment to environmental preservation and represents a key step toward creating a more sustainable and circular fashion industry.

### 3.5 Circularity

Levi's and Kontoor Brands acknowledge the benefits of extending the lifespan of products, reducing waste and pollution, and regenerating the ecosystems. They are therefore working on innovative ideas to give new life to potential waste, moving from the conventional "take-make-waste" approach to a circular one.

Both Brands have jointly embraced the "Jeans Redesign Program" developed by the "Ellen MacArthur Foundation" as a part of their effort to reduce waste and increase the longevity of their products. This program includes a set of guidelines that illustrate how a garment can be designed and produced in a way that is consistent with the principles of a circular economy [9, 11].

Moreover, Levi's and Kontoor Brands are both pioneering innovative methods of reusing and recycling used clothes [9, 11]. For instance, Levi's has launched a program called "SecondChance," which allows customers to exchange their used jeans for a voucher for new one [9]. Similarly, Kontoor Brands has launched the "Take-Back" initiative, which allows consumers to return their old denim for recycling [11]. Levi's even went a step further and established in-store tailor shops where customers can repair, reuse and personalize their jeans [9].

Both Levi's and Kontoor Brands are focused on manufacturing durable and long-lasting jeans that can be worn for many years, even decades. Thus, by offering repairs, reconditioning and warranties, they inspire customers to keep their jeans longer and reduce waste in the fashion industry.

They also encourage consumers to take care of their jeans according to the instructions on care labels, e.g., by washing them less often, using cold water and line drying them, and by giving them away when no longer needed [9, 11].

### *3.6 Materials Sourcing*

Levi's and Kontoor Brands are fully committed to using recycled and renewable materials in their garments as a way to ensure the well-being of people and the environment [9, 11]. In fact, Levi's has set a goal of using only durable fibers by 2030 [9], while Kontoor Brands has committed to 100% use of sustainable cotton by 2025 and 100% use of recycled polyester by 2030 [11].

In line with their ongoing efforts to promote circularity and minimize waste in the fashion industry, they use recycled cotton and recycled polyester to avoid introducing virgin materials into their products and encourage the use of available resources.

They also have switched from conventionally grown cotton to organic cotton and hemp, which require less water and pesticides. Furthermore, these fibers promote biodiversity and minimize soil erosion [9, 11].

Levi's also incorporates into its products fibers such as TENCEL™, Refibra™ and Circulose®, which are produces via closed-loop processes that recycle both water and chemicals [9]. On the other hand, Kontoor Brands uses the highly sustainable fiber "Sorona®" from DuPont, which is produced using plant-based materials [11].

## 4 Discussion

Although both brands have made significant progress in reducing their environmental impact, Levi's is ahead due to its long-standing commitment to sustainability and its accomplishment of numerous sustainability targets throughout the years.

On the other hand, Kontoor, being a newer company, is still in the process of formulating and implementing its sustainability strategy. For instance, Levi's has publicly pledged to reduce its greenhouse gas emissions by 40% by 2025 [9], while Kontoor Brands is still working on developing a similar commitment [11].

In addition, the environmental performance of both brands in China was evaluated by the IPE Green Supply Chain CITI Evaluation. As a result, Levi's received a higher score of 82.62, which ranked them first among all evaluated companies [17], while Kontoor Brands received a score of 62.74 and ranked 13th out of 124 textile brands [18]. To conclude, Levi's outperformed Kontoor in all areas, including energy conservation and emission reduction, transparency and extending green supply chain practices [17, 18].

It should be noted, though, that the IPE Green Supply Chain CITI Evaluation only assesses brands' supply chains in China [19], and thus does not provide a detailed analysis of their overall environmental efforts. In addition, as this assessment focuses on the supply chain rather than the entirety of the company, it may not provide an accurate reflection of its environmental commitments or endeavors.

To conclude, both Brands have made considerable strides in minimizing their environmental footprint. Therefore, a more comprehensive review of their efforts and progress in meeting their sustainability goals would be required to determine which brand is more environmentally committed.

## 5 Conclusion

Becoming a sustainable brand means achieving a compromise between economic success and the demands of consumers, employees as well as the preservation of the environment. Many fashion brands are aware of the challenges that endanger the planet and are constantly searching for solutions that ensure product durability and esthetics while minimizing their impact on the environment.

The objective of this research was to investigate the environmental sustainability practices in the fashion industry. To achieve this, we selected two Jeanswear brands, namely Levi Strauss & Co. and Kontoor Brands, as case studies. This research involved analyzing their sustainability reports which enabled us to identify some of the environmental sustainability initiatives adopted in areas such as greenhouse gas emissions, energy usage, water consumption, waste generation, chemical usage, as well as circularity practices and materials sourcing.

Based on the implementation of various sustainability practices, we can assume that the reviewed labels are committed to reduce their environmental footprint. These practices include the use of environmentally friendly materials, adoption of clean energy sources, responsible water use and management, preferred chemistry initiatives, recycled and recyclable packaging, waste reduction efforts and promotion of circular economy practices. On top of this, both brands outlined some future goals to be achieved by 2025 and beyond.

It is important to remember that each of the examined brand faces unique challenges and opportunities, which can lead to adopting different sustainability measures. Likewise, progress in terms of sustainability may vary depending on factors such as resources availability, technological progress, and their level of commitment to sustainability goals. Regardless of these variations, their ultimate goal remains the same—to minimize their environmental footprint.

Overall, this study provides some important insights that can inspire other brands to boost their environmental sustainability practices and contribute to achieve a more sustainable future. However, an important limitation of this study is the fact that it relied solely on brands' sustainability reports and websites as primary data sources, which may lack relevant information.

Thus, for future research, we propose to include additional sources and to conduct a more in-depth analysis concentrating on the other environmental initiative that we did not cover in this study, such as biodiversity, land use and innovative products. Additionally, we recommend conducting a comparative analysis to assess their progress toward achieving their set targets.

## References

1. "Denim market worldwide," *Statista.* https://www.statista.com/topics/5959/denim-market-worldwide/. (Accessed Mar. 30, 2023).
2. S. S. Muthu, "Sustainability in Denim," 2017.
3. "Global units of denim jeans 2016–2023," *Statista.* https://www.statista.com/statistics/1368034/global-units-of-denim-jeans-market/. (Accessed Mar. 30, 2023).
4. "Global jeanswear market size, 2018," *Statista.* https://www.statista.com/statistics/541304/size-of-the-global-jeans-market/. (Accessed Mar. 30, 2023).
5. "Global units of denim jeans 2021–2026," *Statista.* https://www.statista.com/statistics/1368005/global-units-of-denim-jeans-market/. (Accessed Mar. 30, 2023).
6. "Ellen MacArthur Foundation, The Jeans Redesign Insights report, (2021), https://ellenmacarthurfoundation.org/the-jeans-redesign."
7. F. Asmi, Q. Zhang, M. A. Anwar, K. Linke, and Y. B. Zaied, "Ecological footprint of your denim jeans: production knowledge and green consumerism," *Sustain Sci*, vol. 17, no. 5, pp. 1781–1798, Sep. 2022. https://doi.org/10.1007/s11625-022-01131-0.
8. "Topic: Levi Strauss & Co.," *Statista.* https://www.statista.com/topics/9039/levi-strauss-and-co/. (Accessed Mar. 31, 2023).
9. L. Strauss, "Levi Strauss & Co. 2021 Sustainability Report," 2021.
10. "VF Corporation (VFC), FY2022 MADE FOR CHANGE REPORT, 2022."
11. "Kontoor Brands, Inc. (KTB). 2020 Sustainability Report, 2020."
12. E. Csanák, "DEVELOPING A SUSTAINABLE DENIM COLLECTION: VIEWS, ASPIRATIONS & METHODS OF ETHICAL DENIM," p. 11, 2021.
13. "Kontoor Brands Expands Indigood™ Program to Include Additional Water Saving Technologies," *Kontoor Brands, Inc.*, Mar. 22, 2021. https://www.kontoorbrands.com/investors/news-events/press-releases/detail/76/kontoor-brands-expands-indigood-program-to-include. (Accessed Mar. 31, 2023).
14. Y. Li, Y. Luo, and Q. He, "Chemical footprint of textile and apparel products: an assessment of human and ecological toxicities based on USEtox model," *The Journal of The Textile Institute*, vol. 111, no. 7, pp. 960–971, Jul. 2020. https://doi.org/10.1080/00405000.2019.1710907.

15. T. Hackett, "A Comparative Life Cycle Assessment of Denim Jeans and a Cotton T-Shirt: The Production of Fast Fashion Essential Items From Cradle to Gate".
16. "Safer Chemicals," *Levi Strauss & Co.* https://www.levistrauss.com/sustainability-report/consumption/safer-chemicals/. (Accessed Mar. 31, 2023).
17. "Institute of Public and Environmental Affairs (IPE)." http://wwwen.ipe.org.cn/GreenSupplyChain/CITIDetailV2.aspx?id=27. (Accessed Apr. 02, 2023).
18. "Institute of Public and Environmental Affairs (IPE)." http://wwwen.ipe.org.cn/GreenSupplyChain/CITIDetailV2.aspx?id=581. (Accessed Apr. 02, 2023).
19. "Institute of Public and Environmental Affairs (IPE)." http://wwwen.ipe.org.cn/GreenSupplyChain/CITI.html. (Accessed Apr. 02, 2023).

# Investigation and Optimization of a New Bleaching Process for Sulfur-Dyed Denim Fabric

**Sarra Said, Imed Feki, Mohamed Hamdaoui, Sabri Halaoua, and Walid Sahraoui**

**Abstract** Bleach method is among the most popular washing techniques in the textile industry, especially for washing jeans, to achieve the desired color shade. This process is mostly carried out using hypochlorite and other bleaching agents, which are strongly associated with harmful effects on the environment and human health. Therefore, more environmentally sustainable processes should be created and developed. In this paper, we present an experimental study on the effect of a new bleaching process on the color of a sulfur-dyed black denim fabric. The new process is eco-friendly, because it does not use chlorine. Color degradation (D) was studied to evaluate the color changes. In the first part, a preliminary study was carried out to visualize the field of study of the bleaching treatment parameters (concentration of hydrogen peroxide, temperature, and duration). In the second part, the influence of the chosen parameters on the color degradation was studied using the MINITAB software. Then, a mathematical model was implemented to study the effect of different bleach treatment parameters on color changes, which is required by industrialists in the textile sector. Finally, the best combination of bleaching parameters to obtain an optimal value of color degradation (D) was determined.

**Keywords** New bleach process · Eco-friendly treatment · Color degradation · Full factorial design · optimization diagram

S. Said (✉) · I. Feki · M. Hamdaoui · S. Halaoua
Textile Materials and Processes Research Unit (MPTEX), University of Monastir, National Engineering School of Monastir, Monastir, Tunisia
e-mail: Sarrasaid@yahoo.fr

W. Sahraoui
TANIT Textile Services Company, Monastir, Tunisia



S. B. Abdessalem et al. (eds.), *Proceedings of the Second International Conference of Innovative Textiles and Developed Materials-ITDM'2; 05-06 May 2023; Tunisia*,
https://doi.org/10.1007/978-981-99-7950-9_4

# 1 Introduction

The protection of our environment implies a control of pollutants generated by anthropic activity and, in particular, by industry. Today, due to global awareness, industrial emissions are carefully controlled and subject to increasingly rigorous regulations. Industrial companies are required to select appropriate manufacturing processes in order to minimize toxic waste that threatens the environment. This management aims to ensure the preservation of aquatic ecosystems and the protection of the quality and quantity of surface and underground resources.

Finishing processes are used in the textile industry to offer fabrics and apparel desirable features and an attractive appearance, such as drape, a soft handle, and fashionable aspects. These finishing processes are achieved with chemical or mechanical treatments [1, 2], such as enzymatic wash [3–5], bleach wash [6, 7], acid wash [8], silicone wash [9]. These treatments cause worn appearance and aged look for the garment, thus provoking degradation of the color level.

Chemicals like formic acid, sodium nitrate, sodium chlorite are examples of bleaching agents that are frequently used in the traditional bleaching process. These chemicals undoubtedly have a multitude of disadvantages, such as the release of hazardous emissions, high time and energy requirements, decreased fabric strength, and environmental contamination [10].

Ecological finishing treatments are being frequently applied to treat textile surfaces in order to have better appearance. Hydrogen peroxide is effective and suitable for bleaching some dyed cellulosic fabrics. High levels of whiteness that are stable and don't tend to turn yellow during storage are delivered by hydrogen peroxide. Hydrogen peroxide has also the benefits of being inexpensive, adaptable in application. It provides the degradation and elimination of the dye giving a fashionable effect [11, 12].

Color is a crucial factor that influences how consumers behave. It has an impact on how they evaluate clothing and how they feel about making purchases. The type of finishing treatments used, as well as the interactions between the finishing agents and dyes used, have a significant impact on the final colorimetric values of dyed and finished cotton fabrics [13]. Therefore, the optimization of the parameters influencing the final colorimetric value is required.

The particularity of our work is that we manage a dye known for its high color fastness to any type of test. Changing the shade of a sulfur-dyed fabric is not easy due to the covalent chemical bonds [14, 15].

The aim of this paper is to investigate the change of the color shade of a black sulfur-dyed denim fabric after our new bleaching treatment. Based on experimental results and mathematical modeling, the best combination of bleaching parameters leading to an optimal value of color degradation has been determined.

Table 1 Denim fabric specifications

| Parameter | Description |
|---|---|
| Composition | Pure cotton |
| Weave structure | Z-Twill of 3 warp effect |
| Surface density (g/m$^2$) | 366 |
| Warp density (yarn/cm) | 27 |
| Weft density (yarn/cm) | 23 |

# 2 Materials and Methods

## 2.1 *Materials*

Black sulfur-dyed denim fabric was used in this study. Its basic characteristics are shown in Table 1.

## 2.2 *Methods*

First, the black sulfur-dyed denim was desized at 45 °C for 15 min to remove the adhesives already present in the warp during weaving. The samples were then treated with a two-step eco-friendly bleach wash using hydrogen peroxide:

- A first step where the desized fabric is put in contact with a bath containing a mixture of low concentration of hydrogen peroxide and caustic soda so that the pH of this mixture must have a value between 10 and 12. The bath temperature and contact time are well controlled.
- A second step where the contact time and handling temperature are taken as equal to those of the first step. The fabric is put in contact, a second time, with a new bath containing the same concentrations of hydrogen peroxide and caustic soda as the first bath. A bleach activator is added with a portion of 0.5 g/L. As before, the pH is taken between 10 and 12.

The tests were carried out after pre-conditioning of cotton fabrics in a standard test atmosphere (20 ± 2 °C and 65 ± 4% RH) according to the standard ISO 139:2005.

The SpectroFlash SF600 spectrophotometer with dataMatch 2.0 software (Datacolor International, USA) was used to measure the color intensity of the treated fabrics, expressed as color strength (K/S). The K/S values were measured using the Kubelka–Munk equation:

$$\frac{K}{S} = \frac{(1 - R)^2}{2R} \tag{1}$$

where, $K$ is the absorption coefficient, $S$ is the scattering coefficient, and $R$ is the reflectance.

The degradation $D$ refers to the value of the color strength of the fabric before and after the treatment. This parameter is mathematically defined by the following relationship

$$D = \frac{\left(\frac{K}{S}\right)_{\text{Before}} - \left(\frac{K}{S}\right)_{\text{After}}}{\left(\frac{K}{S}\right)_{\text{Before}}} \tag{2}$$

Color degradation of treated and untreated black sulfur-dyed cotton fabrics was used to analyze and interpret the color data.

## 3 Results and Discussion

### 3.1 *Study of the Experimental Parameters Effect on the Bleach Treatment*

At first, a preliminary study was carried out by selecting the most influential parameters in order to optimize the bleach washing process by the experimental design method (Full Factorial Design). Throughout this study, we fixed the bath ratio at 1/10 and the activator concentration at 0.5 g/L (industrial requirement). Concerning the pH of washing, it is an alkaline pH that varies from 10 to 12, since the bleach treatment with hydrogen peroxide requires an alkaline bath to activate the oxidizing effect of hydrogen peroxide [16, 17].

We chose to study the effect of some experimental parameters on the bleach washing treatment. These parameters are as follows:

- Hydrogen peroxide concentration;
- Washing temperature;
- Washing duration.

In order to better assess the color changes resulting from the bleach wash treatment, the color quality of the treated denim fabrics was reported using the color degradation $D$ value. Each value shown in the graphs, representing the effect of the experimental parameters on color degradation ($D$), is an arithmetic average of three test results.

**Study of the Hydrogen Peroxide Concentration Effect on Color Degradation (D)**. Determining the optimal concentration of hydrogen peroxide for sulfur dye degradation is very practical due to the fact that hydrogen peroxide is easily available and we can buy it at a reasonable price [18].

In this section, we investigated the effect of hydrogen peroxide concentration on color degradation. First, we varied this parameter between 1 and 5 g/L (same variation for each bath, i.e. from 2 to 10 g/L in total) to visualize the effect of the latter on color

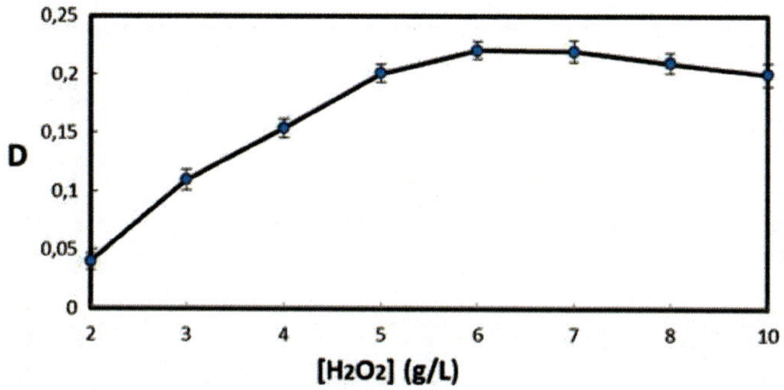


**Fig. 1** Evolution of the color degradation as a function of the hydrogen peroxide concentration (temperature: 80 °C and duration 10 min)

degradation while keeping the temperature at 80 °C and the total washing duration at 10 min (5 min for each bath). We kept the same conditions for both baths. Figure 1 shows the effect of varying the hydrogen peroxide concentration.

We can see that the color degradation increases with the concentration of hydrogen peroxide up to the concentration of 6 g/L (3 g/L in each bath). The reason for this tendency of increasing discoloration is due to the oxidizing ability that was developed and improved from the increasing production of hydroxyl radicals obtained from the increasing concentration of $H_2O_2$ [18]. Then, we obtained an optimal value (equal to 6 g/L) from which the efficiency of degradation decreases due to the inhibitory nature of hydrogen peroxide toward the OH° radical when it occurs at a higher concentration leading to a decrease in the oxidation effect and the recombination of hydroxyl radicals according to the reactions below [18–21].

$$HO^{\cdot} + H_2O_2 \rightarrow H_2O + HO_2^{\cdot}$$

$$HO^{\cdot} + HO^{\cdot} \rightarrow H_2O_2$$

Considering the results obtained by this study, we chose to work with the concentrations of 3 and 6 g/L in the following part (optimization of this bleach process). Indeed,

- Under 3 g/L of hydrogen peroxide, the value of *D* is very low (The industry requires a degradation value higher than 10% i.e. a D value above 0.1).
- Over 6 g/L, there is a decrease in *D*.

**Study of the Temperature Effect on Color Degradation (D).** In addition to the hydrogen peroxide concentration, temperature is a very important factor for the bleach treatment. For this purpose, we studied the effect of the variation of this parameter on the bleach treatment at different temperatures (60–98 °C) while keeping the other washing parameters constant: the concentration of hydrogen peroxide fixed

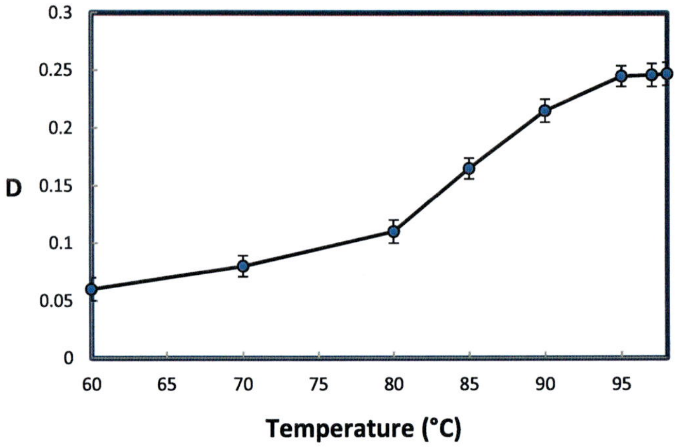


**Fig. 2** Evolution of color degradation as a function of temperature ([$H_2O_2$]:3 g/l and duration: 10 min)

at 3 g/L and the total washing duration at 10 min (5 min for each bath). The effect of temperature variation on color degradation (D) is reported in Fig. 2.

From Fig. 2, we can see that the color degradation (D) increases with the washing temperature up to the temperature of 95 °C. Many research papers indicate that an increase in operating temperature could be beneficial for the oxidation rate [22], which confirms the result found in our study (Fig. 2). It can be seen that from 95 °C, the value of color degradation is almost constant, which means that for a fixed concentration of hydrogen peroxide and for a fixed duration of the bleach treatment, all the hydrogen peroxide has been converted into hydroxyl radicals [23], so that even if the temperature is increased, there would be no color degradation.

Considering the results obtained by this study, we have chosen to work with temperatures 80 and 95 °C in the following part (optimization of this bleach process). Indeed,

- Below 80 °C, the value of *D* is very low (lower than 0.1);
- Over 95 °C, the value of *D* remains stable.

**Study of the Duration Effect on the Color Degradation (D).** The duration of the treatment varied from 5 (2.5 min for each phase) to 30 min (15 min for each phase), while keeping the washing temperature (80 °C) and hydrogen peroxide concentration (3 g/L) constant.

Figure 3 shows that there is an increase in the efficiency of the bleach treatment proportionally with the duration of the treatment up to a duration of 20 min, then there is a constant behavior of the color degradation (D) going from 20 to 30 min, because all the hydrogen peroxide has been converted into hydroxyl radical and their concentration remains constant during the reaction [23].

Considering the results obtained by this study, we chose to work with 10 and 20 min in the following part (optimization of this bleach process). Indeed,

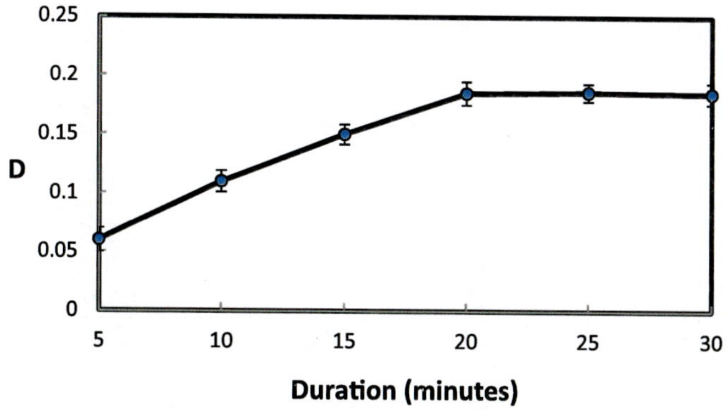


**Fig. 3** Evolution of color degradation as a function of duration (temperature: 80 °C and [$H_2O_2$]: 3 g/l)

- Under 10 min, the value of D is lower than 0.1.
- Over 20 min, the value of D remains stable.

## 3.2 *Optimization of the Bleach Treatment Recipe*

**Realization of the Full Factorial Design.** The present work uses a rigorous statistical methodology, namely experimental design. It allows the results obtained to be analyzed and interpreted in a simple and scientific way. According to the study of the effects already established in the previous section, we have fixed the number of factors (3 factors namely: Hydrogen peroxide concentration, temperature, and treatment duration) and their levels to be tested (two levels for each parameter) as shown in Table 2.

In order to optimize the new bleach recipe, we used a full factorial design using the MINITAB 17 software. The number of experiments required is 8 ($2^3$) experiments as indicated in the Table 3.

From the results summarized in Table 3, we can see that the color degradation of the washed fabric is more or less important depending on the conditions used. They vary from 11 to 38% for more or less severe conditions in terms of bleach

**Table 2** Level and code of variables for experimental design

| Treatment | Variables (inputs) | Symbols | Coded levels | |
|---|---|---|---|---|
| | | | − 1 | 1 |
| Bleach treatment | Concentration of $H_2O_2$ (g/L) | [C] | 3 | 6 |
| | Temperature (°C) | T | 80 | 95 |
| | Duration (min) | D | 10 | 20 |

**Table 3** The matrix of the experimental design with the values of the responses

| Trial number | Variable | | | Response |
|---|---|---|---|---|
| | $[H_2O_2]$ | Duration | Temperature | Degradation (D) |
| Brute | – | – | – | 0 |
| 1 | 3 | 10 | 80 | 0.110 |
| 2 | 6 | 10 | 80 | 0.221 |
| 3 | 3 | 20 | 80 | 0.185 |
| 4 | 6 | 20 | 80 | 0.239 |
| 5 | 3 | 10 | 95 | 0.245 |
| 6 | 6 | 10 | 95 | 0.322 |
| 7 | 3 | 20 | 95 | 0.295 |
| 8 | 6 | 20 | 95 | 0.381 |

concentration, duration, and temperature of the treatment. These different values indicate the presence of sulfur dye removal.

**Diagram of the Main Effects on Color Degradation (D).** Factors influencing color degradation after bleach wash treatment were evaluated using factorial plots: main effects and interactions.

Using MINITAB software, we plot the effects of the different factors selected (hydrogen peroxide concentration, temperature, and treatment time) on color degradation after wash treatment (Fig. 4).

We can conclude that the concentration of hydrogen peroxide, the temperature, and the duration of the treatment have a positive effect on the response (D), as shown in the main effects diagram in Fig. 4. In fact, as shown by the positive slopes of

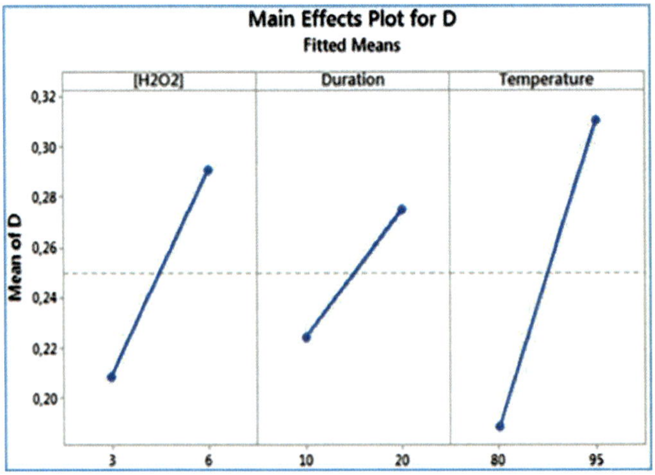


**Fig. 4** Color degradation (D) main effects diagram

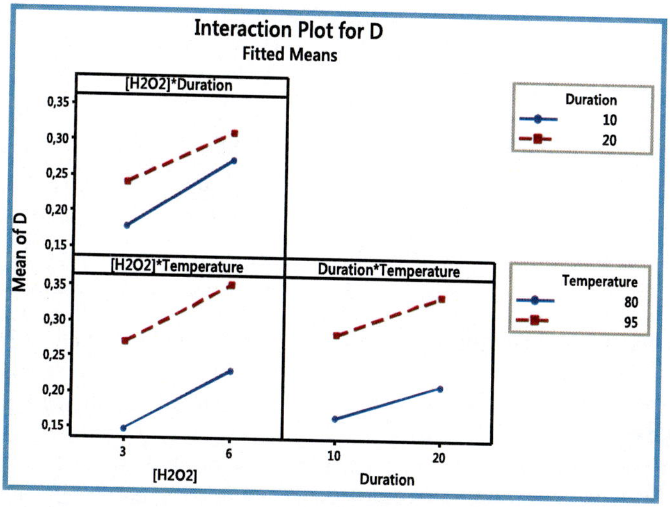


**Fig. 5** Interaction diagram for color degradation (D)

the curves related to these three parameters, increasing these three selected factors improves the bleaching result (better degradation). We can also see that the slopes of their effect curves are important, and that both the temperature and the concentration of hydrogen peroxide have a great influence on the color degradation.

**Interaction Diagram for Color Degradation (D)**. The aim of this study is to investigate the degree of interaction between the various parameters studied and how they affect the response, namely color degradation following bleaching. The interaction diagram of the factors influencing the response (D) is shown in Fig. 5.

From Fig. 5 of the interaction diagram for color degradation (D), we can see that there are no interactions between the different parameters studied. This is because the transition of the different factors from one level to another is represented by almost parallel lines. Therefore, in the following part, we can neglect the effect of interactions.

**Establishment of a Mathematical Model**. Modeling the experimental design allowed us to determine the relationship between the input variables and the chosen response. The equation which represents the model obtained using the MINITAB software (excluding the effect of interactions) is given below, with encoded variables:

$$\mathbf{D} = 0.24975 + 0.04100[\mathbf{H_2O_2}] + 0.02525\mathbf{Duration} + 0.06100\,\mathbf{Temperature} \tag{3}$$

The equation in uncoded variables is:

$$\mathbf{D} = -0.661 + 0.027[\mathbf{H_2O_2}] + 0.005\mathbf{Duration} + 0.008\mathbf{Temperature} \tag{4}$$

Table 4 Color degradation model variance analysis (D)

| Source | F | P |
|---|---|---|
| Model | 74.48 | 0.001 |
| Linear | 74.48 | 0.001 |
| $[H_2O_2]$ | 62.19 | 0.001 |
| Duration | 23.59 | 0.008 |
| Temperature | 137.66 | 0.000 |
| $R^2 = 98.24\%$ | | |

The coefficient of determination $R^2$ indicates the degree of predicting the model, a value of 100% implies a perfect prediction, while a value of 0% means that the model fails to predict. Based on the results, the R2 is 98.24%, indicating that the proposed model is highly predictive.

All the parameters studied have a positive effect on the response, as shown by the coded variable equation. Indeed, temperature has the greatest influence on the bleaching wash, followed by hydrogen peroxide concentration and duration.

**Analysis of Variance**. The sources of variation that are statistically significant are shown in the ANOVA table (Table 4).

Indeed, the *P* value is an indication of the importance of each parameter in the regression equation.

- If $P > 0.05$, the target parameter is not significant in the equation.
- If $0.01 < P < 0.05$, the target parameter is significant in the equation.
- If $P < 0.05$, the target parameter is highly significant in the equation.

It can be seen that $P < 0.05$ for all the terms studied. This allows us to conclude that the contributions of "temperature", "hydrogen peroxide concentration", and "duration" have high significance.

F is the Fisher ratio: It is the mean square of the factor divided by the mean square of the error. The value of F reflects the importance of the factors studied; a large value of F means that this factor has the greatest influence on the response under investigation. In fact, in our case, the temperature factor is the term with the greatest influence on the degradation of the color, followed by the $[H_2O_2]$ factor and then the duration.

**Study of the Contour Plot**. The contour plot represents a two-dimensional view of the surface, demonstrating the different points that have the same response. These points are connected to form contour lines of constant response. Contour plots are therefore used to show desirable response values and operating conditions. Studying contour plots is a simple method of optimizing the color degradation value (D). In our case, the contour plot consists of three curves, each representing an infinite number of combinations between two variables when the third variable is held constant (Fig. 6).

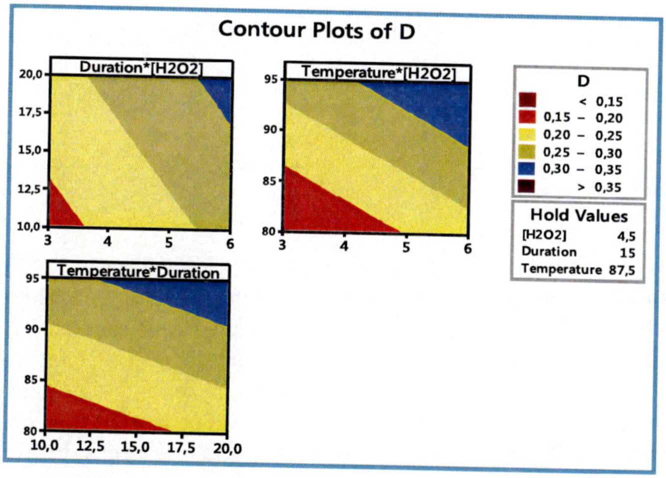


**Fig. 6** Contour plot of color degradation (D)

For each variable studied, these constant levels are the central levels. Figure 6 shows that:

- Color degradation (D) increases as the values of the parameters studied increase.
- Degradation (D) reaches its maximum when the value of the three factors studied is high.

**The Optimization Diagram.** This section is aimed at determining a combination of bleaching parameters which results in optimum degradation (D). Figure 7 shows the results of the optimization.

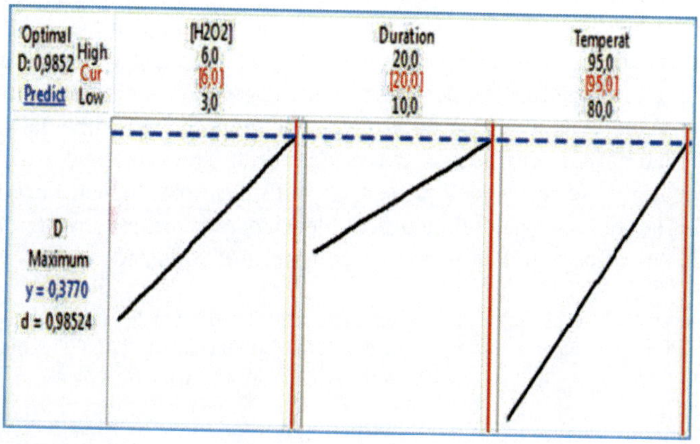


**Fig. 7** Optimization curves of the color degradation (D)

Table 5 Experimental and theoretical values of model validation

| | Theoretical value of D | Experimental value of D |
|---|---|---|
| Temperature = 80 (°C)<br>$[H_2O_2] = 4$ g/L<br>Duration = 16 min | 0.167 | $0.174 \pm 0.011$[a] |

[a] Mean deviation value

MINITAB software is a tool that allows us to use mathematical modeling to distinguish the combination of input factors that optimizes the output. The command used is "Response Optimization", which provides an optimal solution (in our case, the solution is D) for the combinations of input variables ($[H_2O_2]$, temperature, and duration), as well as the optimization plot. The full factorial design method yields the following combination to identify the bleach wash parameters that optimize the selected response: $[H_2O_2]$ is equal to 6 (g/L), temperature is equal to 95 (°C), and duration is equal to 20 (minutes), as shown in Fig. 7.

The expected response for the optimum combination is D equal to 0.377. Experimental validation of the optimal conditions was carried out (Table 5).

Indeed, the experimental result obtained is very close to the theoretical optimum (Table 5), hence the validation of our model.

## 4 Conclusion

In this work, we investigated the influence of a new bleaching process for sulfur-dyed denim fabric. Our treatment can attract the attention of industrialists, thanks to its green effects on the environment as well as the new range of shades that this treatment gives. The effect of temperature, hydrogen peroxide concentration, and duration of the bleach treatment was studied. The color change of the treated fabric is evaluated according to the color degradation values. The experimental data were analyzed using the experimental design method. Regression equations for the color degradation of the treated fabrics were developed from the experimental data and solved using MINITAB statistical software. The results showed that the predictions of the linear model were in good agreement with the experimental observations. Based on this study, the best combination of bleach parameters required to obtain an optimal value of color degradation was determined and evaluated.

**Acknowledgements** This project is carried out under the MOBIDOC scheme, funded by the EU through the EMORI program, and managed by the ANPR. My thanks also to TANIT Textile Services Company for providing us chemicals and for welcoming us to realize the treatments.

## References


1. Dalbaşi, E.S. and G. Özçelik Kayseri, *A research on the comfort properties of linen fabrics subjected to various finishing treatments*. Journal of Natural Fibers, 2019: p. 1–14.
2. Feki, I., et al., *Features selection of stonewashed denim to a sensory descriptor.* Fibers and Polymers, 2015. **16**(9): p. 2066-2076.
3. Mondal, M.I.H. and M.M.R. Khan, *Characterization and process optimization of indigo dyed cotton denim garments by enzymatic wash.* Fashion and Textiles, 2014. **1**(1): p. 1-12.
4. Khan, M.M.R., M.I.H. Mondal, and M.Z. Uddin, *Sustainable washing for denim garments by enzymatic treatment.* Journal of Chemical Engineering, 2012. **27**: p. 27-31.
5. Sarkar, J. and M. Elias Khalil, *Effect of enzyme washing combined with pumice stone on the physical, mechanical and color properties of denim garments.* 2014.
6. Khan, M.M.R., M.I.H. Mondal, and M.Z. Uddin. *Effect of bleach wash on the physical and mechanical properties of denim Garments.* in *International Conference on Mechanical Engineering.* 2011.
7. Sarkar, J. and E. Khalil, *Effect of industrial bleach wash and softening on the physical, mechanical and color properties of denim garments.* IOSR Journal of Polymer and Textile Engineering, 2014. **1**(3): p. 46-49.
8. Haq1, U.N. and M.M.R. Khan, *Technology of acid wash on woven denim apparel with damp pumice stone.* 2014.
9. Mazumder, S., *Effects of sand blasting with industrial enzyme silicon wash on denim apparel characteristics.* Daffodil International University Journal of Science and Technology, 2010. **5**(1): p. 6-9.
10. Farooq, A., et al., *Comparative performance evaluation of conventional bleaching and enzymatic bleaching with glucose oxidase on knitted cotton fabric.* Journal of Cleaner Production, 2013. **42**: p. 167-171.
11. Varadarajan, G. and P. Venkatachalam, *Sustainable textile dyeing processes.* Environmental chemistry letters, 2016. **14**(1): p. 113-122.
12. Said, S., et al. *The Effect of Bleach Washing Treatment on the Comfort Properties of Dyed Cotton Garments.* in *Advances in Applied Research on Textile and Materials-IX: Proceedings of the 9th International Conference of Applied Research on Textile and Materials (CIRATM).* 2022. Springer.
13. Yavas, A., O.O. Avinc, and E. Kalayci, *THE EFFECTS OF DIFFERENT FINISHING APPLICATIONS ON THE COLORIMETRIC AND COLOR FASTNESS PROPERTIES OF REACTIVE DYED COTTON FABRICS.* TEXTILE SCIENCE AND ECONOMY: p. 193.
14. Aspland, J., *Sulfur Dyes and Their Application.* Textile Chemist & Colorist, 1992. **24**(3).
15. Aspland, J., *Part 2: practical application of sulfur dyes.* Text Chem Color, 1992. **24**(4): p. 27-31.
16. Du, W., et al., *Comparative study on the effects of laser bleaching and conventional bleaching on the physical properties of indigo kapok/cotton denim fabrics.* Applied Sciences, 2019. **9**(21): p. 4662.
17. Oliveira, B.P.d., L.T. Moriyama, and V.S. Bagnato, *Colorimetric Analysis of Cotton Textile Bleaching through H 2 O 2 Activated by UV Light.* Journal of the Brazilian Chemical Society, 2018. **29**: p. 1360–1365.
18. Bhatti, H.N., M. Iqbal, and A. Nazir, *Decolorization of basic turquise blue X-GB and basic blue X-GRRL by the fenton's process and its kinetics.* Zeitschrift für Physikalische Chemie, 2019. **233**(3): p. 361-373.
19. Dhahir, S.A., K.A. Al-Saade, and I.S. Al-Jobouri, *Degradation studies of rhodamine B in the presence of UV/H 2 O 2/Fe 2.* Int. J. Tech. Res. Appl, 2014. **2**: p. 123-127.
20. Lucas, M.S. and J.A. Peres, *Decolorization of the azo dye Reactive Black 5 by Fenton and photo-Fenton oxidation.* Dyes and Pigments, 2006. **71**(3): p. 236-244.
21. Lucas, M.S. and J.A. Peres, *Degradation of Reactive Black 5 by Fenton/UV-C and ferrioxalate/ H2O2/solar light processes.* Dyes and pigments, 2007. **74**(3): p. 622-629.

22. Javaid, R. and U.Y. Qazi, *Catalytic oxidation process for the degradation of synthetic dyes: an overview.* International journal of environmental research and public health, 2019. **16**(11): p. 2066.
23. Buthiyappan, A., A.R.A. Aziz, and W.M.A.W. Daud, *Recent advances and prospects of catalytic advanced oxidation process in treating textile effluents.* Reviews in Chemical Engineering, 2016. **32**(1): p. 1-47.

# Investigation of an Eco-Friendly Acid Dyeing Process of Cotton Fabric

**Adel Elamri, Imed Feki, Mohamed Amine Touati, Mohamed Hamdaoui, and Omar Harzallah**

**Abstract** This research work is a new investigation to develop a chemical free and eco-friendly textile dyeing process to reduce environmental negative impact of textile industry. To improve cotton fabric affinity to acid dye, the former was pre-cationized with a quaternary amine salt. Preliminary tests showed that dyeing time, temperature and unison agent concentration were the main parameters influencing the acid dyeing process of pre-cationized cotton fabric. The optimal dyeing conditions were determined using the Response Surface Method (RSM). Factors that influence Color strength (K/S) and dyebath exhaustion were evaluated using factorial plots. Then, ANOVA system was used for variance experimental analysis and the response surface methodology was performed with statistical software tool Minitab 17. The fitting extend of the proposed mathematical model was evaluated according to the obtained coefficient $R^2$, the adjusted $R^2$ and the $p$-values. To assess the properties of free salts acid dyed cotton fabrics, morphological (MEB), mechanical and thermal (KAWABATA KES) characterization tests were carried out. Dyed samples showed an improved surface morphology compared to greige cotton one. Also, free chemicals dyed cotton samples had higher thermo-physiological comfort properties.

**Keywords** Cationized · Cotton · Ecofriendly · Dyeing

A. Elamri (✉) · I. Feki · M. Hamdaoui
MPTex UR17ES33, ENIM, Monastir University, 5000 Monastir, Tunisia
e-mail: amri.adel2201@gmail.com

M. A. Touati
Textile Department, ENIM, Monastir University, 5000 Monastir, Tunisia

O. Harzallah
LPMT UR 4365, ENSISA, UHA, 68093 Mulhouse Cedex, France



S. B. Abdessalem et al. (eds.), *Proceedings of the Second International Conference of Innovative Textiles and Developed Materials-ITDM'2; 05-06 May 2023; Tunisia*,
https://doi.org/10.1007/978-981-99-7950-9_5

# 1 Introduction

The textile dyeing and finishing industries induce water pollution and environmental threat around the globe [1]. The textile wastewater loaded with hazardous chemicals, such as dyes, acids, pigments, etc. is considered as a major pollution, serious danger to organisms and environmental ecosystem [2]. Recently, the increased awareness of eco-safety practices has push textile industrials to develop eco-friendly production techniques, instead of the implementation of wastewater treatment facilities. In fact, the effluent treatment technologies are not cost-effective and their efficiency is, in most cases, moderate [3].

Cotton fibers are widely used in textile industries due to their high properties, such as high breathability, good softness and moisture absorption. Cellulose fabrics present excellent characteristics of comfort and environment safety. Dyeing cotton fabrics relies mainly to the high affinity of cellulose to several classes of dyes [4]. The conventional main dyes for cotton fibers are water soluble anionic dyes (reactive and direct). The classic dyeing systems for cotton fabrics with these dyes need the addition of large quantities of chemical auxiliary agents, in particular electrolytes, to avoid the negative charges of the fiber surface and increase dye adsorption [5]. These incorporated chemicals are primordial to improve the dye exhaustion and fixation rates and unison efficiency. The amount of added chemical agents can, in some dyeing cases, reach 100 g/L [6]. At the end of the dyeing processes, the obtained wastewater effluent containing important quantities of salt and dye molecules is spread in nature without any treatments, leading to serious ecological threats. Also, the dyeing procedures are generally time and energy consuming to allow the dye molecules to diffuse into the fiber.

To overcome these drawbacks, eco-friendly cotton dyeing strategies were investigated [7]. Dyeing cotton with acid dyes is one of the promising processes [8]. Nevertheless, interactions between cotton fibers and acid dye are generally weak [9]. An efficient solution to improve acid dye-cotton bonds and decrease electrostatic repulsions between cellulose and dyes is the chemical modification of cotton fabrics with cationic groups [10–12].

Cotton cationization consists of creating cationic sites by reaction with hydroxyl groups of cellulose via esterification, etherification, grafting or crosslinking reaction [13]. Quaternary or tertiary $NH_2$ groups are incorporated into cotton to create nucleophilic group to cotton fabric which shows greater affinity to anionic dyes resulting in formation of strong chemical bond without salts. Among several cationization techniques, the usage of polymers with cationic or amino groups is the most efficient due to high reactivity with cotton fibers and homogeneous dyeing, which are very promising characteristics in industrial field [14].

The main scope of this study is to develop an eco-friendly free auxiliaries dyeing process of cotton fabrics with acid dye. This non-polluting dyeing of pre-cationized cellulose presents a promising revolutionary procedure of safe process with reduced emissions into the environment.

# 2 Materials and Methods

## 2.1 *Fabric*

100% cotton pique knit fabric (160 g $m^{-2}$ mass per area) was used in all experiments. A preliminary study was carried out to determine the optimal preparation process to adopt on our cotton fabric prior to cationization and dyeing. The obtained optimal pretreatment procedure consisted of scouring, caustification and bleaching, respectively.

Scouring treatment was carried out at 100 °C in a sodium carbonate alkali bath. Then, cotton fabric was kept under tension in a caustic soda solution (150 g/L) for 15 min at room temperature to allow its caustification. After that, scoured and caustified fabric was bleached for 30 min in hydrogen peroxide (35%) at 90 °C in presence of sodium carbonate (to obtain alkali medium with pH 10, 5 to 12). In all preparation operations, deionized (DI) water was used.

## 2.2 *Cotton Cationization and Dyeing*

The pretreated cotton fabric was cationized using synthetic resin at 60 °C for 30 min in alkali medium (pH = 11 is maintained by using caustic soda). The cationization resin is a commercially available epoxy compound derivative that contains cationic reactive groups. The OH group of cotton fabric is related to the epoxy group in an alkali bath by strong covalent bond [15]. Dyeing of pre-cationized cotton was carried out at pH range 5–6. By introduction of cationic groups into cotton fibers, the affinity of acid dye for cotton can be significantly improved. The electrostatic attractions (Fig. 1) between cationized cotton and acid dye result in increased dye uptake.

## 2.3 *Color Strength and Dye Exhaustion Measurement*

The color strength of dyed fabrics was evaluated by a reflectance spectrophotometer (Color Eye 7000). The absorbance of acid dye was determined in the visible range of the spectra from 400–700 nm. The reflectance (*R*) of dyed fabric in the maximal wavelength of absorption (λmax) was depicted to calculate the color strength (K/S) of dyed fabric by the Kubelka–Munk [16] relation:

$$\frac{K}{S} = \frac{(1-R)^2}{2R} \tag{1}$$

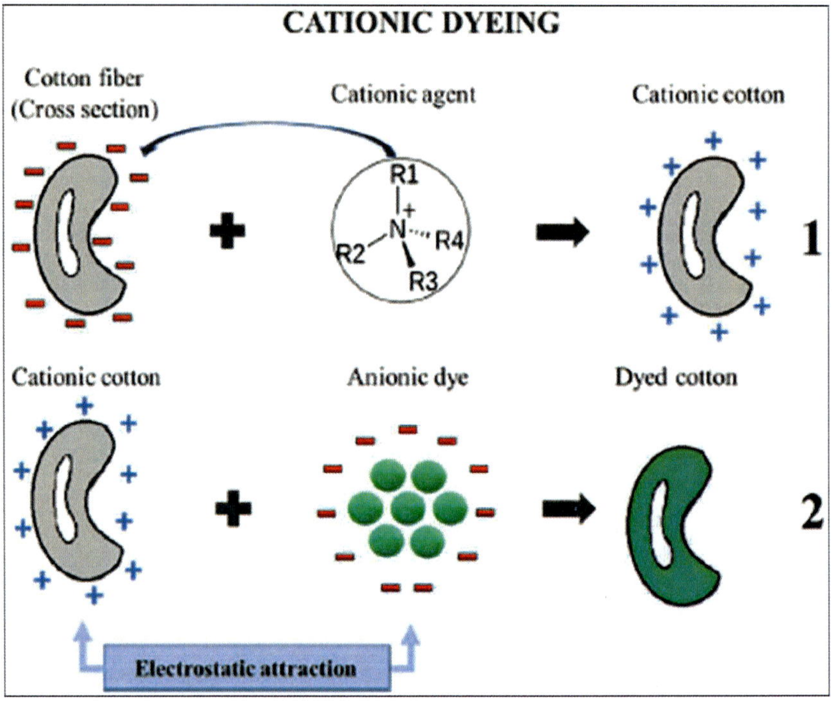


**Fig. 1** Mechanism of cationized cotton dyeing

where $K$ is the absorption coefficient and $S$ is the scattering coefficient of the substrate. For each sample, the measurements were made on both sides of the fabric and the average value was reported.

Dye exhaustion rate noted "$E$" was also used to evaluate dyeing performance. The exhaustion rate is defined as the proportion of dye absorbed by the fiber in the relation to that remaining in the dye bath at the end of dyeing test [16]. It is calculated according to the following relation:

$$\% E = \frac{C_0 - C_s}{C_0} \times 100 \quad (2)$$

where $C_0$ and $C_s$ are the initial and final concentrations of dye in the bath, respectively.

## 2.4 Dye Optimization and Prediction Models

The first objective in this investigation is to depict the optimal conditions (time, temperature and sodium sulfate concentration) of pre-cationized cotton acid dyeing by using response surface methodology (RSM). The response surface methodology data are obtained from experimental dyeing color values. It implies the measure of exhaustion rate ($E$) and color strength (K/S) values previously defined. Dyeing tests were carried out by varying lengths of time duration called "Dur" (15, 30 and 45 min), temperature named "Temp" (60, 75 and 95 °C) and sodium sulfate concentration called "Sulfate" (0, 1.5 and 3%).

Minitab 17 software with Taguchi method was used for variance analysis (ANOVA) and corresponding response surfaces. Linear regression analysis of experimental values and the design of graph contours were carried out on the same tool (Minitab 17). The extend of fitting of the obtained mathematical models was estimated via $R^2$ and adjusted $R^2$ coefficients, and $p$-values.

## 2.5 Scanning Electron Microscopic Analysis

Scanning Electron Microscopy (SEM) provides quickly high-resolution and high-depth-of-field images of the sample surface and near-surface useful for evaluating various material surfaces. This technique was used to assess surface morphological properties of crude, cationized and dyed cotton fabrics. Jeol JSM 6360LV SEM apparatus was used to obtain images at two different magnifications (×40 for fabric surface level; and ×2k for fiber surface level). The SEM samples were sputtered with gold before the observation.

## 2.6 Compression and Thermal Properties Analysis

Kawabata Evaluation System (KES) was used to evaluate the sensorial comfort properties of fabrics. Two properties of crude, cationized and dyed cotton fabrics were evaluated: compressibility (EMC) and constant thermal conductivity (K). The compression tests of cotton specimens were evaluated by Automatic Compression Tester, FB3 instrument of KES (Fig. 2.a). The compression tests were performed by exerting maximum force (50gf/cm2) on the samples using a pressure plate which acts as a probe. The compressibility EMC was calculated from the equation below:

$$\text{EMC}\ (\%) = \frac{h_0 - h_\text{m}}{h_0} \times 100 \tag{3}$$

where $h_0$ represents the thickness of the cotton sample relative to the initial pressure 0.5 gf/cm$^2$ and $h_\text{m}$ is the thickness of the sample at the maximal pressure 50 gf/cm$^2$.

Thermal analysis of cotton fabrics was carried out by Automatic Compression Tester, FB7 instrument of KES (Fig. 2.b). The constant thermal conductivity (K) has been measured as the heat transmitted from a heat plate with a constant temperature (30 °C) through a sample to a heat plate with a separate constant temperature (20 °C)

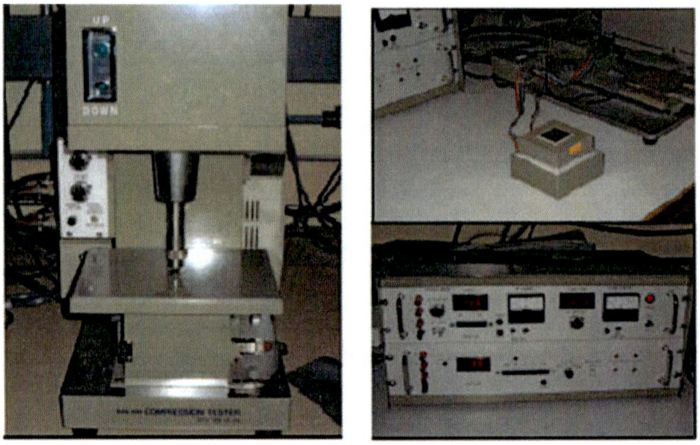

(a) (b)

**Fig. 2** **a** KES-FB3 and **b** KES-FB7 instruments of Kawabata evaluation system

by the following relation:

$$K = \frac{W \times h}{S \times \Delta T} \tag{4}$$

where $h$ is fabric thickness, $S$ is fabric surface and $T$ is temperature.

# 3 Results and Discussion

## 3.1 *Dyeing Optimization and Modelization*

Factorial experimental design was used to decrease the number of tests and minimize the possible combinaisons of parameters required to study the effect of the main factors (Temp, Dur and Sulfate) on the dyeing system responses (K/S and E). The effects of each factor on the dye exhaustion and color strength are respectively reported on Figs. 3 and 4.

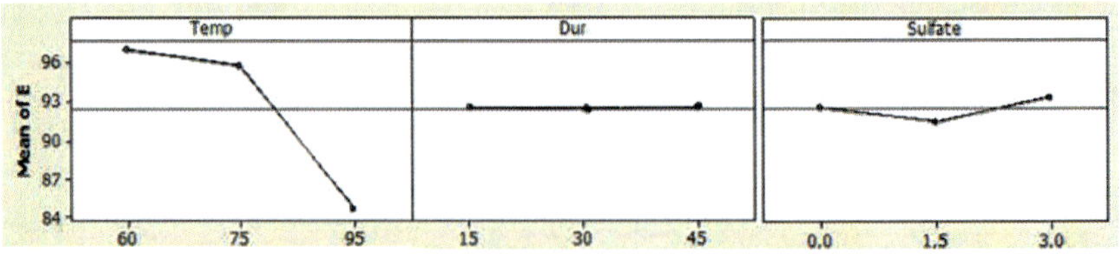


**Fig. 3** Main effects diagram for dye exhaustion (E)

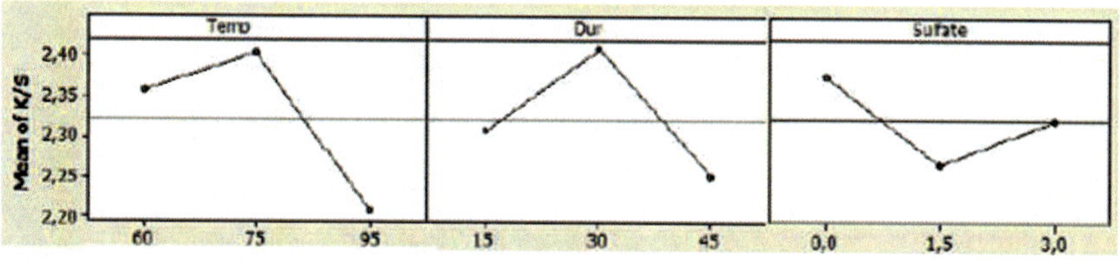


**Fig. 4** Main effects diagram for color strength (K/S)

We remark that temperature has an important negative effect on exhaustion rate of the acid dyeing bath of cationized cotton. In addition, unison agent (sodium sulfate) concentration has less influence on dye exhaustion. The highest dye uptake percentage is obtained at 3 g/L of sodium sulfate.

Analysis of main effects plots on Fig. 4 showed an important effect of temperature on K/S evolution, with a positive slope in the range [60 °C; 75 °C] and a negative slope in the range [75 °C; 95 °C]. Similarly, the dyeing time has an influence on color strength with a positive slope in the range [15 min; 30 min] and a negative slope in the range [30 min; 45 min].

On the contrary, the sodium sulfate concentration effect on K/S is opposite to the two previous factors (temperature and time). In fact, when sodium sulfate concentration rises from 0 to 1.5 g/L, we obtain a negative slope of K/S variation. Then, in the range [1.5 g/L; 3 g/L], color strength evolution becomes proportional to sulfate concentration.

Interaction diagrams between the main factors and their effects on K/S and dye exhaustion of dyed fabrics are reported on Figs. 5 and 6. In fact, an interaction between two of the three factors is considered as effective when their lines on diagrams do not run parallel.

The results were analyzed by the statistical software Minitab version 17. The relation between the input variables and the response data is estimated as linear regression model that depends on the main effects of selected factors (Temp, Dur and Sulfate). The variance experimental analysis (ANOVA) was set up according to the employed model which considers the three main variables and corresponding significant interaction coefficients. The effect of each prediction model term on response data, the contribution of each term in the form of regression coefficients and the significance of each term under *t*-test as *p*-value show the significance of the data. Details of the regression equation for exhaustion are presented in Table 1.

For dye exhaustion E, the high square correlation coefficient ($R^2$) proves that the linear regression model fits well. So, exhaustion rate E could be expressed using the following equation:

$$E = 119.787 - 0.363 \text{ Temp} + 0.002 \text{ Dur} + 0.264 \text{ Sulfate} \tag{5}$$

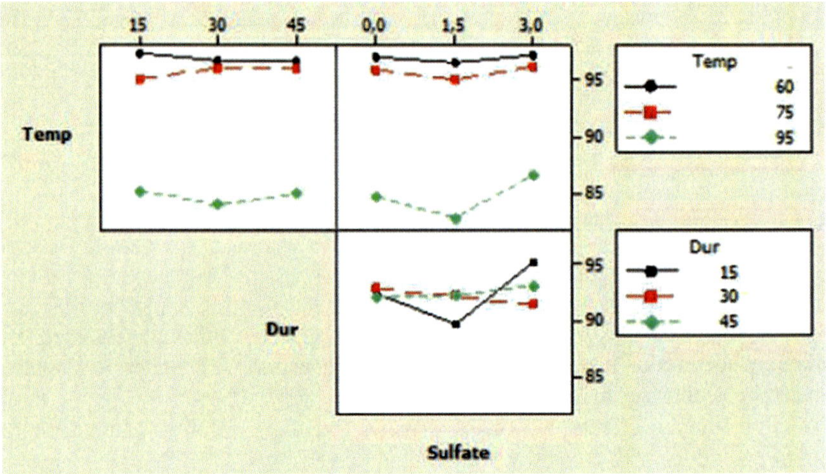


**Fig. 5** Interaction plots for main factors of dye exhaustion E

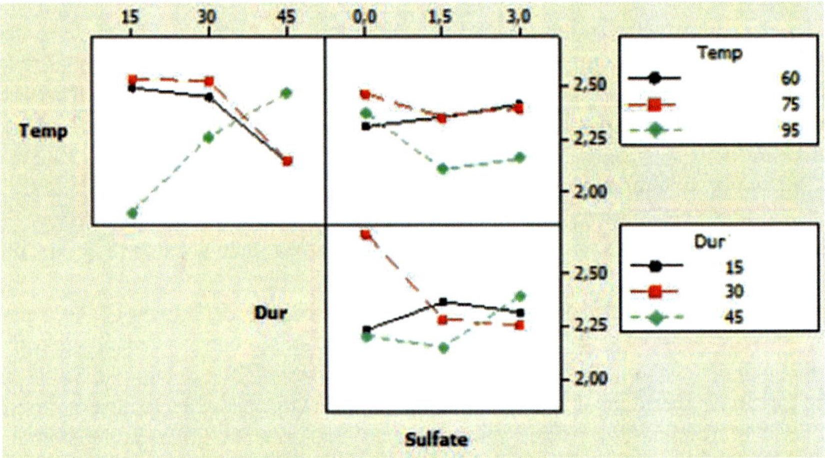


**Fig. 6** Interaction plots for main factors of K/S

**Table 1** Estimated regression coefficients for exhaustion E

| Term | Coef | SE Coef | $T$-value | $p$-value |
|---|---|---|---|---|
| Constant | 119.787 | 3.47207 | 34.5 | 0.000 |
| Temperature | – 0.363 | 0.03976 | – 9.125 | 0.000 |
| Duration | 0.002 | 0.04655 | 0.052 | 0.048 |
| Sulfate | 0.264 | 0.46547 | 0.568 | 0.075 |

**Table. 2** Estimated regression coefficients for exhaustion K/S

| Term | Coef | SE Coef | *T*-value | *p*-value |
|---|---|---|---|---|
| Constant | 2.7555 | 0.4090 | 6.739 | 0.000 |
| Temperature | – 0.0045 | 0.0046 | – 0.972 | 0.341 |
| Duration | – 0.0018 | 0.0054 | – 0.344 | 0.734 |
| Sulfate | – 0.0181 | 0.0546 | – 0.331 | 0.744 |

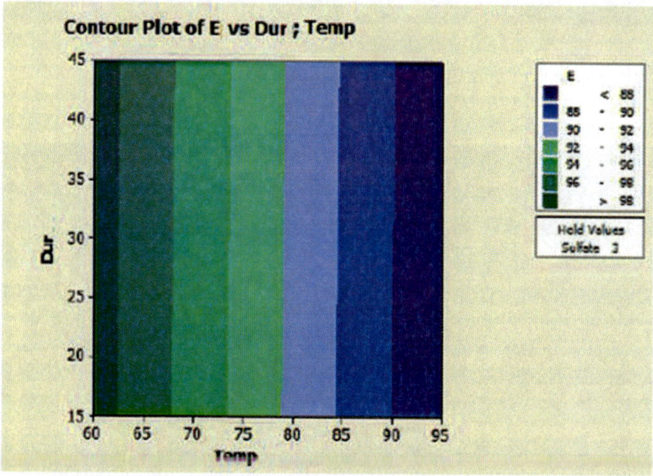


**Fig. 7** Contour plot of exhaustion response function

Nevertheless, color strength ANOVA analysis (Table 2) shows that the effects of linear terms were not significant. The model suffers lack-of-fit, because the *p*-values are larger than the level of significance (0.05).

The main output of RSM study is to depict the values of the input variables (Temp, Dur and Sulfate concentration) that lead to maximal response. As illustrated by Fig. 7, when the pre-cationized cotton is dyed at 95 °C for 45 min with 3 g/L of sulfate, the dye exhaustion attains its highest value (near 98%).

## 3.2 *Surface Morphological Analysis*

The changes in the cotton fabric surface, by cationization and acid dyeing, were evaluated using scanning electron microscopy (SEM). Micrographs at two different magnifications of greige cotton surface, cationized fabric surface and dyed cotton fabric are reported on Figs. 8, 9 and 10, respectively.

According to micrographs of Figs. 8 and 9, the surface of cationized fabric is rougher than that of greige one, due to cationization agent deposits. It could also be

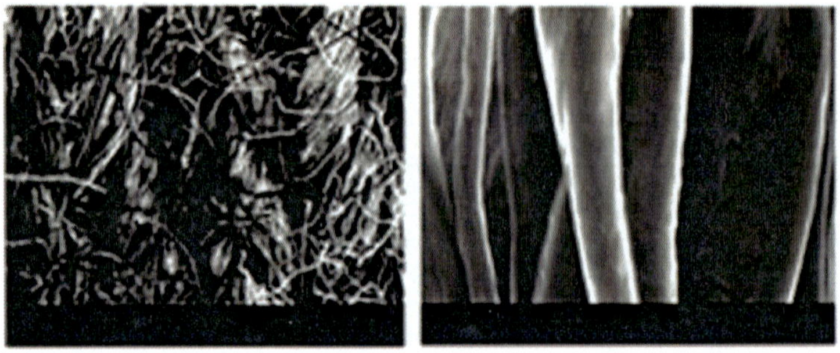

**Fig. 8** SEM micrographs of untreated cotton surface (×40 and ×2K)

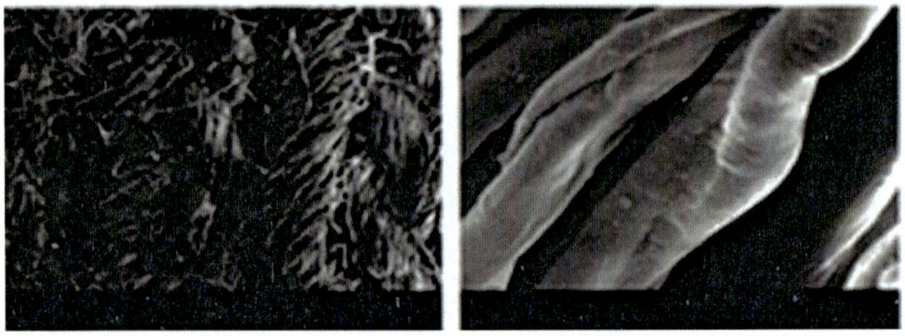

**Fig. 9** SEM micrographs of cationized cotton fabric (×40 and ×2K)

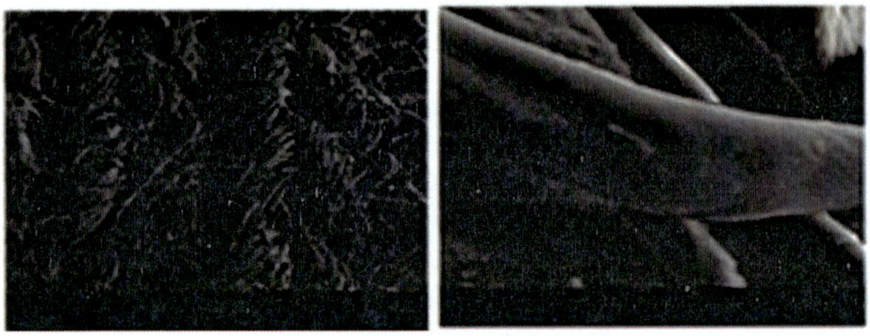

**Fig. 10** SEM images of dyed cationized cotton (×40 and ×2K)

noted that the extent of cationization decreases slightly the fibers surface pilosity. Nevertheless, the overall hand of cationized samples remained unaltered. This is an advantage of the applied resin compared to other cationic agents which promote fabric stiffness.

On SEM micrographs of dyed fabric surface (Fig. 10), we can see that the dyed fibers are no longer smooth and become rough and slightly degraded. These changes could be due to the creation of voids in the fiber structure that leads to a physical

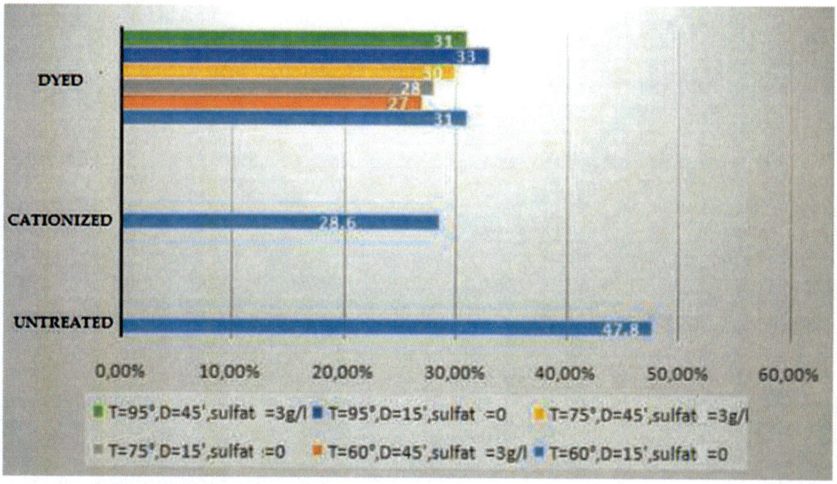


**Fig. 11** Compressibility EMC (%) of cotton fabrics

weakness of the dyed fiber structure with an improved capacity of penetration of dye molecules into the fibers.

## 3.3 *KES Characterization*

*Compressibility EMC*

The obtained results of compressibility evaluation with KES-FB3 instrument for untreated, cationized and dyed cotton fabrics are reported on Fig. 11.

The compressibility rate of untreated cotton fabric is about 47.8%. After preparation and cationization treatments, this rate is decreased to 28.6%. This decrease of compressibility is due to the rise of the superficial surface accentuated by the fibrils and the high surface pilosity. The increase of pilosity during preparation and finishing processes leads to non-reversible deformations during the compression test.

Moreover, if compressibility value EMC is low, the fabric compressibility is low; or if compressibility value EMC is high, the fabric compressibility is high. If a material is harder and denser, it can be compressed less [17]. This is also confirmed by the compressibility values of acid dyed cotton samples.

*Thermal Conductivity K*

In order to evaluate the thermo-physiological comfort of cotton fabric after different treatments, measure of thermal conductivity K, which refers to heat being transferred from the skin to the fabric, has been performed. Thermal conductivity values for tested samples are reported on Fig. 12.

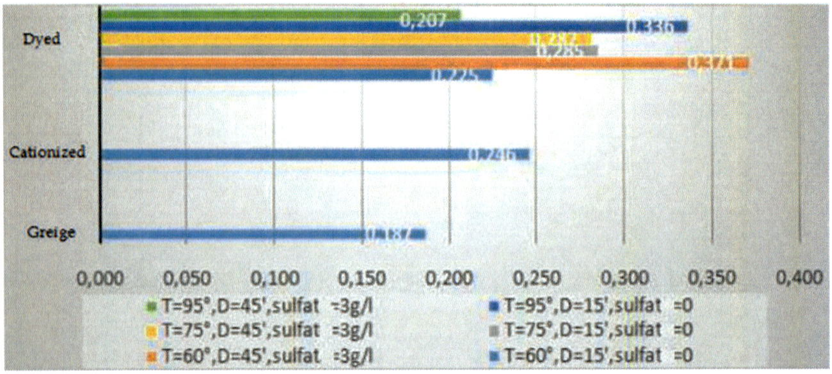


**Fig. 12** Thermal conductivity K (W $m^{-1}$ $K^{-1}$) of cotton fabrics

The thermal conductivity increased for cationized and dyed fabrics, improving the comfort properties compared to the greige fabric, which allows heat transfer rapidly. This could be caused to the increase of cotton fabric thickness after preparation, cationization and dyeing processes. These treatments increased the thickness of cotton fabric and decreased the air gap between the fabric and the skin, leading to increased K values, which is also confirmed in the study of Dalbaşi et al. [18].

## 4 Conclusion

In this research work, the potential of non-conventional chemical auxiliaries free acid dyeing of cotton fabrics as an environmentally friendly alternative has been explored. Prior to dyeing, cotton samples were cationized to create positive sites on their surface and increase their affinity to anionic dye. The influence of different experimental conditions (temperature, duration and sulfate concentration) on dyeing properties was examined. A regression equation of the acid dye exhaustion rate was established. KES tests revealed improved compression and thermal behavior of acid dyed fabric cotton.

## References


1. Khan, A., Hussain, M. T., Jiang, H., and Gul, S.: Development of Functional Wool Fabric by Treatment with Aqueous and Alkaline Extracts of Cinnamomum Camphora Plant Leaves. Journal of Natural Fibers 17(4), 472–481 (2020).
2. Dasgupta, J., Sikder, J., Chakraborty, S., Curcio, S., and Drioli, E.: Remediation of Textile Effluents by Membrane Based Treatment Techniques: a State of the Art Review. Journal of Environmental Management 147, 55–72 (2015).

3. Correia, J., Rainert, K.T., Oliveira, F.R, Rita, V., Valle, J.: Cationization of cotton fiber: an integrated view of cationic agents, processes variables, properties, market and future prospects. Cellulose 27, 8527–8550 (2020).
4. Aysha, T.S., Ahmed, N.S., El-Sedik, M.S., Youssef, Y.A., El-Shishtawy, R.M.: Eco-friendly salt/alkali free exhaustion dyeing of cotton fabric with reactive dyes. Sci Rep. 12(1), 22339 (2022).
5. Hamdaoui, M., Romdhani, Z., Djebbi, I., Harzallah, O., Harzallah, J.: Chemical auxiliaries free dyeing of cationized cotton with 1:2 metal complex dye. Indian Journal of Fiber and Textile Research 43(3), 357–362 (2017).
6. Pei, L.J., Gu, X.M., Wang, J.P.: Sustainable dyeing of cotton fabric with reactive dye in silicone oil emulsion for improving dye uptake and reducing waste water. Cellulose 28, 2537–2550 (2021).
7. Fu, S., Farrell, M.J., Hauser, P.J.: Optimising the anionisation of cotton with 3-chloro – 2-hydroxy-1-propanesulphonic acid sodium salt for dyeing with basic dyes. Coloration Technology 133, 194-199 (2017).
8. Lada, Z.G., Mathioudakis, G.N., Pavlidou, S., Goulas, G., Anastasopoulos, C., Bokias, G., Andrikopoulos, K.S., Voyiatzis, G.A.: Comparative Assessment of the Dyeing Process for Pristine and Modified Cotton Fabrics towards the Reduction of the Environmental Fingerprint. Sustainability 15, 3144 (2023).
9. Kamel, M., Youssef, B., Shokry, G.: Dyeing of cationised cotton with acid dyes. J. Soc. Dye. Colour. 114, 101–104 (2008).
10. Yang, H., Fang, K., Liu, X., An, F.: High-Quality Images Inkjetted on Different Woven Cotton Fabrics Cationized with P(St-BA-VBT) Copolymer Nanospheres. ACS Appl. Mater. Interfaces 11, 29218–29230 (2019).
11. Zhai, S., Li, Y., Dong,W., Zhao, H., Ma, K., Zhang, H., Cai, Z.: Cationic cotton modified by 3-chloro-2-hydroxypropyl trimethyl ammonium chloride for salt-free dyeing with high levelling performance. Cellulose 29, 633–646 (2022).
12. Niu, T., Wang, X., Wu, C., Sun, D., Zhang, X., Chen, Z., Fang, L.: Chemical Modification of Cotton Fabrics by a Bifunctional Cationic Polymer for Salt-Free Reactive Dyeing. ACS Omega 5, 15409–15416 (2020).
13. Zhang, F., Chen, Y., Lin, H., Lu, Y.: Synthesis of an amino-terminated hyperbranched polymer and its application in reactive dyeing on cotton as a salt-free dyeing auxiliary. Color. Technol. 123, 351–357 (2007).
14. Shindhal, T., Rakholiya, P., Varjani, S., Pandey, A., Ngo, H.H., Guo,W., Ng, H.Y., Taherzadeh, M.J.: A critical review on advancesin the practices and perspectives for the treatment of dye industry wastewater. Bioengineered 12, 70–87 (2121).
15. Rehman A., Irfan M., Hameed A., Saif M.J., Qayyum M.A., Farooq T.: Chemical-Free Dyeing of Cotton With Functional Natural Dye: A Pollution-Free and Cleaner Production Approach. Front. Environ. Sci. 10, 848245 (2022).
16. Broadbent, A.D.: Basic principles of textile coloration. : Society of Dyers and Colorists, West Yorkshire, United Kingdom (2001).
17. Utkun, E.: Comparison of sensorial comfort properties of different cotton fabrics using the Kawabata Evaluation System; Industria Textila 72(6), 587–593 (2021).
18. Dalbaşi, E.S., Özçelik K.G.: A research on the comfort properties of linen fabrics subjected to various finishing treatments. Journal of Natural Fibers 18, 909–922 (2021).

# Dyeing of the Polyester Fabric with the Fluorescein Molecule

**Fredj Saad, Ayda Baffoun, and Mohamed Hamdaoui**

**Abstract** This study examined the effects of dyeing polyester fabrics with a fluorescent dye under different conditions (temperature, pH and fluorophore concentration) on the color intensity and fluorescence quality of the dyed fabrics. Exhaustion dyeing method was applied in this experiment for textile fiber. The dependence of fluorescein on pH has been studied, and a bathochromic and hyperchromic effect of the absorption spectra of its fluorescent solution by going to higher pHs has been shown. In addition, images taken under UV excitation at 365 nm gave different emissions from the treated polyester at different pH values.

**Keywords** Fluorescein · Polyester fabric · Colorimetric measurements

## 1 Introduction

The literature mentions many attempts to develop fluorescent dyes for textile dyeing and finishing. These fluorescent dyes occupy a unique niche in textiles, used primarily where the high visibility they provide is of paramount importance. Fluorescent fabrics can be used for specific aesthetic and fashion purposes, with eye-catching bright colors offering unique creative possibilities for textile and garment developers. Fluorescent dyes are much more important when applied to synthetic fibers. The classic example of a fluorescent dye, fluorescein, is an example of a rigid but non-planar molecular structure, which results in high fluorescence intensity [1]. Fluorescein is an organic compound widely used as a synthetic dye. The fluorophore is easily obtained by mixing phthalic anhydride, resorcinol and $ZnCl_2$ or methane sulfonic acid [2]. This agent emits an intense green-yellow fluorescence at alkaline pH. This emission is better recognized by the human eye than any other color under various

F. Saad (✉) · A. Baffoun · M. Hamdaoui
Textile Materials and Processes Research Unit, National Engineering School of Monastir, University of Monastir, Monastir, Tunisia
e-mail: fredj93.saad@gmail.com

 
S. B. Abdessalem et al. (eds.), *Proceedings of the Second International Conference of Innovative Textiles and Developed Materials-ITDM'2; 05-06 May 2023; Tunisia*,
https://doi.org/10.1007/978-981-99-7950-9_6

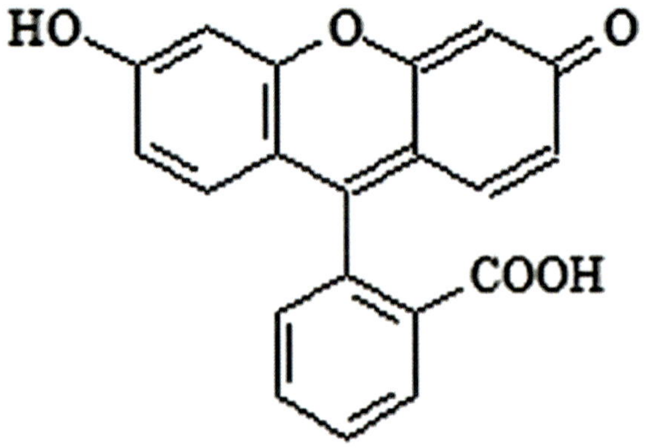


**Fig. 1** Chemical structure of fluorescein

lighting conditions. It is rare in nature and has the ability to contrast with most backgrounds [3]. In the field of textiles, fluorescein has been used by our research group as an anti-counterfeiting tracer [4, 5]. Its presence in the textile substrate can only be detected after excitation at a very specific wavelength under UV light.

The aim of this study was to investigate the application of fluorescein to a polyester fabric without any additives, in order to develop a fluorescent textile. The influence of pH, fluorescent dye concentration and application temperature on the fluorescence of polyester fabrics was determined and analyzed. Images were taken under UV excitation to highlight the fluorescent effect of the developed samples.

## 2 Materials and Methods

### *2.1 Polyester and Fluorescein Dye*

A weight plain weave polyester fabric (154 g/m$^2$) was used. An aqueous solution of fluorescein with a concentration of $5.5 \times 10^{-5}$ mol/L was prepared and evaluated by measuring its reflectance using a spectrophotometer. The fluorescein dye has an absorption maximum ranging from 460 to 490 nm and an emission maximum ranging from 520 to 560 nm, respectively, and the color of its aqueous solution varies from green to orange depending on the pH of the bath (Fig. 1).

### *2.2 Dyeing Procedure*

According to literature, fluorescein is sparingly soluble in water. In practice, before dyeing began, a stock solution of the dye was prepared. For this, a large quantity of 1 g of the dye molecule is placed in an aqueous bath, which is highly insoluble. By

**Table 1** Control factors and their levels

| Factor levels | Concentration of FL (%) | pH | Temperature |
|---|---|---|---|
| 1 | 0.5 | 4.5 | 100 |
| 2 | 1 | 5.5 | 110 |
| 3 | 1.5 | 6.5 | 120 |
| 4 | 2 | 7.5 | 130 |

adding a few drops of dispersing agent, a good dispersion of the stock solution is obtained. In this case, fluorescein behaves like a dispersed dye. Table 1 shows the experimental parameters used and their levels.

The dyeing processes were achieved in a laboratory apparatus (AHIBA Nuance speed, DataColor), with a liquor-to-fiber ratio of 10:1. After immersing polyester fabric in the dye solutions at 40 °C, the temperature was increased to 130 °C at the rate of 1 °C/min with a holding time of 60 min and finally cooled to 80 °C at 2 °C/ min. At the end of dyeing, the dyed sample was rinsed thoroughly in distilled water and allowed to dry in the open air [6].

For each pH, we plot the fluorescein calibration line, which relates absorbance to tracer concentration, based on Beer-Lambert's law. The absorbance of the aqueous fluorescein solution is determined at the dye's wavelength of maximum absorption at each pH. Using a UV–vis spectrophotometer, the absorbance of the dye solutions before and after depletion was measured, and the percentage depletion (E) was calculated using Eq. (1):

$$\%E = 100 \times \frac{A_0 - A_i}{A_0} \tag{1}$$

where $A_0$ and $A_i$ are the absorbance of dye solutions before and after exhaustion, respectively. The coloring strength (K/S) and the reflectance ($R$) at the fluorescein emission wavelength of the treated samples were measured on a datacolor spectrocolorimeter. All the measurements were carried out under the conditions of a D65 illuminant and an observer at 10°.

# 3 Results and Discussions

## 3.1 *pH Quenching of Fluorescein Absorbance*

The ionic charge and the chemical structure of fluorescein evolve according to the surrounding pH. This is explained by the presence of phenolic fragments and the state of equilibrium between a carboxylic function and a lactone, leading to fluctuations in the photophysical properties [7]. The fluorescence quantum yield is very high under

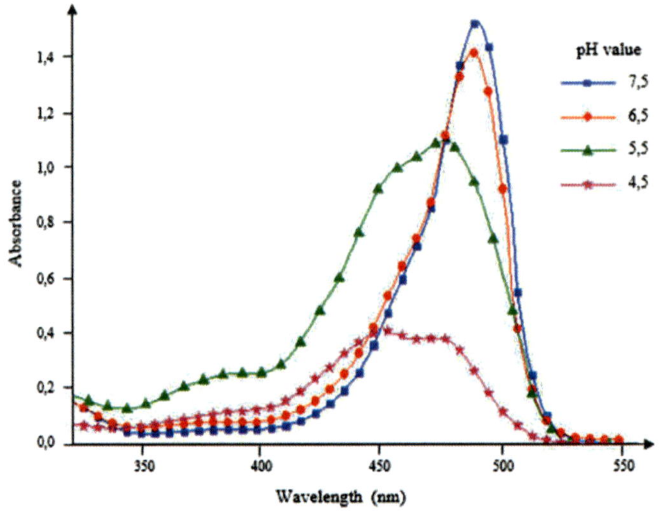


**Fig. 2** pH-dependent changes in absorption spectra of fluorescein (dye concentration: 5.5 × $10^{-5}$ mol/L)

basic conditions (ΦF: 0.95 in 0.1 M NaOH) under excitation at 490 nm, and the acidification of the solution gradually leads to the extinction of the fluorescence.

In the current study, solutions of known concentrations of fluorescein were prepared in solutions from pH 4.5 to 7.5 at 0.5 pH increments (Fig. 2).

The absorption spectra of the fluorescein dye as a function of pH (Figs. 3 and 4) are in line with data published in the literature. In a basic solvent, dianion species form with a maximum wavelength of 490 nm and a curvature of the curve near 478 nm. The relative fraction of dianions decreases with decreasing pH and the simultaneous increase in the fraction of other proteolytic species—anion (with two maxima near 472 and 450 nm), neutral species including three tautomers (zwitterion 22%, quinoid 11% and lactone 67%) and cation [8, 9]. Thus, a batochromic and hyperchromic effect was observed following an increase in pH.

The basic pH causes the fluorescein ring to open, becoming highly emissive and displaying a greenish-yellow color (Fig. 3). In the pH range of existence of the fluorescein monoanion (at around pH 5.5), the fluorescence quality is very close to that of the neutral molecule, and the emission color tends toward yellow-orange at pH 4.5 (with a slight emission contribution from the dianion).

## 3.2 *Effect of pH and Temperature on Dye Exhaustion*

Due to the sensitivity of fluorescein to bath pH, the effect of this parameter on bath exhaustion at different temperatures will be studied in this section.

**Fig. 3** Effect of pH on fluorescein emission color under UV excitation at 365 nm

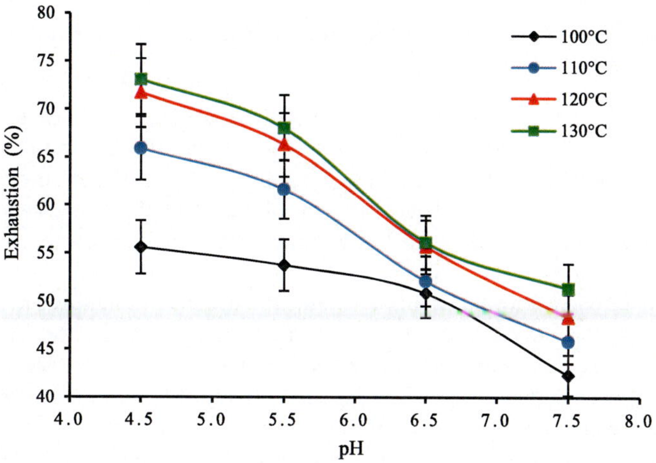


**Fig. 4** Effect of pH and temperature on dye exhaustion (dye concentration: 1% owf)

Standard curves for each pH solution were constructed with reference to the linear relationship between fluorescein concentration and absorbance. At a fixed temperature, a decrease in bath depletion was observed when moving toward a basic pH.

## *3.3 Effect of pH and Fluorescein Concentration on the Reflectance*

In this review, the measurement of the fluorescence of samples treated with fluorescein is made in a qualitative way by measuring the reflectance at the emission wavelength of the dye. At this wavelength, the maximum reflectance value of the

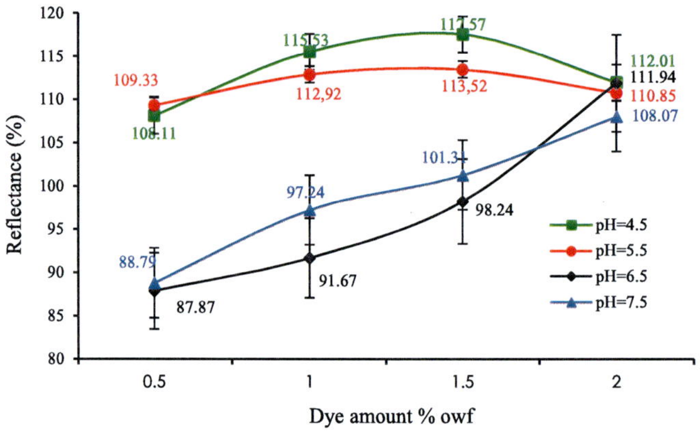


**Fig. 5** Effect of pH and concentration of fluorescein on reflectance at temperature 100 °C

fluorescent polyester is greater than the reflectance of the standard bleached sample. However, the emitted fluorescent light adds to the light reflected from the treated material, producing an apparent increase in reflectance in the emitting region of the fluorescent molecule [5]. Therefore, the judgment of the fluorescence of the treated textile samples is determined by measuring their reflectance at the emission wavelength of the dye.

Many researchers have shown that changing the structure of fluorescein by protonation or deprotonation also has a significant influence on the quantum yield of the molecule [10]. To this extent, the influence of pH and fluorophore concentration on fluorescence was studied by plotting the curve giving the reflectance value at the emission wavelength of fluorescein (Fig. 5).

By Comparing bath exhaustion and reflectance of a sample treated at pH 4.5 and 100 °C, exhaustion is 55%, while reflectance is maximum at 115%. This means that dye fixation on polyester is maximum, without quenching. So the return to the fundamental state of excited molecules is totally radiative-fluorescent. For information, maximum exhaustion doesn't mean maximum fluorescence. This is because the probability of collision between molecules in the ground state (static quenching) and those in the excited state (dynamic quenching) increases, resulting in a decrease in fluorescence.

In general, polyethylene terephthalate is dyed in an acidic environment at a pH of 4.5–5.5 [11] giving maximum coloring strength and therefore more molecules have the chance to become excited and fluoresce once they are irradiated with ultraviolet light. Going toward higher pH, the emission color of the treated samples varies with a decrease in fluorescence (Table 2).

At a strongly acidic pH, the reflectance of fluorescent tissue exceeds 100%, gradually increases to a threshold and then decreases from a fluorophore concentration of 0.5–2%. A self-extinction phenomenon may be the cause of the decrease in fluorescence. Indeed, by increasing the concentration of the xanthene dye, the collision

**Table 2** Visualization of samples treated at a temperature of 100 °C and a fluorescein concentration of 1%, under daylight and under a UV lamp at 365 nm

| Under daylight | | Under UV light at 365 nm | |
|---|---|---|---|
| pH = 4.5 | pH = 5.5 | pH = 4.5 | pH = 5.5 |
|  | |  | |
| pH = 6.5 | pH = 7.5 | pH = 6.5 | pH = 7.5 |

between the molecules is favored thus forming a non-fluorescent complex in the ground state. In addition, excited-state energy transfer processes may take place between the fluorescent molecules [12].

## 4 Conclusion

In summary, a fluorescent dye has been exploited in the dyeing of polyester fabrics for high visibility and to obtain fluorescent tints. Different concentrations of fluorescein were applied to the textile fiber by changing the temperature and pH of the bath, via the exhaustion technique. The measurement of the absorbance of aqueous solutions of fluorescein at different pH shows a bathochromic and hyperchromic shift toward a basic bath. The low solubility of fluorescein in water favors the dispersed behavior of this dye and thus reflects its application on polyester.

## References

1. Clark, M., Handbook of textile and industrial dyeing: principles, processes and types of dyes. Elsevier (2011).
2. Shabir, G., A. Saeed, and P. Ali Channar, A review on the recent trends in synthetic strategies and applications of xanthene dyes. Mini-Reviews in Organic Chemistry, 15(3): p. 166–197 (2018).
3. Arenas, L.M., J. Troscianko, and M. Stevens, Color contrast and stability as key elements for effective warning signals. Frontiers in Ecology and Evolution, 2: p. 25 (2014).
4. Baatout, K., and al., Luminescent cotton fibers coated with fluorescein dye for anti-counterfeiting applications. Materials Chemistry and Physics, 234: p. 304–310 (2019).
5. Saad, F., et al., Polyester Fabric with Fluorescent Properties Using Microwave Technology for Anti-Counterfeiting Applications. Journal of Fluorescence, 32(1): p. 327–345 (2022)
6. Rashaduzzaman, M., and al., A study on the effects of dispersing agent concentrations on the colorfastness of synolon yellow exw fluorescent disperse dye. International Journal of Materials Science and Applications 4(5):216–221 (2016).
7. Le Guern, F., and al., Fluorescein derivatives as fluorescent probes for ph monitoring along recent biological applications. International Journal of Molecular Sciences, 21(23): p. 9217 (2020).
8. Doughty, M.J., pH dependent spectral properties of sodium fluorescein ophthalmic solutions revisited. Ophthalmic and Physiological Optics, 30(2): p. 167–174 (2010).
9. Kotásková, M., Synthesis of new xanthene derivatives. Mainz, Univ., Diss (2012).
10. McLoughlin, C.K., and al., Oxygen-and pH-Dependent Photophysics of Fluorinated Fluorescein Derivatives: Non-Symmetrical vs. Symmetrical Fluorination. Sensors, 20(18): p. 5172 (2020).
11. Casetta, M., Procédé de teinture du polyester par les colorants dispersés: étude physico-chimique et modélisation. Lille 1 (2000)
12. Lakowicz, J.R., Principles of fluorescence spectroscopy. Springer (2006).

# Study of a New Dyeing Process Based on the Capillary Rise Technique

**Sarra Said, Imed Feki, Mohamed Hamdaoui, and Walid Sahraoui**

**Abstract** This paper presents a new dyeing process of knitted fabric. This process consists to using a capillary rise technique with the goal to obtain a local dyeing with single color, degraded from dark to light. The C.I. Direct Red 79 dye was used in this dyeing process. The effects of dye concentration, temperature and salt (NaCl) concentration on the surface tension of the dye bath were initially investigated in order to identify the optimized parameters, required for the dyeing process. The result shows that the surface tension of the bath is almost constant when the dye concentration is varied from 0 to 12 g/L. The experimental capillarity result of salt-free bath reveals the appearance of two distinctive progression fronts. The first one, which progress rapidly through the knitted fabric, reflects the capillary progression of water. The second front, however, advances very slowly. It presents the capillary rise of the dye. When the salt is added to the bath, the capillarity result shows almost one progression front, indicating a homogeneous rise of water and dye. Likewise, color strength (K/S) investigation reveals a gradual decrease of the K/S value as a function of the fabric height. The aforementioned results highlight the use of the capillary rise technique as a simple and nondestructive method to get a dyed knitted fabric with one color, degraded from dark to light.

**Keywords** Jersey knitted fabric · Direct dye · Dyeing capillary rise · Color strength

S. Said (✉) · I. Feki · M. Hamdaoui
Textile Materials and Processes Research Unit (MPTEX), National Engineering School of Monastir, University of Monastir, Monastir, Tunisia
e-mail: Sarrasaid@yahoo.fr

W. Sahraoui
TANIT Textile Services Company, Monastir, Tunisia



S. B. Abdessalem et al. (eds.), *Proceedings of the Second International Conference of Innovative Textiles and Developed Materials-ITDM'2; 05-06 May 2023; Tunisia*,
https://doi.org/10.1007/978-981-99-7950-9_7

## 1 Introduction

Dyeing process is one of the most important aspects in the textile commerce. Indeed, the global textile dyeing market reached 9.68 billion USD in 2023, and is expected to raise by a 6.8% during the short upcoming period [1]. In this regard, several efforts have been expanded by concentrating the research activities on the improvement of the textile industry and on providing a creativity in the dyeing process and methods [2–4].

The process of dyeing involves the imbuing of a textile with a color by immersing it in a dye bath or by applying dye to the material using several methods, in order to achieve color with desired fastness. Generally, textile materials can be dyed using different procedures and techniques, including piece dyeing, continuous dyeing, pad-batch dyeing and tie-dye dyeing [2, 3, 5]. Several factors influence the type of the used dyeing procedures and techniques, such as the type of material, the basic type of fiber, the type of dye and the intended outcomes of the dyed fabric [3, 5].

Piece dyeing is a procedure of coloring fabric after it has been woven or knitted. This process is mostly used for linen, cotton and rayon materials. Continuous dyeing is based on a continuous process of dyeing by using big machines, which allow the treatment of enormous amounts of material at one time. However, pad-batch technique is often used for dyeing little amounts of fabrics with a pad dyeing equipment. From other hand, the tie-dye approach, which is frequently used for cotton and silk materials, involves various procedures of folding, tying, binding and twisting of fabric before dyeing it [2].

Dyeing process is mostly carried out in aqueous solution, which involves enormous quantity of water. To optimize the water consumption, from one hand, and to ensure the quality requirements in the dyed fabric, from other hand, a deep understanding of liquid transport mechanism along fabrics, including water diffusion and capillary wicking, is necessary.

Owing to its simplicity, effectiveness and its nondestructive nature, experimental method based on the capillary rise is extensively used to control the capillary flow through textile materials and liquid adsorption into textile fabrics [6–11]. The aim is to measure the evolution of the height penetration and/or the mass gain of wetting liquids versus time.

In line with the topic related to design, fashion and creativity in the textile industry, the present work presents a new process of local dyeing using the capillary rise technique. The goal of this process is to obtain a knitted fabric dyed with one color, degraded from dark to light shade.

**Table 1** Characteristics of the knitted fabric

| Composition | 100% cotton | Norms |
|---|---|---|
| Structure | Single jersey | ISO 23606 |
| Area density ($g/m^2$) | 175 $g/m^2$ | EN 12127 |
| Thickness | 1.1 mm | ISO 5084 |
| Absorbed yarn length | 0.452 cm /stitch | EN 14790 |
| Number of rows per cm | 20 | EN 14971 |
| Number of columns per cm | 13 | |

**Table 2** Characteristics of the dye

| Dye | C.I. Direct Red 79 |
|---|---|
| Chemical formula | $C_{37}H_{28}N_6Na_4O_{17}S_4$ |
| Class (SDC) | Di-azo |
| Macromolecular weight | 1048.87 |
| Solubility in water | 15 g/L (60 °C)<br>24 g/L (97 °C) |

## 2 Materials and Methods

### *2.1 Materials*

A bleached jersey knitted sample was used in this investigation and its basic characteristics are shown in Table 1.

The commercial dye used in this study was a C.I. Direct Red 79 (Table 2).

### *2.2 Methods*

A device assuring the capillary rise of the dye bath in the jersey knitted fabric was inspired from the experimental device developed during the work of Hamdaoui et al. [6–9]. The fabric is held so that the columns are perpendicular to the surface of dyeing bath by means of a metal rod system and permit the capillary rise of colored solution in the textile structure (Fig. 1).

The knitted fabric (25 cm × 35 cm) is attached to a sensitive electronic balance (AND GX-2000 precision balance) with an accuracy of 0.01 g. The balance has the ability to record the weight of the absorbed dye bath (g) as a function of time (s).

The height reached by the liquid is evaluated every 5 min by fixing a rule next to the fabric.

Tests were realized following cotton fabrics preconditioning at standard testing atmosphere (20 ± 2 °C and 65 ± 4% RH) according to ISO 139:2005.

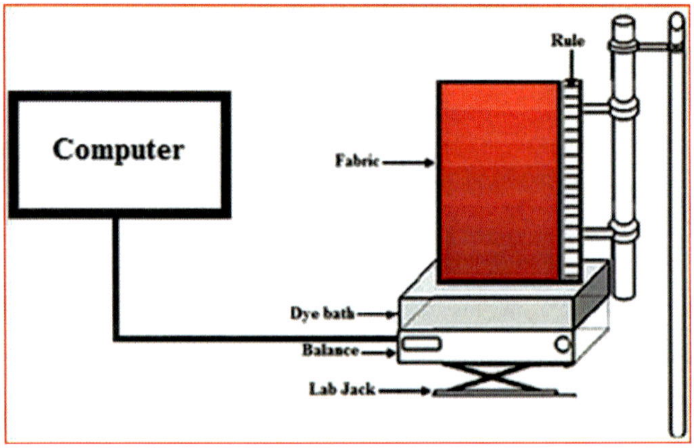


**Fig. 1** Experimental device for vertical suspension of the knit fabric

SpectroFlash SF600 spectrophotometer with DataMatch 2.0 software (Datacolor International, USA) was used to measure the color intensity of the treated fabrics expressed as color strength (K/S). The K/S values were measured using the Kubelka–Munk equation:

$$\frac{K}{S} = \frac{(1 - R)^2}{2R} \tag{1}$$

where, $K$ is the absorption coefficient, $S$ is the scattering coefficient and $R$ is the reflectance.

The tensiometer is used to measure the surface tension of a liquid at equilibrium, where a thin plate connected to a precision balance is involved.

# 3 Results and Discussion

## *3.1 Physicochemical Study of the Dyeing Process Based on the Capillary Rise Technique*

In this part, we investigate the effect of the different factors of the dyeing process (the concentration of the dye and the temperature of the bath) on the physicochemical properties of the bath (the surface tension of the liquid).

**Effects of Dye Concentration on the Surface Tension of the Bath.** Figure 2a shows the variation of the surface tension of the bath as a function of the C.I. Direct Red 79 dye concentration. It is clearly observed that the bath surface tension slightly increases with the rise of the dye concentration. Moreover, as observed from Fig. 2a,

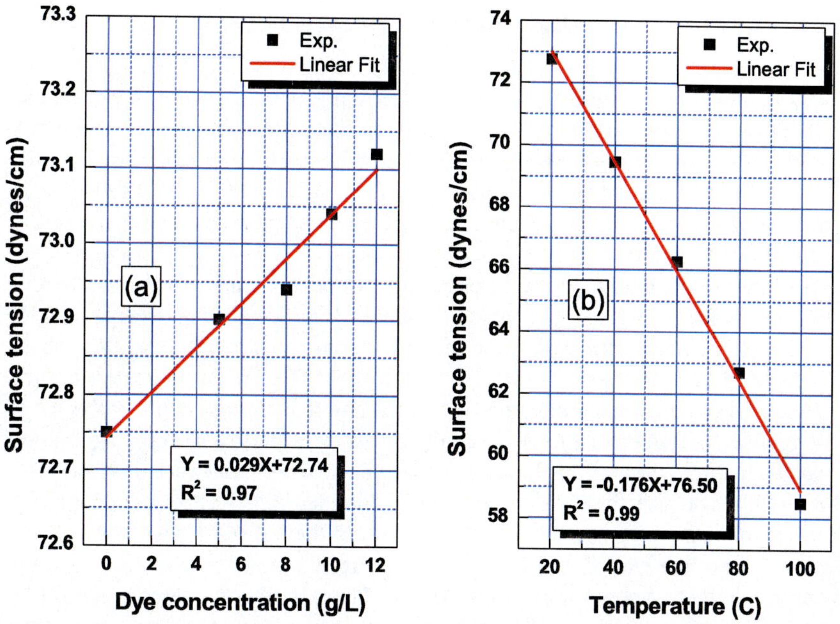


**Fig. 2** Effect of [C.I. Direct Red 79] (**a**), and temperature (**b**), on the surface tension of the bath

this variation is linear, with a slope of about $0.03\,\frac{\text{dynes/cm}}{\text{g/L}}$. When the dye concentration is varied from zero to 12 g/L, the surface tension of the bath is increased from 72.7 to 73.1 dynes/cm. This rise is very small (less than 1%), which means that the surface tension of the bath is almost constant in this dye concentration range (from 0 to 12 g/L).

**Effect of Temperature on Bath-Textile Interface Phenomena.** In order to study the effect of temperature on the surface tension, the bath temperature is varied from 20 to 100 °C. The evolution of the surface tension as a function of temperature is presented in Fig. 2b. Obviously, when the temperature increases, the surface tension of the bath decreases significantly. Furthermore, the surface tension of the bath shows a linear dependence on the bath temperature with a negative slope of about $-0.17\,\frac{\text{dynes/cm}}{\text{C}}$. When the bath temperature is raised from 20 to 100 °C, the surface tension of the water is dropped by about 19%, from 72.8 to 58.6 dynes/cm. From industrial point of view, this reduction of the surface tension is considered as useless compared to the enormous gain in energy, if we work at 20 °C instead of 100 °C. For this reason, in the following part, the direct dyeing process was carried out at room temperature using the capillary rise method.

## 3.2 Study of the Capillary Rise in a Knitted Fabric

**Study of the Water Capillary Rise in a Knitted Fabric.** In order to understand and analyze the process of capillary rise within the knitted fabric, we have firstly investigated the kinetics of water capillary rise within a knitted fabric, without the addition of dye. The fabric is maintained in vertical contact with an infinite reservoir. The evolution of the water height absorbed by the knitted fabric over time, ($h = f(t)$), is presented in Fig. 3a.

We can observe that the curve has a positive slope. It clearly proves that there is a phenomenon of capillary rise, which decreases with time and reaches a value of zero at equilibrium (at full saturation of the material and when the capillary phase is completed to have a greater effect of gravity) [6]. This result indicates that the process of water capillary rise could be divided into two regimes, namely the short-time regime (the first phase), where capillary rise of water occurs rapidly, and the long-time regime (the second phase), where capillary rise of water occurs slowly [6]. During the first phase, the water vertical rise is dominated by the effect of capillarity. In this regime, the rise process is manifested through the diffusion of water within the macro-pores of the knitted structure [6, 8]. As a result, the water rises very quickly to reach a height of about 8 cm after 10 min. Then, during the second phase (the long regime), when the rise of the liquid is established within the micro-pores of the

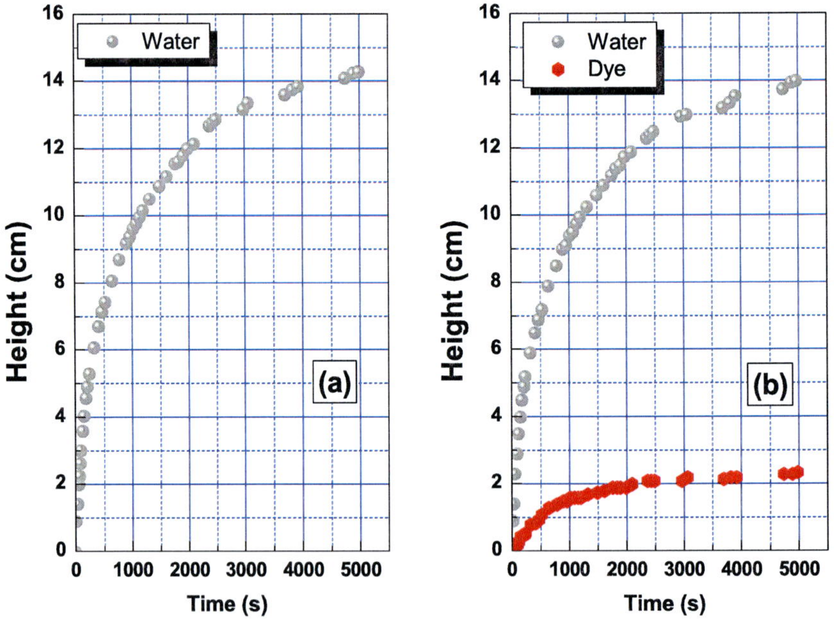


**Fig. 3** The height reached by the liquid within the knitted fabric: **a** case of the reservoir containing only water, **b** case of the reservoir containing water and dye

knitted structure, the capillarity decreases and the gravity begins to give an effect. At a certain height, called the equilibrium or saturation height, the effect of gravity stops completely the capillarity rise, and the curve tends toward an asymptote.

**Study of the Dye Addition Effect on the Capillary Rise Process**. To study the effect of dye addition on the capillary rise process within a knitted fabric, we add 1 g/L of Direct Red 79 dye to a reservoir containing water. As shown in Fig. 3b, adding the Direct Red 79 dye to the bath leads to the appearance of two capillary progression fronts. The first front advances very quickly. It reflects the capillary progression of water. However, the second front advances very slowly. It reflects the capillary progression of the dye. This front reaches a maximum height at equilibrium less than 2.3 cm.

The height reached by the water at equilibrium is almost constant in both test cases. It is slightly influenced by the presence of the Direct Red Dye 79. Indeed, in the first case (water alone in the bath), the value of the water height at equilibrium within the knitted fabric is estimated to be 14.5 cm. In the second case (bath containing water and dye), this value of the water height (First front) at equilibrium slightly decreases to have a value of 14 cm.

The addition of the dye to the bath does not affect the volume of liquid absorbed by the textile. However, it influences the overall process of capillary progression and leads to the appearance of two progression fronts (Fig. 4). The water retains its ability to progress rapidly vertically through the textile. However, the dye was unable to increase in the same way. It rises slightly within the textile with a very slow kinetics. The aforementioned observation could be attributed to the electrostatic repulsion occurred between the anion dye and the cotton fiber [12–14]. Indeed, from one hand, the direct dyes are anionic. From other hand, cotton fibers present a negative surface charge when immersed in water [15]. As depicted in Fig. 4, the electrostatic repulsion phenomenon affects the capillary rise of the dye bath, leading consequently to the appearance of double wicking front.

## 3.3 *Study of the Salt Effect on the Capillary Rise Process*

**Effect of NaCl Concentration on the Surface Tension**. To overcome the electrostatic repulsion between the dye and the cotton fiber, sodium chloride (NaCl) will be added to the bath [12, 16, 17]. However, high concentration of NaCl could increase the surface tension of the bath, causing subsequently a reduction in the capillary rise. Therefore, an investigation of the effect of NaCl concentration of the surface tension of the bath is needed to identify the optimum concentration that will be added to the dye bath.

Figure 5 shows the effect of salt concentration on the surface tension of the bath. It can be seen that the bath surface tension value increases linearly with increasing the concentration of salt added to the bath. In fact, when the NaCl concentration is varied from 0 to 70 g/L, the surface tension is raised from 72.7 to 75.4 dynes/cm, respectively. Based on the above-mentioned results, the designated NaCl concentration, which will

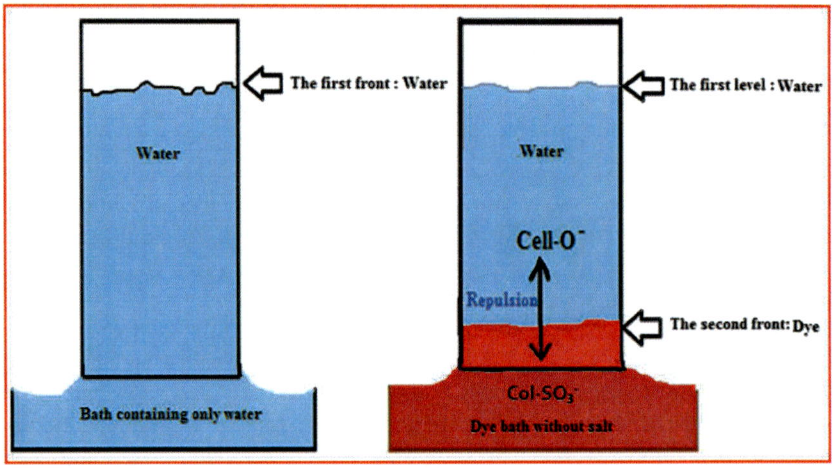


**Fig. 4** Effect of the dye addition on the capillary rise within the knitted fabric

be added to the bath, is around 30 g/L. Within this concentration, the salt can act as an electrolyte in the dye without having a great effect on the surface tension (the surface tension will increase by about 1%, when the salt concentration increases from 0 to 30 g/L).

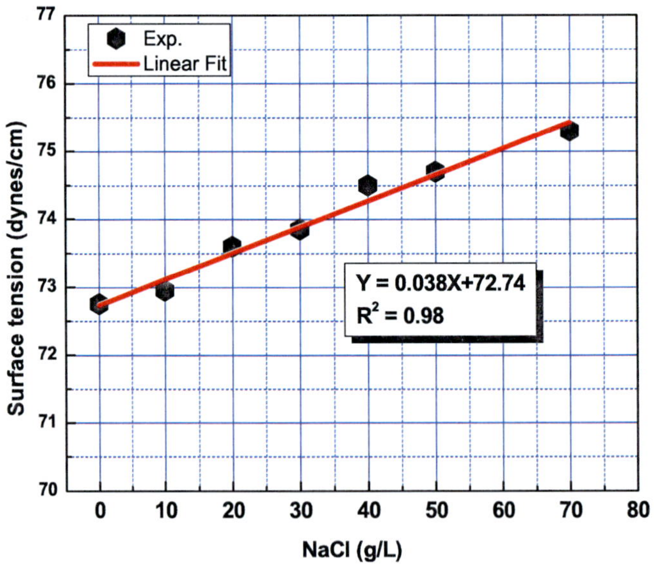


**Fig. 5** Effect of NaCl concentration on the surface tension of the bath

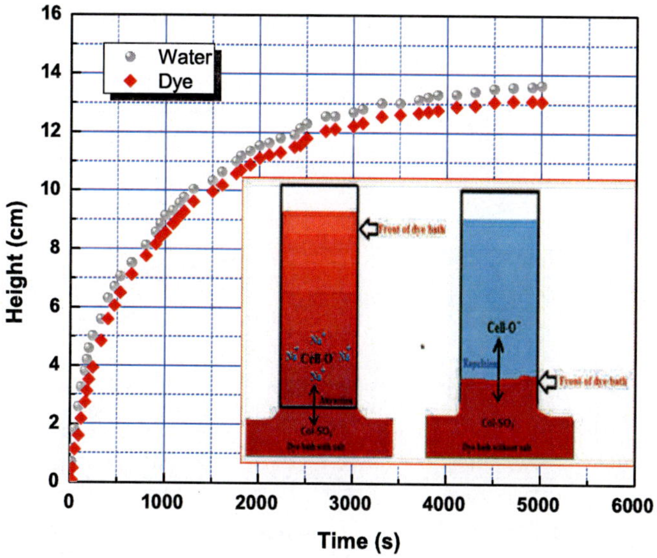


**Fig. 6** The height reached by the liquid within the knitted fabric after salt addition. The insert graph the role of salt as an electrolyte in the dye bath

**Effect of Salt Addition on the Capillary Rise Process**. To study the effect of the salt addition (sodium chloride) on the capillary progression behavior of the dye bath, we have added 30 g/L of salt in the dye solution. Moreover, to increase the wettability of the cotton and promote the diffusion of the dye inside the fiber, 1 g/L of sodium carbonate and 8 g/L of commercial wetting agent have been added to the bath [12, 18].

The curve $h = f(t)$, given in Fig. 6, describes the evolution over time of the height reached by the liquid within the chosen fabric.

It can be seen from Fig. 6 that the height reached by the dye bath at different times is influenced by the addition of salt. Indeed, the existence of salt in the dye reduces the electrostatic repulsion between the dye and the cotton, leading consequently to an increase in the bath volume absorbed by the knitted fabric. Moreover, it avoids the appearance of the double front, reflecting a nearly homogeneous rise of water and dye. The height reached by the dye is about 13 cm, very close to the value reached by the water (13.7 cm). The insert graph in Fig. 6 explains more clearly the phenomenon, visually observed, and illustrates the role of salt as an electrolyte in the dye bath.

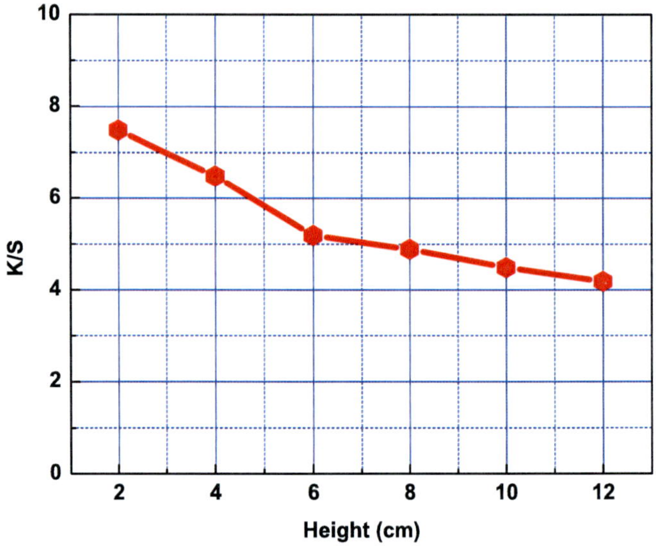


**Fig. 7** Behavior of the color strength with the increase of height

### *3.4 Behavior of the Color Strength with the Increase of Height*

Figure 7 shows the evolution of color strength (K/S) as a function of the height. At the bottom, i.e. at height value around 2 cm, the K/S value is about 7.5. When the height is increased, the K/S value is gradually dropped to reach 4.2 for a fabric height of about 12 cm, i.e. in the top the knitted fabric. The above-mentioned results highlight the usefulness of the capillary rise technique as a simple and nondestructive method to get a dyed knitted fabric with one color, degraded from dark to light shade.

## 4 Conclusion

In this work, we have successfully developed a new manner of one-color dyeing based on the capillary rise technique to create gradation of color from dark to light within knitted fabric. For this purpose, C.I. Direct Red 79 dye has been used in this dyeing process. A preliminary study has been performed to investigate the impact of the different dyeing factors, such as concentration of dye, temperature and concentration of salt, on the bath-textile interface phenomena. When salt-free bath is used, the capillarity rise showed two distinctive fronts, one with fast progression and the other with slow evolution. These fronts present the capillary rise of water and dye, respectively, through the knitted fabric. The aforementioned observation was attributed to the electrostatic repulsion occurred between the anion dye and the cotton fiber, which inhibits the dye to follow the water and diffuse within the fabric. Yet, when the salt was added to the bath, i.e., the electrostatic repulsion between the dye and the cotton is reduced, the capillarity result showed almost one progression front, reflecting a similar rise of water and dye. Moreover, color strength measurement revealed a gradual decrease of the K/S value as a function of the fabric height, indicating the obtaining of dyed fabric with one color, degraded from dark to light shade.

## References


1. Textile Dyes Global Market Report 2023, https://finance.yahoo.com/news/textile-dyes-global-market-report-200300555.html
2. Said, S., Hamdaoui, M. and Sahraoui, W.: Mathematical Models toFollow theDyeing ofCotton Knitted Garment Using the Capillary Rise Technique. Fibers and Polymers 24, 45–56 (2023)
3. Shang, S. M.: 13 - Process control in dyeing of textiles, Editor(s): Abhijit M., Apurba D., Alagirusamy, R., Kothari, V. K., In Woodhead Publishing Series in Textiles, Process Control in Textile Manufacturing, pages 300–338, (2013)
4. Wang, J., Gao, Y., Zhu, L., Gu, X., Dou, H. and Pei, L.: Dyeing Property and Adsorption Kinetics of Reactive Dyes for Cotton Textiles in Salt-Free Non-Aqueous Dyeing Systems. Polymers 10 (9), 1030 (2018)
5. Farah, M., Gisele A., Elisa R., Cardoso, J., Zanoni, M. and Oliveira, D.: Textile Dyes: Dyeing Process and Environmental Impact, From the edited volume Eco-Friendly Textile Dyeing and Finishing, Edited by Melih G., (2013) DOI: https://doi.org/10.5772/53659
6. Hamdaoui, M. and Ben Nasrallah, S.: Capillary rise kinetics on woven fabrics – Experimental and theoretical studies. Indian Journal of Fibre & Textile Research 40, 150–156 (2015)
7. Hamdaoui, M., Fayala, F. and Ben Nasrallah, S.: Dynamics of Capillary Rise in Yarns: Influence of Fiber and Liquid Characteristics. Journal of Applied Polymer Science 104, 3050–3056 (2007)
8. Hamdaoui, M., Fayala, F., Perré, P. and Ben Nasrallah, S : Experimental study of capillary rise in fabric using an electrical resistance technique. AUTEX Research Journal, 8 (2), 44–48 (2008)
9. Hamdaoui, M., Fayala, F. and Ben Nasrallah, S. :Experimental apparatus and mathematical model for determination of parameters of capillary rise in fabrics. Journal of Porous Media, 9 (4), 381–392 (2006)
10. Fayala, F., Hamdaoui, M., Perré, P., and Ben Nasrallah, S. :Study of liquid distribution during capillary rise in fabrics using an electrical resistivity technique: influence of structure and composition. Journal of Porous Media, 11 (3), 231–240 (2008)
11. Perwuelz, A., Casetta M. and Caze, C.: Liquid organisation during capillary rise in yarns—influence of yarn torsion. Polymer Testing 20 553–561(2001)
12. Ghulam, M., Mahar, F.,Khatri, A., Ali, S., Ali, A., Ali, Az., Pirzado, A. and Jiang. Q.: Potential of reactive cationization of cotton over mercerization for dyeing characteristics. North American Academic Research 3(10), 139–153 (2020)
13. Correia, V., Stephenson, T. and Judd, S.: Characterisation of textile wastewaters-A review. Environ. Technol. 15, 917–929 (1994).
14. Arivithamani,N. and Dev, V.: Industrial scale salt-free reactive dyeing of cationized cotton fabric with different reactive dye chemistry. Carbohydrate Polymers. 174, 137–145 (2017)
15. Ribitsch, V. and Kleinscheck, K.: Characterizing textile fibre surfaces with streaming potential measurements. J. Text. Res. 68 (10), 701–707 (1998).
16. Ibrahim, N., El-Sayed, W. and Ameen, N.: A novel technique to minimise energy and pollution in the dyeing of linen fabric. Color. Technol. 126, 289–295 (2010).

17. Cid, M., Spronsen, J., Kraan, M., Veugelers, W., Woerlee, G. and Witkamp, G.: Excellent dye fixation on cotton dyed in supercritical carbon dioxide using fluorotriazine reactive dyes. Green Chem. 7, 609–616 (2005)
18. Montazer, M.,Malek, R. and Rahimi, A.: Salt free reactive dyeing of cationized cotton. Fibers and Polymers, 8(6), 608–612 (2007).

# Innovative Materials & Polymers

# Ballistic Impact Response of Multi-Layers Armors Against Piercing Projectile: Finite Element Modeling

**Hassouna Amira and Salah Mezlini**

**Abstract** The development of reliable bulletproof vests requires the study of the factors that affect the ballistic resistance of the body armor. The numerical simulation is a fundamental tool adopted in this process. This work is focused on the ballistic performance of a multi-layer armor against an ogival nosed hardened steel projectile. The multi-layers armors are constituted of two composite layers and a ceramic layer. The Johnson-Holmquist-2 (JH-2) model was employed for modeling the behavior and damage of ceramic material and the Hashin criterion for simulating the damage of the composite material. The projectile was considered as a deformable part and its behavior is governed by the Johnson–Cook (JC) model. The results reveal that the residual velocity is highly influenced by the impact velocity. The effectiveness of the body armor is reduced by increasing the impact velocity. Furthermore, it is found that adding a thin composite layer in front of the ceramic layer can reduce the damage of the back layer made of composite material. The ballistic resistance is improved by increasing the thickness of both ceramic and composite plates. The residual velocity is more influenced by the thickness of the ceramic plate than by the thickness of the composite plate.

**Keywords** Ballistic performance · Body armors · Ceramic and composite thicknesses

## 1 Introduction

The development of high-performance bulletproof vests with a high level of protection is a great challenge for researchers and manufacturers. To contribute to this development, it is necessary to study and better understand the effects induced by the ballistic impact of a projectile against ceramic armor [1, 2], composite armor [3,

H. Amira (✉) · S. Mezlini
Laboratoire de Génie Mécanique, Ecole Nationale d'Ingénieurs de Monastir, Université de Monastir, Monastir, Tunisia
e-mail: Hassounaamira0@gmail.com



S. B. Abdessalem et al. (eds.), *Proceedings of the Second International Conference of Innovative Textiles and Developed Materials-ITDM'2; 05-06 May 2023; Tunisia*,
https://doi.org/10.1007/978-981-99-7950-9_8

4] and bilayers armors consisting of ceramic and composite plates [5]. Based on Artificial Neural Networks, Malik et al. [6] evaluated the absorbed energy of composite laminate subjected to low velocity when the thickness, the stacking sequence, and the number of layers are varied. The ballistic performance of the composite increases with the thickness and the number of laminate layers. Khan et al. [1] performed an experimental and numerical study to explore the ballistic resistance of a ceramic target against ogival nosed projectile. It was found that the absorbed energy decreases with the increase of the impact velocity. Moreover, the damage in the target increases with the angle of obliquity. Feli et al. [7] analyzed the ballistic perforation of ceramic composite armors against cylindrical tungsten projectiles. The results show that the delamination of composite layers and the angle of the ceramic cone decrease with increasing the impact velocity.

Motivated by the previously mentioned research, a new design of body armor was developed. This body armor was composed of two layers of composite materials and a ceramic layer inserted in the middle. The objective of this work is to study the ballistic resistance of multiple layer armor under single hit using finite element method. Furthermore, the effect of design parameters (ceramic and composite thicknesses) as well as the effect of the impact velocity on the ballistic performance (residual velocity and target damage) were examined.

## 2 Finite Element Modeling

### *2.1 Description of the Finite Element Model*

To simulate the ballistic impact of ceramic and composite materials against ogival nosed projectile, a Finite Element (FE) model is developed through the Abaqus / explicit code. Figure 1 shows a description of the finite element model and meshing. The target is composed of three layers, two composite layers, and a ceramic plate inserted in the middle of these two layers. The thickness of the forward composite layer, the ceramic layer, and the back composite layer are 1.5 mm, 5 mm, and 3 mm, respectively. The dimension of the target is 120*120*9,5 $mm^3$. The projectile is in hard steel, with a diameter of 7.62 mm and a mass of 30 g. The lateral sides of the target are embedded, and an impact velocity of 275 m/s is applied to the impactor, which is assumed to be deformable. The ceramic layer and the projectile are meshed with C3D8R three-dimensional elements, while the composite layer is meshed with SC8R elements. A general contact algorithm is adopted between the projectile and the target and between the target layers. The coefficient of friction between the projectile/ ceramic plate and the projectile/ composite plate is assumed to be negligible; this assumption was adopted based on the high impact speed.

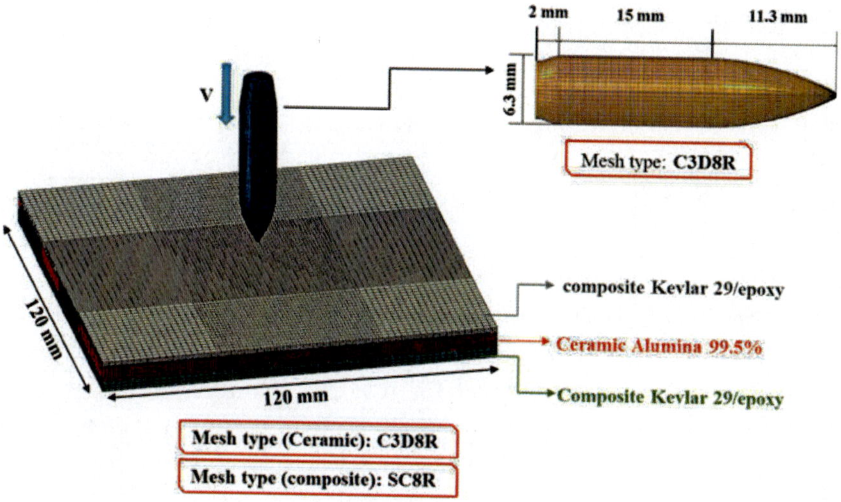


**Fig. 1** Description of the finite element model and meshing

## 2.2 *Material Properties*

In this study, two behavior laws are used for modeling the target material. The composite Kevlar/ epoxy is modeled as an anisotropic material, and the Hashin criterion is adopted to describe the damage initiation and evolution of this material. The Johnson-Holmquist model [Holmquist et al. 1993] (JH-2) is employed to describe the behavior and the damage of the ceramic plate. The material parameters are summarized in the following tables (Tables 1 and 2). The Johnson–Cook (JC) behavior law and failure model are applied to the projectile, which is made of steel 4340. The Johnson cook parameters are illustrated in the Table 3.

**Table 1** Properties of Kevlar/epoxy [8]

| $E_1$ (GPa) | E2 (GPa) | E3 (GPa) | $\upsilon_{12}$ | $\upsilon_{13}$ | $\upsilon_{23}$ |
|---|---|---|---|---|---|
| 25.63 | 25.63 | 7.5 | 0.2 | 0.3 | 0.3 |
| **$G_{12}$ (GPa)** | **$G_{13}$ (GPa)** | **$G_{23}$ (GPa)** | **$X^T$ (MPa)** | **$X^C$ (MPa)** | **$Y^T$ (MPa)** |
| 2.12 | 5.43 | 5.43 | 586 | 112 | 586 |
| **$Y^C$ (MPa)** | **$S^L$ (MPa)** | **$S^T$ (MPa)** | **$\rho$(Kg/m$^3$)** | | |
| 112 | 66 | 66 | 1251 | | |

**Table 2** Properties of Ceramic (Alumin 99%) [8]

| G (GPa) | HEL (GPa) | A | N | C | $\rho$ (Kg/m$^3$) |
|---|---|---|---|---|---|
| 90.16 | 19 | 0.93 | 0.6 | 0 | 3890 |
| **B** | **M** | $\sigma_{fmax}$ | **PHEL (GPa)** | $D_1$ | $D_2$ |
| 0.31 | 0.6 | 0.2 | 1.46 | 0.005 | 1 |
| **β** | $K_1$ **(GPa)** | $K_2$ **(GPa)** | $K_3$ **(GPa)** | | |
| 1 | 130.95 | 0 | 0 | | |

**Table 3** Properties of projectile material (steel 4340) [8]

| E (GPa) | υ | ρ(Kg/m$^3$) | A (GPa) | B (GPa) | n |
|---|---|---|---|---|---|
| 210 | 0.3 | 7850 | 0 0.95 | 0 0.725 | 0.375 |
| **C** | **m** | **D1** | **D2** | **D3** | **D4** |
| 0.015 | 0.625 | −0.8 | 2.1 | 0.5 | 0.002 |
| **D5** | $T_f$ **(K)** | $T_a$ **(K)** | $\varepsilon^{\cdot 0}_p$ ($s^{-1}$) | | |
| 0.61 | 1793 | 293 | 1 | | |

# 3 Results and Discussion

## *3.1 Validation of the Finite Element*

The validation of the finite element model is carried out for each component of the target, independently. Firstly, the composite model is validated by the trials performed by Millan et al. [4] (Table 4). Secondly, the results' issues from the ceramic model are validated by those obtained by Khan et al. [1] (Fig. 2). The FE model made of a composite plate is vetted by comparing the residual velocity of the numerical model to the experimental one from the literature. Moreover, the FE model of the ceramic is validated in terms of damage on the backside of the target. It can be noticed that a strong fit is obtained between our numerical results and the results of the literature for both models. Both numerical and experimental results show that radial cracks form on the back surface of the ceramic plate.

**Table 4** Comparison between our numerical results and the results of the literature

| Our numerical model | Impact velocity = 300 m/s | Residual velocity = 243 m/s |
|---|---|---|
| Literature [4] | Impact velocity = 300 m/s | Residual velocity = 245 m/s |

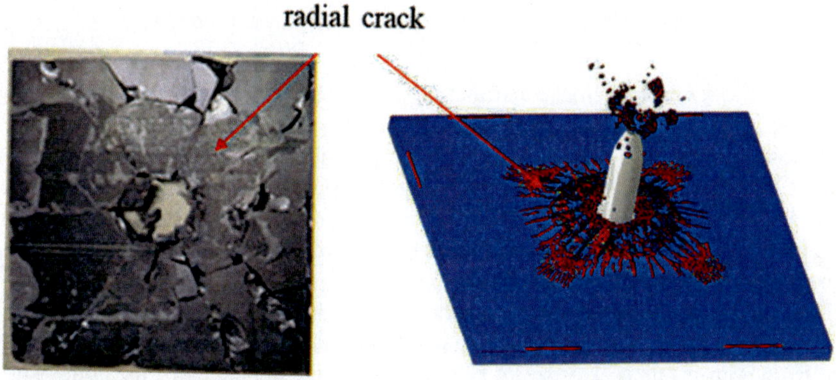


**Fig. 2** Comparison of numerical and experimental ceramic failure modes

## 3.2 *Effect of Impact Velocity*

To highlight the effect of the impact velocity on the residual velocity, three impact velocities 275, 650, and 850 m/s were tested. Figure 3 shows the evolution of the residual velocity versus the impact velocity. It can be noted that the residual velocity is highly influenced by the impact velocity. The increase in the impact velocity results in a decrease in the average relative difference between the impact velocity and the residual velocity. For an impact speed of 275 m/s, the observed residual velocity is 164 m/s, i.e., an average relative error of 40%. For an impact velocity of 850 m/s, the residual velocity is 792 m/s, so the average relative error is 6%. It can be concluded that the performance of the body armor, meaning the energy absorbed, decreases as the impact velocity increases.

**Fig. 3** Effect of the impact velocity on the residual velocity

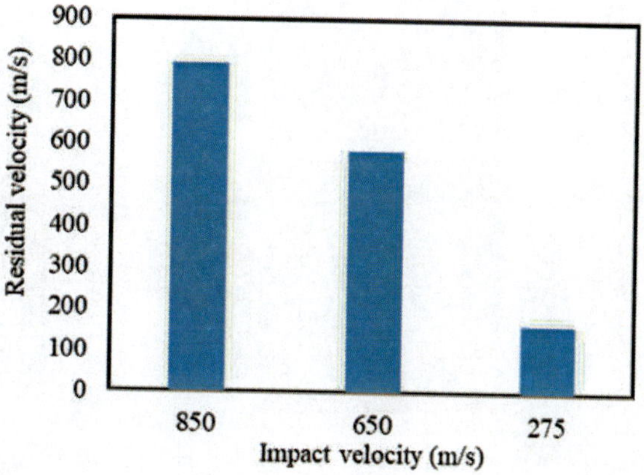

### 3.3 Comparison Between Two- and Three-Layer Targets

Figure 4 illustrates the damage distribution in the ceramic plate and composite back layer for bilayer target and three-layer target. It is noted that the damage to the composite part for the bilayer target is more pronounced than that of the tri-layer target. In fact, for the three-layer target, the absorbed energy is spread over three layers instead of two, minimizing the damage to the rear composite layer. It is observed that the fracture in the composite plate is oriented toward the transverse direction of the fiber, which is explained by the low strength of the matrix compared to that of the fiber. Concerning the damage of the ceramic plate, it can be seen that radial cracks form in the ceramic plate, and these are more important in the three-layer target than in the two-layer target. This result is explained by the deflection of the composite plate in front of the ceramic plate, which provides the propagation of cracks in the ceramic plate.

To examine the effect of adding a composite layer in front of the ceramic layer, a comparison between bilayer target formed with ceramic layer and composite backing layer and three-layer target, in which the ceramic is inserted in the middle of two composite layers, is shown in Fig. 5. It is found that the addition of a composite layer in front of the ceramic plate has a negligible influence on the residual velocity. This can be explained by the small thickness of the added composite layer. Therefore, the objective of the following section is to investigate the effect of the composite and ceramic thicknesses on the residual velocity.

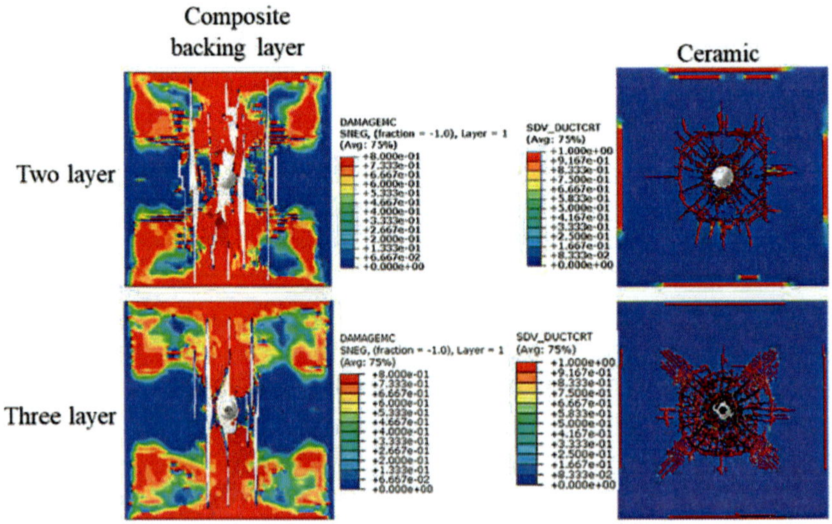


**Fig. 4** Comparison of damage between two- and three-layer targets

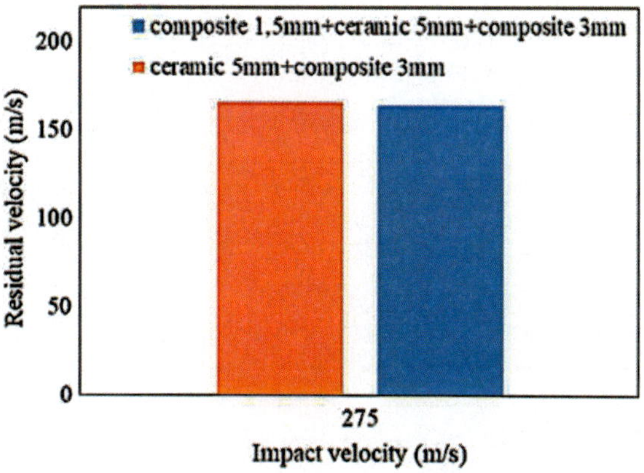


**Fig. 5** Comparison of residual velocity between two- and three-layer targets

### 3.4 *Effect of the Design Parameters*

To evaluate the effect of composite and ceramic thicknesses on the residual velocity, different tests were performed (Table 5). Figure 6 illustrates the evolution of residual velocity as a function of composite and ceramic plate thicknesses. It can be seen that increasing the thickness of the back composite layer (from test 1 to test 2) and the front composite layer (from test 2 to test 3) has a little influence on the residual velocity. Indeed, increasing the thickness of the composite from 3 to 5 mm generates a decrease in the residual velocity from 164 to 156 m/s, resulting in an average relative difference of 4%. Moreover, when the composite thickness passes from 3 to 10 mm (from test 3 to test 4), an average relative error of 16% is detected. This proves that the thickness of the composite has an influence on the residual velocity, if it undergoes a very large variation. The effect of the ceramic thickness on the residual velocity is observed in tests 5 and 6. We can see that when the ceramic thickness increases from 5 to 8 mm, the residual velocity decreases of about 65% and the impactor is totally penetrated. However, when the ceramic thickness reaches 10 mm, the impactor is completely blocked, and the kinetic energy of the projectile is totally absorbed by the target. It can be inferred that the ceramic thickness has a great influence on the ballistic resistance.

Figure 7 shows the target penetration depth (pd) for different ceramic and composite thicknesses. It is observed that the ceramic and composite thicknesses have a significant influence on the target deformation. When the composite plate thickness varies (from test 2 to test 3), the depth of penetration varies from 20 to 11.9 mm. A decrease of about 40% was detected. As the thickness of the ceramic plate increases (Trial 3 to Trial 5), the depth of penetration decreases from 11.9 to 2.6 mm. A decrease of about 78% was observed. It can be deduced that increasing the thickness of the composite and ceramic plates leads to a decrease in the target deformation. The low penetration depth is observed for high ceramic thickness.

**Table 5** Performed tests

| Tests | Composite front layer (mm) | Ceramic plate (mm) | Composite back layer (mm) |
|---|---|---|---|
| Test 1 | 1.5 | 5 | 3 |
| Test 2 | 3 | 5 | 3 |
| Test 3 | 3 | 5 | 5 |
| Test 4 | 10 | 5 | 5 |
| Test 5 | 3 | 8 | 5 |
| Test 6 | 3 | 10 | 5 |

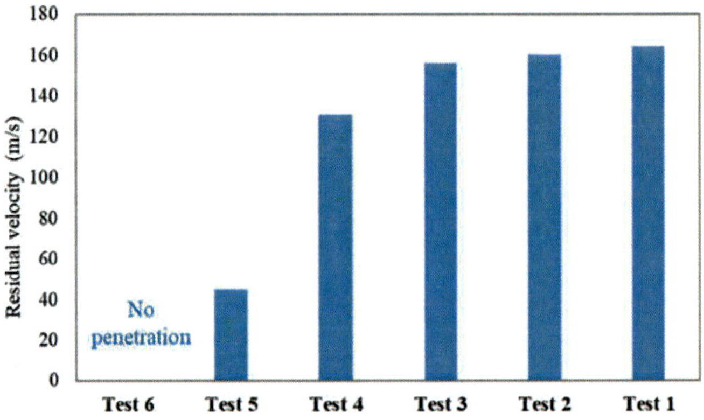


**Fig. 6** Effect of the composite and ceramic thicknesses on the residual velocity

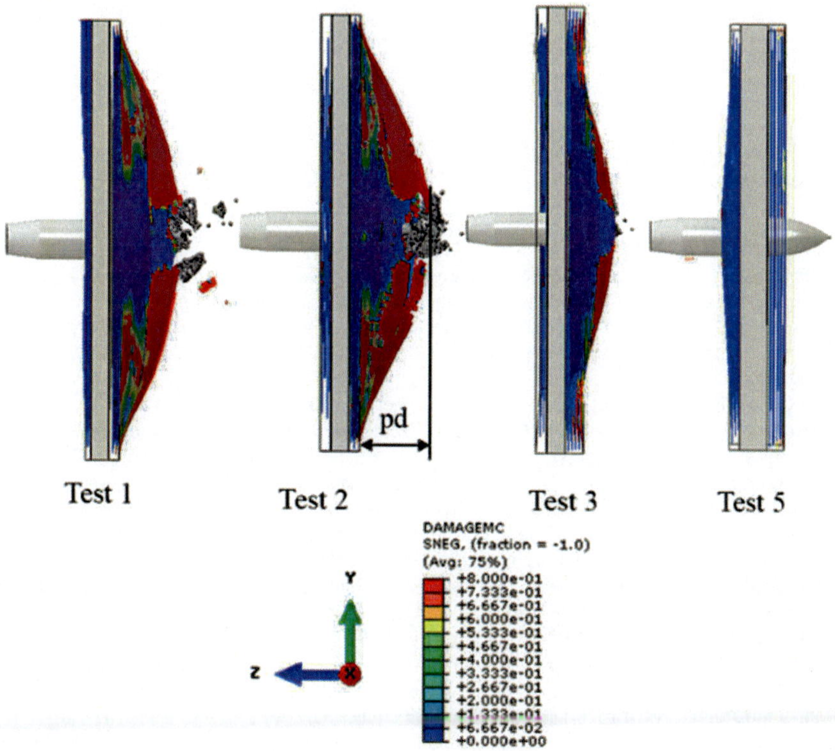


**Fig. 7** Effect of composite and ceramic thicknesses on the depth of penetration

# 4 Conclusion

This paper studied the ballistic performance of multiple layer armor formed with two composite layers and one ceramic layer inserted in the middle against ogival nosed projectile. The influence of the impact velocity as well as the influence of ceramic and composite thicknesses on the residual velocity and the target damage are examined using numerical simulation. It was found that the ballistic performance of body armor decreases with increasing the impact velocity. In addition, ceramic and composite thicknesses have a considerable influence on the residual velocity. The residual velocity is more affected by the ceramic thickness than the composite thickness. The target deformation decreases with increasing both ceramic and composite thicknesses.

The perspectives envisaged in this study are the investigation of the effects of the projectile obliquity angle and its diameter on the ballistic performance of the bulletproof vest.

**Acknowledgements** I would like to thank the Applied Mechanics and Engineering laboratory (LMAI) of the National Engineering School of Tunis for the Abaqus license.

## References

1. M. K. Khan, M. A. Iqbal, V. Bratov, N. F. Morozov, and N. K. Gupta, "An investigation of the ballistic performance of independent ceramic target," *Thin-Walled Structures*, vol. 154, no. January, p. 106784, 2020, doi: https://doi.org/10.1016/j.tws.2020.106784.
2. R. Zaera and V. Sánchez-Gálvez, "Modelling of fracture processes in the ballistic impact on ceramic armours," *Journal De Physique.*, vol. 7, no. 3, 1997, doi: https://doi.org/10.1051/jp4:19973117.
3. G. Gopinath, J. Q. Zheng, and R. C. Batra, "Effect of matrix on ballistic performance of soft body armor," *Compos Struct*, vol. 94, no. 9, pp. 2690–2696, Sep. 2012.https://doi.org/10.1016/j.compstruct.2012.03.038
4. M. Rodríguez Millán, C. E. Moreno, M. Marco, C. Santiuste, and H. Miguélez, "Numerical analysis of the ballistic behaviour of Kevlar® composite under impact of double-nosed stepped cylindrical projectiles," *Journal of Reinforced Plastics and Composites*, vol. 35, no. 2, pp. 124–137, Jan. 2016, doi: https://doi.org/10.1177/0731684415608004.
5. Z. Fawaz, W. Zheng, and K. Behdinan, "Numerical simulation of normal and oblique ballistic impact on ceramic composite armours," *Compos Struct*, vol. 63, no. 3–4, pp. 387–395, 2004.https://doi.org/10.1016/S0263-8223(03)00187-9
6. M. H. Malik and A. F. M. Arif, "ANN prediction model for composite plates against low velocity impact loads using finite element analysis," *Compos Struct*, vol. 101, pp. 290–300, Jul. 2013.https://doi.org/10.1016/j.compstruct.2013.02.020
7. S. Feli and M. R. Asgari, "Finite element simulation of ceramic/composite armor under ballistic impact," *Compos B Eng*, vol. 42, no. 4, pp. 771–780, Jun. 2011.https://doi.org/10.1016/j.compositesb.2011.01.024
8. G. Guo, S. Alam, and L. D. Peel, "Numerical analysis of ballistic impact performance of two ceramic-based armor structures," *Composites Part C: Open Access*, vol. 3, no. October, p. 100061, 2020, doi: https://doi.org/10.1016/j.jcomc.2020.100061.

# Effect of the Cultivar on the Development Hemp Stem and Textile Fiber in the East of France

**Aurélie Decker, Adrien Tritter, Vivien Sarazin, Jean-Yves Drean, and Omar Anis Harzallah**

**Abstract** Historically, the textile hemp fiber was cultivated in the East of France until the 1930s. Since a decade, textile hemp cultivation has been re-examined because hemp is a great water and fertilizer consumer. So, SADEF, specialist in the plant nutrition and analysis, produces and studies hemp growing in the East of France. The hemp genotype has evolved, so four cultivars (Férimon 12, Santhica 27, Santhica 70 and Fibror 79), whom three are used for the fiber production, are studied at different development stage (development, before and after flowering and seed maturity). The study is the stem characterization (color, height, fresh mass and yield) carried out by SADEF and fiber characterization (apparent diameter and mechanical properties) carried out by LPMT, laboratory of physical and mechanical textile. During the hemp cultivation, the fiber stress, strain and Young's modulus increased like the stem height. However, the fiber diameter decreased between June and October. The fresh stem mass decreased at the cultivation end. The optimal cultivation duration was October in the East of France. The cultivar Fibor 79 has stunted growth and a yellow stem color. But, there is no significant difference in morphological and mechanical properties between the different cultivar along the cultivation. Cultivar selected for the dual-production (Férimon 12) produced the same fiber quality than cultivar selected for the fiber production (Santhica 27, Santhica 70 and Fibror 79). So, the four cultivar were adapted to the East of France hemp cultivation.

**Keywords** Hemp · Cultivation · Fiber

A. Decker (✉) · J.-Y. Drean · O. A. Harzallah
Université de Haute-Alsace, LPMT UR 4365, 68100 Mulhouse, France
e-mail: aurelie.decker@uha.fr

Université de Strasbourg, Strasbourg, France

A. Decker · A. Tritter · V. Sarazin
SADEF, 30, Rue de La Station, Aspach-Le-Bas, France



S. B. Abdessalem et al. (eds.), *Proceedings of the Second International Conference of Innovative Textiles and Developed Materials-ITDM'2; 05-06 May 2023; Tunisia*,
https://doi.org/10.1007/978-981-99-7950-9_9

# 1 Introduction

According to the cultivar catalogue of Hemp-it (French hemp seed) [1], each cultivar has a specific use: production of stem, seed and/or fiber. Santhica and Fibror are two cultivars used for textile fiber production. In particular, Fibror 79 was created to absorb less azote for improve scutching [1, 2]. Hemp production in France for the textile industry is mainly carried out in Normandy with the cultivar Santhica 27. Other regions of France are interested in hemp fiber production, such as the East of France, in order to recreate a historic textile industry [3, 4]. Currently, the hemp production in the East of France is used mainly for the paper and building industry. Cultivar specificity changes with the localization due to climate and sunshine duration [5–7]. So, a cultivar intended for fiber production in Le Havre, West of France (Standard localization of Hemp-it) [1] may change specificity in the other part of France. By definition, the development duration depends on the genotype, an early cultivar is in flowering faster than a late cultivar.

The stem and fiber characteristics vary according to the localization, the cultivar and the cultivation duration [8–14]. According to the study of Vandepitte et al. on hemp cultivation in Belgium in 2018, the height was impacted by the cultivar. The height of Santhica 70 was higher that of Santhica 27 [11]. The stem hemp height increases all along the cultivation [5, 13, 15]. In particular, in the first month, the height increase is quick, then after flowering, it's more slow [5, 6, 15]. The stem height increase depends on the cultivar [5, 15]. The cultivars Uniko B and Kompoiti Hybrid TC studied by Sankari and Mela in Finland were growing faster than the cultivar Uso 11 and Uso 31 [15].

The cultivar affects not only the stem characteristics but also that of the hemp fiber [16]. The cultivar impacts the fiber fineness [8, 17]. The study of Amaducci et al., shows that the Tibor cultivar produces finer fibers than Futura cultivar grown in Italy in 2004 [8]. Moreover, cultivar impacts the mechanical properties too, in particular, the breaking stress [11, 14, 17–19]. The cultivar effect on the breaking strain and the Young's modulus is weak [11, 14, 17]. Fiber characteristics change all along the hemp cultivation [13, 20–22]. The breaking stress increases significantly during the first cultivation weeks [21, 22], then after flowering, the breaking stress appears to decrease [20, 21]. The impact of the cultivation duration is very small on the breaking strain and the Young's modulus. Finally, the hemp fiber diameter decreases between the end of flowering and the maturity of seed [21].

For the production of fiber in the East of France, the cultivar coupled to the harvest time is studied. The analysis of the impact of the cultivar on the optimal harvest time is performed. Firstly, the stem characteristic (color, height, weight and yield) was studied to understand the hemp plant development. Then, the stems were peeling to extract the fiber bundle. The fiber bundles were combined to obtain technical fiber. Secondly, the fiber characteristic, like diameter and mechanical properties, was studied to understand the impact of the cultivation duration and the cultivar on the hemp fiber quality. Moreover, a link between the stem and the fiber characteristics was realized.

# 2 Material and Methods

## 2.1 *Raw Material*

In 2018, four hemp monoecious (Canabis sativa L.) cultivars were carried out in the East of France by SADEF laboratory. The selected cultivars have a different genotype. The seeds came from the French seed producing, Hemp-it [1]. Férimon 12 is early cultivar and is used for seed and stem production. Whereas, the three other cultivars are used for textile fiber production. Santhica 70 and Fibror 79 are late cultivars, unlike Santhica 27 is early cultivar. Moreover, Fibror 79 was selected to make scutching easier and to be adapted to the fiber textile industry.

The soil was clay-sandy silt, it's proportionally composed of clay, silt and sand. The pH was 6. But, the soil contains a lot of stones that facilitate soil dryness. Tillage and mechanical weeding were carried out to prepare the soil before sowing. The soil was fertilized with chemical fertilizers to boost the growth of hemp plant. The applied doses are one hundred units per hectare of nitrogen and potassium and fifty units per hectare of phosphorus. Moreover, the soil was limed to reduce soil acidity. No herbicide and no fungicide were used during the hemp cultivation. The seeding was carried out on 28 may with a seed density of 50 kg ha$^{-1}$.

The data of temperature and rain were collected from a metrological station and shown at Fig. 1. The meteorological data came from the website: www.meteociel.fr.

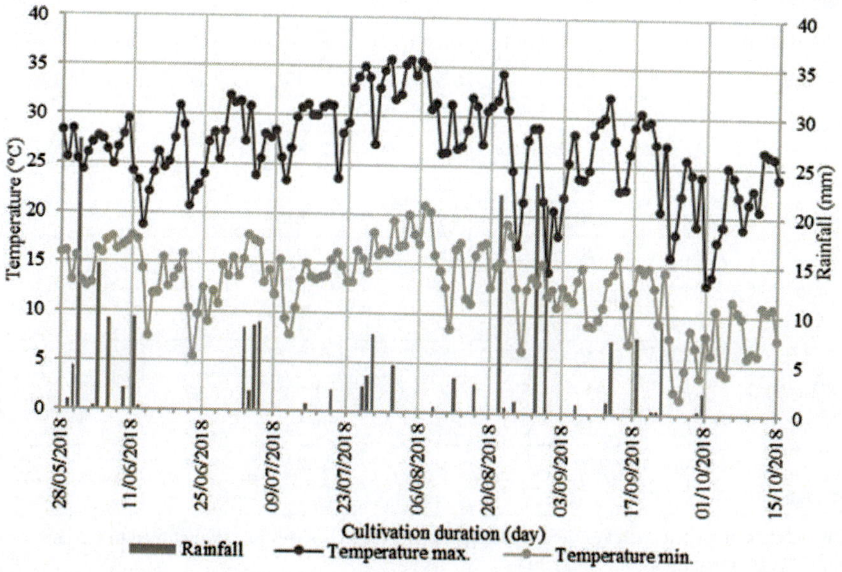


Fig. 1 Daily weather conditions (minimum and maximum temperature and rainfall) during the 2018 hemp cultivation

## 2.2 Stem Characterization

The cultivation duration impacts the hemp plant. Table 1 shows the stem characteristics (cultivation color, height, fresh mass and yield) estimated for the four cultivars and the cultivation duration. The three harvest time corresponded to the plant development stage (June), the flowering stage (July) and the after-flowering stage (August).

The color defines the plant development. The darker green was the plant, the more development the cultivation was considered. Four measurement was performed using a predefined color scheme from 1 (yellow) to 10 (very dark green) in field. The appreciation of the color was dependent on the operator and the weather.

The stem height was measured with a ruler in field expressed in cm. Twenty stems are measured randomly by cultivar and by harvest time. The height was taken from the ground to the top of the hemp stem.

The stem fresh mass was measured with Mettler Toledo PC400 weighing scales expressed in g. The stems were taken in the field and measured in the laboratory. The bundle of five stems was measured for each cultivar and harvest time.

Finally, the stem yield was calculated from the fresh mass and the plant density. The plant density corresponded to the plant number in a 0.25 $m^2$-square measured in June. The stem yield was expressed in kg $m^{-2}$.

**Table 1** Stem characteristics and confidence interval for the height estimated for different cultivar and cultivation duration

| Cultivar | Month | Height (cm) | Color(/10) | Mass (g) | Stem Yield (kg $m^{-2}$) |
|---|---|---|---|---|---|
| Férimon 12 | June | 65 ± 3 d | 6.0 | 14.36 | 1.84 |
| | July | 133 ± 12 bc | 6.7 | 30.48 | 3.90 |
| | August | 154 ± 18 ab | 7.5 | 25.83 | 3.31 |
| Santhica 27 | June | 63 ± 4 d | 5.5 | 15.12 | 1.51 |
| | July | 130 ± 8 bc | 6.9 | 22.84 | 2.28 |
| | August | 140 ± 8 ab | 6.5 | 17.93 | 1.79 |
| Santhica 70 | June | 66 ± 3 d | 5.8 | 9.56 | 1.15 |
| | July | 140 ± 7 ab | 6.5 | 26.22 | 3.15 |
| | August | 140 ± 6 ab | 7.5 | 15.73 | 1.89 |
| Fibror 79 | June | 53 ± 3 e | 4.6 | 11.00 | 0.67 |
| | July | 113 ± 7 c | 5.9 | 21.34 | 2.48 |
| | August | 160 ± 17 a | 2.0 | 32.23 | 3.74 |
| p-value | | < 0,001 | / | / | / |

The letters a to e correspond to the comparison of the impact of the cultivar coupled to the harvest time in the same column according to an ANOVA analysis

**Fig. 2** Photo of a specific cardstock framework and a fiber

## 2.3 Fiber Characterization

The fiber development changes according to the stem development as shown previously. Ten hemp stems of each cultivar were randomly taken in June, July and October to observe the fiber apparent diameter and mechanical properties. The stems were manually peeling to separate fiber bundles and shives. Then, fiber bundles were brushing to extract the technical fiber. Fifty fibers were randomly selected and prepared on a specific cardstock framework with a 20 mm-gauge length, Fig. 2. The environment room conditions were a temperature of 20°C and a relative humidity of 30%.

According to the adaptation of NF EN ISO 137 standard, fiber apparent diameter was measured with a projection microscope, Projectina®. Five diameters were measured along the fiber.

According to NF EN ISO 2062 standard, the tensile test was performed with MTS® to obtain the fiber mechanical properties like breaking force, strain and stress and the Young's modulus. The specific cardstock framework was placed in the clamp. MTS' setup was 20mm gauge, a 2N cell and a 1mm $min^{-1}$ constant crosshead displacement rate.

## 2.4 Statistical Analysis

The data analysis of the stem height and the fiber characteristic (diameter and mechanical properties) is performed according to a one-way analysis of variance (ANOVA) with the software R®.

# 3 Results and Discussion

## *3.1 Weather*

The weather data is shown Fig. 1. The weather conditions were considered as normal. The maximum temperature was about 35 °C and the minimum temperature was close to 5 °C at the end of the cultivation.

## *3.2 Stem Characterization*

Hemp stems changed along the cultivation [5, 13, 15]. The stem characterization (color, height, weight and yield) is shown in Table 1. The stems of Férimon 12, Santhica 27 and Santhica 70 cultivars darkened along the cultivation from light green (5.5–6) to green (7–7.5), Table 1. Hemp stem height increased significantly, especially from June (62 ± 2) cm to July (129 ± 5) cm on average whatever the cultivar [5, 15]. The stem mass almost doubled between June and July for the four cultivars. Compared to Férimon 12, Santhica 27 and Santhica 70, the cultivar Fibror 79 stunted their growth, Table 1. Férimon 12, selected for the production of heep seed and straw, produced the best stem yield, 1.84 kg $m^{-2}$ in June and 3.90 kg $m^{-2}$ in July. In contrast with Santhica 27 and 70 and Fibror 79, which were selected for the production of fiber. From July to August, the height didn't change for the cultivars Férimon 12, Santhica 27 and Santhica 70 like the stem color, whereas the Fibror 79 height increased significantly from (130 ± 8) to (140 ± 8) cm like the mass from 21.34 to 32.23g. But, the stem color changed significantly from a green to a yellow. The color difference between Fibror 79 and the other cultivar is shown in Fig. 3: Fibror 79 in the left was yellower than Santhica 70 in the right in August. The cultivar Fibror 79 produced the best stem yield, 3.74 kg $m^{-2}$, Table 1.

In contrast, the mass of Férimon 12, Santhica 27 and Santhica 70 decreased slightly. After flowering, differences in height between cultivar were smaller according to ANOVA analysis. The height of the cultivar Fibror 79, (160 ± 17) cm, was the highest followed by Férimon 12, (154 ± 18) cm, then Santhica 27, (140 ± 8) cm and Santhica 70, (140 ± 6) cm [11].

The difference between the cultivar Fibror 79 and the other cultivar came from a lack of nitrogen predefined by the genetics [1, 2].

So, the hemp development depended on the cultivar. The cultivar impacted the hemp stem color, height and mass during the cultivation.

**Fig. 3** Photo of hemp plot comparing Fibror 79 (left) and Santhica 70 (right) in August 2018

### 3.3 *Fiber Characterization*

As the stem changed according to cultivar and harvest time, the fiber could change. Table 2 represents the fiber diameter and mechanical properties. All along the hemp cultivation, there was no difference in diameter, Young's modulus, breaking force and stress between cultivars according to an ANOVA analysis, Table 2. In contrast in July, there was a significant difference in breaking strain between Santhica 70 (1.58 $\pm$ 0.20) % and Fibror 79 (2.24 $\pm$ 0.25) %.

As fiber characteristics were similar between cultivars, Fig. 4 represents the typical strain stress curve of the Férimon 12 cultivar. At the beginning of the hemp cultivation, the fiber diameter was similar around 100 μm whatever the cultivar. Then, the hemp fiber diameter decreased along the cultivation [19, 23, 24]. The fibers were refined during the plant growth, as well as the efficiency of the manual peeling was improved. This decrease was significant for Fibror 79 cultivar, whose diameter decreased from (94.48 $\pm$ 8.11) μm in June to (58.9 $\pm$ 6.15) μm in October. Whereas Férimon 12 decreased from (105.27 $\pm$ 11.08) μm in June to (70.75 $\pm$ 6,15) μm in October. So, the diameter of Fibror 79 cultivar decreased faster than that of Férimon 12. The diameter of Fibror 79 was slightly thinner than the other cultivar corresponding to the stunted growth. The breaking force increased between June and October in particular for the cultivar Férimon 12 from (1.15 $\pm$ 0.24) N to (1.83 $\pm$ 0.28) N. The increase of the breaking force was not significant for Santhica 27, Santhica 70 and Fibror 79. The Young's modulus and the breaking strain and stress increased significantly from June to October whatever the cultivar, Table 2 [23]. It was corresponding to the

**Table 2** Fiber characteristics and confidence interval at 95% estimated for different cultivar and cultivation duration

| Cultivar | Month | Diameter (μm) | Breaking force (N) | Breaking stress (MPa) | Breaking strain (%) | Young's modulus (GPa) |
|---|---|---|---|---|---|---|
| Férimon 12 | June | 105.27 ± 11.08 a | 1.15 ± 0.24 b | 156 ± 34 f | 1.88 ± 0.29 bc | 7.95 ± 1.55 e |
| | July | 78.43 ± 6.77 bd | 1.42 ± 0.25 ab | 310 ± 50 c | 1.88 ± 0.22 bc | 15.7 ± 1.85 ac |
| | October | 70.75 ± 6.14 cde | 1.83 ± 0.28 a | 495 ± 76 ab | 2.47 ± 0.22 a | 19.31 ± 2.32 ab |
| Santhica 27 | June | 107.81 ± 12.14 a | 1.35 ± 0.22 ab | 198 ± 42 ef | 1.92 ± 0.21 bc | 8.87 ± 1.78 e |
| | July | 88.15 ± 9.02 abc | 1.58 ± 0.31 ab | 276 ± 48 ce | 1.93 ± 0.25 bc | 12.36 ± 1.64 cd |
| | October | 66.79 ± 7.06 de | 1.85 ± 0.37 a | 536 ± 71 a | 2.61 ± 0.24 a | 19.04 ± 2.36 ab |
| Santhica 70 | June | 92.44 ± 10.24 ab | 1.19 ± 0.21 b | 209 ± 36 def | 1.87 ± 0.16 bc | 9.61 ± 1.34 de |
| | July | 85.43 ± 8.97 abc | 1.44 ± 0.24 ab | 297 ± 58 cd | 1.58 ± 0.2 c | 16.43 ± 1.87 ac |
| | October | 65.56 ± 5.96 de | 1.58 ± 0.26 ab | 500 ± 74 ab | 2.22 ± 0.23 ab | 20.98 ± 2.61 a |
| Fibror 79 | June | 94.48 ± 8.11 ab | 1.06 ± 0.15 b | 172 ± 28 f | 2.16 ± 0.23 ab | 6.89 ± 0.98 e |
| | July | 71.35 ± 7.06 cde | 1.33 ± 0.26 ab | 363 ± 54 bc | 2.24 ± 0.25 ab | 14.61 ± 1.71 bc |
| | October | 58.9 ± 6.15 e | 1.38 ± 0.28 ab | 549 ± 96 a | 2.63 ± 0.31 a | 18.65 ± 2.2 ab |
| p-value | | < 0.001 | < 0.001 | < 0.001 | < 0.001 | < 0.001 |

The letters a to f correspond to the comparison of the impact of the cultivar coupled to the harvest time in the same column according to an ANOVA analysis

increase of the stem height, mass and yield, Table 1. Finally, the fiber diameter and mechanical properties increased along the stem and the fiber development whatever the cultivar [20–22].

According to ANOVA analysis, the difference between the four cultivars was not significantly. Cultivar selected for the dual-production (Férimon 12) produced the same fiber quality than cultivar selected for the fiber production (Santhica 27, Santhica 70 and Fibror 79). So, all cultivar, early and late cultivars, could be cultivated in the East of France.

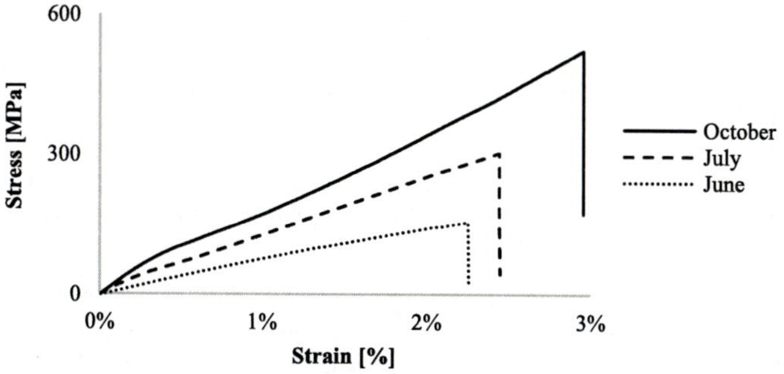


**Fig. 4** Typical strain stress curve case of Férimon 12

# 4 Conclusion

The harvest time was primordial for the hemp cultivation according to previous study. In the East of France, the optimal cultivation duration was October according to the diameter and the mechanical properties whatever the cultivar. From June to October, the stem characteristic changed like the height increase and the color modification. In parallel, the fiber characteristic changed too, like the diameter decrease and the stress and Young's modulus increase.

However, there is no significant difference in morphological and mechanical properties between the different genotypes along the cultivation. In contrast, Fibror 79 cultivar producing fiber has stunted. All cultivar can be grown in the East of France whatever the utilization. To optimize the hemp cultivation, early cultivar, like Férimon 12 and Santhica 27, is favored to have a mowing in August and a retting until September in the East of France.

# References

1. Hemp-it, https://www.hemp-it.coop/, 2022/03/01
2. Allard, L.-M., Augereau, L.: Guide de culture Chanvre, (2017)
3. Kapp, E., Schaefer, R.: La culture du chanvre dans la plaine rhénane: Importance économique et rôle de cette fibre textile dans la vie rurale de l'Alsace de jadis. Arts et traditions populaires. 13e Année, 35–52 (1965)
4. Stoskopf, N.: L'industrie en Alsace de la fin du XVII e siècle au début du XIX e siècle. Fédération des sociétés d'histoire et d'archéologie d'Alsace. 1129–1137 (2017)
5. Cosentino, S.L., Testa, G., Scordia, D., Copani, V.: Sowing time and prediction of flowering of different hemp (Cannabis sativa L.) genotypes in southern Europe. Industrial Crops and Products. 37, 20–33 (2012). https://doi.org/10.1016/j.indcrop.2011.11.017
6. Struik, P.C., Amaducci, S., Bullard, M.J., Stutterheim, N.C., Venturi, G., Cromack, H.T.H.: Agronomy of fibre hemp (Cannabis sativa L.) in Europe. Industrial Crops and Products. 11, 107–118 (2000).
7. Tang, K., Struik, P.C., Yin, X., Thouminot, C., Bjelková, M., Stramkale, V., Amaducci, S.: Comparing hemp (Cannabis sativa L.) cultivars for dual-purpose production under contrasting

environments. Industrial Crops and Products. 87, 33–44 (2016). https://doi.org/10.1016/j.indcrop.2016.04.026
8. Amaducci, S., Zatta, A., Pelatti, F., Venturi, G.: Influence of agronomic factors on yield and quality of hemp (Cannabis sativa L.) fibre and implication for an innovative production system. Field Crops Research. 107, 161–169 (2008).
9. Campiglia, E., Radicetti, E., Mancinelli, R.: Plant density and nitrogen fertilization affect agronomic performance of industrial hemp (Cannabis sativa L.) in Mediterranean environment. Industrial Crops and Products. 100, 246–254 (2017).
10. Cromack, H.T.H.: The effect of cultivar and seed density on the production and fibre content of Cannabis sativa in southern England. Industrial Crops and Products. 7, 205–210 (1998).
11. Vandepitte, K., Vasile, S., Vermeire, S., Vanderhoeven, M., Van der Borght, W., Latré, J., De Raeve, A., Troch, V.: Hemp (Cannabis sativa L.) for high-value textile applications: The effective long fiber yield and quality of different hemp varieties, processed using industrial flax equipment. Industrial Crops and Products. 158, 112969 (2020).
12. Vera, C.L., Malhi, S.S., Raney, J.P., Wang, Z.H.: The effect of N and P fertilization on growth, seed yield and quality of industrial hemp in the Parkland region of Saskatchewan. Can. J. Plant Sci. 84, 939–947 (2004).
13. Decker, A., Tritter, A., Sarazin, V., Harzallah, O., Drean, J.-Y.: Comparison of Hemp Cultivation and Fiber Quality in Two Different Types of Soil in the East of France in 2020. Presented at the The Fiber Society 2022 Spring Conference, Leuven, Belgium (2022)
14. Decker, A., Sarazin, V., Harzallah, O., Drean, J.-Y.: Evolution of the mechanical and morphological properties of two hemp cultivars (santhica 27 and félina 34) during the dew-retted. Presented at the AUTEX2022, Lodz, Poland (2022)
15. Sankari, H., Mela, T.J.N.: Plant Development and Stem Yield of Non-domestic Fibre Hemp (Cannabis sativa L.) Cultivars in Long-day Growth Conditions in Finland. Journal of Agronomy and Crop Science. 181, 153–159 (1998).
16. Zimniewska, M.: Hemp Fibre Properties and Processing Target Textile: A Review. Materials. 15, 1901 (2022).
17. Marrot, L., Lefeuvre, A., Pontoire, B., Bourmaud, A., Baley, C.: Analysis of the hemp fiber mechanical properties and their scattering (Fedora 17). Industrial Crops and Products. 51, 317–327 (2013).
18. Amarasinghe, P., Pierre, C., Moussavi, M., Geremew, A., Woldesenbet, S., Weerasooriya, A.: The morphological and anatomical variability of the stems of an industrial hemp collection and the properties of its fibres. Heliyon. 8, e09276 (2022).
19. Sankari, H.: Comparison of bast fibre yield and mechanical fibre properties of hemp (Cannabis sativa L.) cultivars. Industrial Crops and Products. 11, 73–84 (2000).
20. Keller, A., Leupin, M., Mediavilla, V., Wintermantel, E.: Influence of the growth stage of industrial hemp on chemical and physical properties of the fibres. Industrial Crops and Products. 13, 35–48 (2001).
21. Mazian, B., Bergeret, A., Benezet, J.-C., Malhautier, L.: A Comparative Study of the Effect of Field Retting Time on the Properties of Hemp Fibres Harvested at Different Growth Stages. Fibers. 7, 108 (2019).
22. Pickering, K.L., Beckermann, G.W., Alam, S.N., Foreman, N.J.: Optimising industrial hemp fibre for composites. Composites Part A: Applied Science and Manufacturing. 38, 461–468 (2007).
23. Duval, A., Bourmaud, A., Augier, L., Baley, C.: Influence of the sampling area of the stem on the mechanical properties of hemp fibers. Materials Letters. 65, 797–800 (2011).
24. Rahman Khan, Md.M., Chen, Y., Belsham, T., Laguë, C., Landry, H., Peng, Q., Zhong, W.: Fineness and tensile properties of hemp (Cannabis sativa L.) fibres. Biosystems Engineering. 108, 9–17 (2011).

# Energetic Valorization of Cutting Textile Waste by Transforming It into Thermal Insulation Products

**Nour Sghaier and Nourhene Dibej**

**Abstract** The quantity of textile cutting waste in Tunisia has been increasing in recent years. Tax-exempt raw materials have been exported, which are essential for reducing consumption through reusing and recycling textile waste. These textile scraps can serve as thermal insulation, effectively reducing heat loss through thermal conduction when placed in wall gaps, ceilings, and floors. Properly installed cuttings can improve a building's energy efficiency, leading to reduced heating and cooling costs. This study focuses on valorizing textile cutting waste as a source of thermal insulation products, presenting a unique approach to sustainable waste management and energy efficiency in construction. It offers a potential solution to decrease raw material consumption by promoting the reuse and recycling of textile products. The study examines the technical properties, performance, and suitability of insulation products for use in the construction industry. The first stage involves a chemical analysis of the textile cutting waste to identify its composition and characteristics. Subsequently, a process for designing and optimizing a textile thermal insulation sample is developed. Polyester binds are used to combine textile cuttings and fiber/PET in various ratios to produce different samples. These samples, including cotton (white and color) and polyester (white), undergo testing and evaluation using standard methods. The findings indicate that the insulation products demonstrate good thermal performance, making them highly suitable for construction applications.

**Keywords** Frayed of cotton cutting textile waste/polyester fibers · Polyester bonding · Conductivity · Diffusivity · Valorization

N. Dibej
The Higher Institute of Biotechnology of Monastir Road Taher Hadded, B.P 74 Monastir, Tunisia

N. Sghaier (✉)
University of Monastir, ENIM, Laboratory for the Study of Thermal and Energy Systems LESTE, LR99ES31, ENIM, 5000 Monastir, Tunisia
e-mail: nour.sghaier@enim.rnu.tn



S. B. Abdessalem et al. (eds.), *Proceedings of the Second International Conference of Innovative Textiles and Developed Materials-ITDM'2; 05-06 May 2023; Tunisia*,
https://doi.org/10.1007/978-981-99-7950-9_10

## 1 Introduction

Textile waste is a growing environmental concern, as it contributes to landfills and pollution. In recent years [1], there has been an increasing interest in recycling old clothes and other textile waste into new products [2]. In fact, several studies have explored the fact of recycling old clothes. For instance, a study conducted by Eder and colleagues (2018) [3] investigated the feasibility to use them as a raw material for textile recycling. The authors found that recycled textile products made from old clothes can have similar properties and qualities as virgin textile products. In addition, another study by Shafiqul Islam and Gajanan Bhat showed the recycled textiles to be environmental friendly thermal isolation [4] added to another study by Ankang Kan and colleagues [5] related to vacuum insulation panels prepared with recyclable fibrous cotton core. In Tunisia, textile cutting waste is a significant issue. More specifically, considerable amount of textile waste is exported without being utilized for its full potential. According to the National Institute of Statistics (2021) [6], the country exports around 70% of its textile waste, despite having the potential to transform it into value-added products.

To address this issue, we decided to valorize textile waste by transforming it into thermal insulation products as other authors have demonstrated in their studies [7]. Insulation products made from recycled textile waste, including insulation panels offer several benefits, including improved energy efficiency and reduced environmental impact.

In conclusion, the recycling of old clothes and textile cutting waste is a promising approach to reduce environmental impacts and create value-added products. By valorizing textile waste, Tunisian companies do not only address the issue of textile waste, but also create new economic opportunities.

## 2 Chemical Characterization of Textile

Before recycling textile waste, it is essential to conduct a thorough textile characterization to identify the materials that will be recycled to ensure that the recycled products have the needed properties and qualities. In the case of recycling cotton waste, for instance, it is important to confirm that the material is, indeed, cotton. For that, chemical tests, involving the use of various reagents that react with specific fibers to produce a color change or other observable changes, are usually performed. For instance, the solubility of cotton in acid can be confirmed by using Sulfuric Acid 70% [8]. In case of Sulfuric Acid solution added to a sample of cotton, it will turn Soluble.

Textile characterization is a crucial step in textile recycling. By identifying, the materials that will be recycled and confirming their properties, the process can be optimized to produce high-quality and value-added products. Alike, the use of chemical tests to confirm the presence of cotton in textile waste is commonly used to ensure the quality of recycled cotton products.

## 3 Sample Design

In addition to using recycled cotton and recycled polyester edge binding waste fibers, this study also adhered to the principles of eco-conception, as outlined in ISO 14026 [9]. It is a design approach that considers the environmental impact of a product or process throughout its entire life cycle, from raw materials extraction to end-of-life disposal. By incorporating eco-conception principles into the manufacturing process of these non-woven fabric bricks. Raw materials for this study were cotton and polyester scrap collected by a textile industry.

After analyzing textile cutting waste samples, we select cotton textile one. Then we unravel the textile. After that, we weighed the used textile, while respecting the desired proportion (cotton/polyester) and the homogenized mixture with the two components.

The process involves taking a frame of 27/27cm size and 5 cm thickness (see Fig. 1a). We, then, placed the fabric mixture onto the surface of the frame where the proportions of polyester and cotton were verified by weight and then homogenized (see Fig. 1b). We use baking paper over it to melt the polyester and connect it with the cotton (Fig. 1c). Then, we iron the fabric mixture. Also, we use a clamp until reaching 3 cm thickness (see Fig. 1d) and e). The resulting bricks (see Fig. 1f) are useful in the construction of buildings, providing effective thermal insulation. This collaboration between the textile and construction industries has resulted in a sustainable and innovative solution for thermal insulation in building construction.

Two textile samples (textile Brick) are prepared and composed of a mixture of polyester and cotton. The first one is 20% polyester used for connection and 80% cotton whereas the second is composed of 30% polyester used for connection and 70% cotton.

In all this publication, the first sample will be qualified by brick 1 and the second by brick 2.

## 4 Thermal Characterization

### 4.1 Conductivity

In the process of recycling textile waste into thermal insulation products, it is essential to ensure that the recycled samples have the required thermal insulation characteristics. This is typically achieved through various types of analysis, including thermal

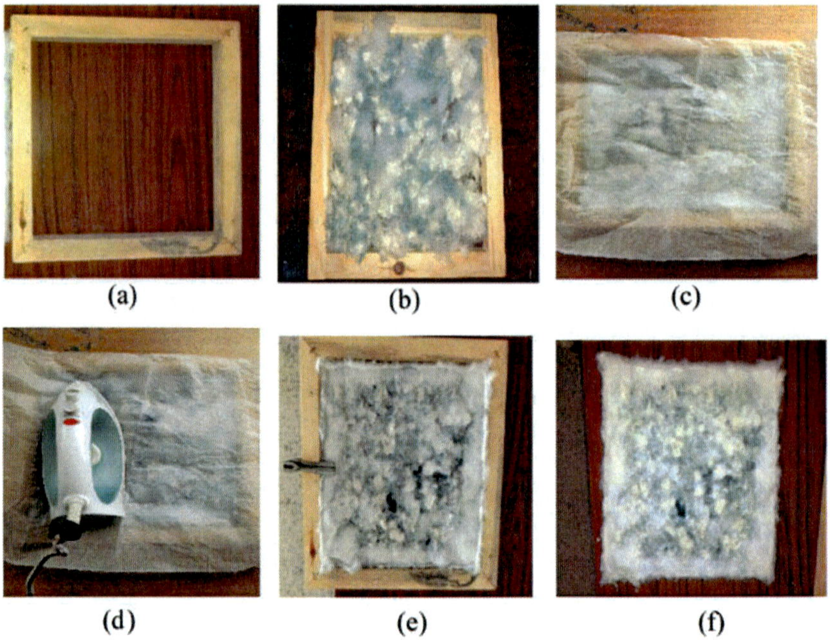

Fig. 1 Stages of sample preparation

conductivity measurements. In fact, it is considered to be one of the key thermal parameters determining the capacity of a material to provide thermal insulation. To measure thermal conductivity as shown in his study [10], one common method used, the box method (See Fig. 2), which involves placing a sample of the material (brick) in a copper box between two temperature-controlled surfaces and measuring the temperature difference across the sample. The heat flow through the sample can then be calculated from the temperature difference and the known dimensions of the sample.

The dispositive is composed of two volumes (A) and (B). More precisely, Volume (A) is the low-temperature room fixed with a cryostat and an exchanger whereas volume $B$ is the high-temperature room fixed with a rheostat. The sample placed between $A$ and $B$ and his surface temperature are measured and qualified respectively as ($T_f$) for the cold surface near to $B$ and ($T_c$) for the hot surface near to $A$.

The acquisition of different temperatures in function of time is established with acquisition chain Agilent 34970A allowing the storage and recording of the different temperatures ($T_a$, $T_B$, $T_c$, $T_f$) every 10 s in °C.

The thermal total conductivity of the sample can be determined as follows (Eq. 1):

$$\lambda_{\text{th.Total}} = \frac{e}{S\left(T_c - T_f\right)}\left[\frac{U_e^2}{R_e} - C_{th}(T_B - T_a)\right] \quad (1)$$

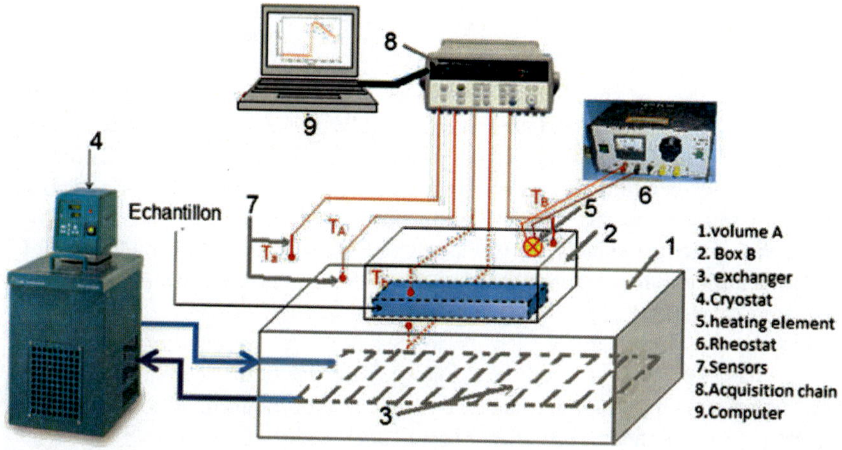


**Fig. 2** Box method for evaluating conductivity [13]

where $\boldsymbol{S}$ is the sample surface (m$^2$), $\boldsymbol{Re}$: electrical resistant of the heating element ($\Omega$) $\boldsymbol{e}$: the sample thickness (m) $\boldsymbol{Tc}$: temperature of the face of the copper box near the hot zone (°C), $\boldsymbol{Tf}$: temperature of the face of the copper box near the cold zone (°C), $\boldsymbol{T_B}$ : of the hot zone (°C), $\boldsymbol{Tc}$: temperature of the cold zone (°C). $\boldsymbol{Ue}$: tension generated by the heating element (V) and $\boldsymbol{C_{th}}$ is the heat loss coefficient, which is evaluated with a reference material equal to $C_{th} = 0.17$ (W/K).

The total thermal resistance is evaluated as follows:

$$R_{thT} = \frac{2e_{\text{copper}}}{\lambda_{\text{th copper}}} + \frac{e_{\text{sample}}}{\lambda_{\text{th sample}}} = \frac{e_t}{\lambda_{\text{th. Total}}} \quad (2)$$

where $\boldsymbol{e}_{\textbf{copper}} = 0.005$ m thickness of the box of copper, $\boldsymbol{\lambda}_{\textbf{th copper}} = 380$ w/k is the conductivity of the copper

$\boldsymbol{e_t}$ the total thickness of the sample and the box $\boldsymbol{e_t} = 0.04$ m, $\boldsymbol{e}_{\textbf{sample}} = 0.03$ m.

We can then deduce the thermal sample conductivity as:

$$\lambda_{\text{th Sample}} = \frac{e_{\text{sample}}}{\frac{e_t}{\lambda_{\text{th.Total}}} - \frac{2e_{\text{copper}}}{\lambda_{\text{thcopper}}}} \quad (3)$$

We can deduce the thermal resistance of de specimens as:

$$R_{\text{th}} = \frac{e}{\lambda_{\text{th}}} \quad (4)$$

## 4.2 Diffusivity

In order to determine the thermal diffusivity of our specimens, we modified the method boxes apparatus [10]. Our approach involved placing an incandescent lamp of 1000 W on the lower surface of the sample, specifically on zone A. This lamp provided a brief burst of thermal energy, which was then measured on the opposite face of the sample, located on side B of the box.

The diffusivity is the behavior of the sample to store heat and to conduct it.

The identification of the thermal diffusivity of the sample is performed by applying the method of Parker [11].

Assuming that the sample is well isolated from the surrounding atmosphere, Parker's model consists of briefly exciting the front face of the sample. The temperature evolution is then captured on the opposite side of the form, (See Fig. 3). This process produces a thermogram (see Fig. 4).

Parker established the method for determining the effective thermal diffusivity of the sample.

The diffusivity is then evaluated as:

$$\alpha_{th} = \frac{1.38e^2}{\pi^2 t_{\frac{1}{2}}} \quad (5)$$

where $(e)$ represents the thickness of the sample $(m)$, $(\pi)$ is 3.14 and $(t_{1/2})$ is the time needed to reduce the temperature of the back side with the halt of its maximum value.

The time required for the temperature at the back of a sample to drop by half of its maximum value is represented by (t1/2), while (e) stands for the sample thickness.

Having the conductivity and the diffusivity, we can estimate the massic heat by the relation:

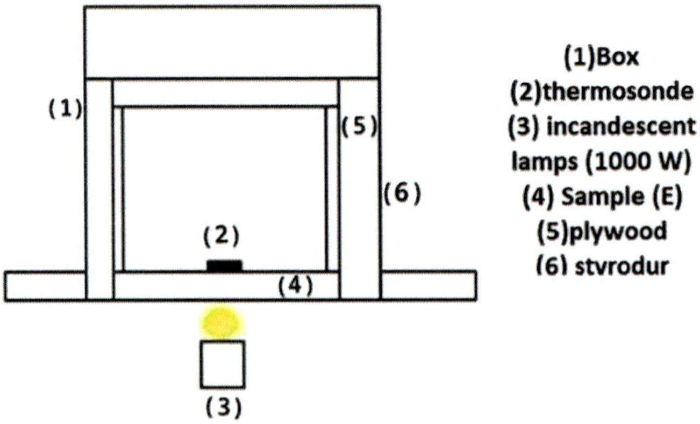


**Fig. 3** Box method to calculate the diffusivity

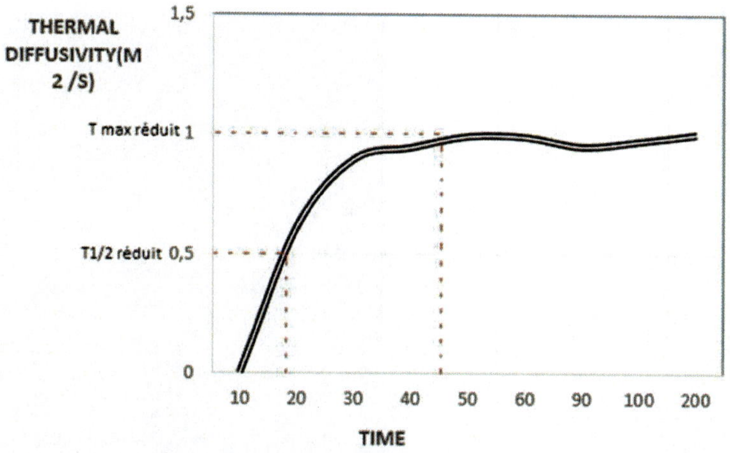


**Fig. 4** Parker thermogram [11]

$$C_p = \frac{\lambda_{th}}{\rho\alpha_{th}} \tag{6}$$

$\rho$ is the volumic mass (kg/m$^{3)}$.

# 5 Results

## *5.1 Textile Characterization*

Sulfuric acid ($H_2SO_4$) is a highly corrosive and reactive substance commonly used in laboratory experiments to identify the composition of materials [12]. Alike, it has been used to determine the various tested samples composition. After careful analysis of the results, we found that all the samples were made of cotton.

The process of testing involved adding a small amount of sulfuric acid to each one and observing any changes. Cotton is a natural fiber highly susceptible to the effects of sulfuric acid. When added, it breaks down the fibers and causes them to dissolve (See Fig. 5), leaving a brownish-yellow residue.

In our experiment, we observed that all the samples similarly reacted to the sulfuric acid, indicating that they were all made of cotton. This result was confirmed through further analysis, including microscopic examination and chemical testing.

**Fig. 5** Yellou residue after adding sulfuric acid to sample tissue

## 5.2 *Conductivity*

We measure the conductivity for different applicated electrical tension for the two samples brick 1 and brick 2 to evaluate the effects of the applicated heat and the sample temperature.

For each sample, five company measures are done and the average and standard deviations are estimated repetitions.

The conductivity calculated for each sample is the average of five done repetition for each temperature of the sample evaluated as $T_{sample} = (T_c + T_F)/2$.

For the brick 2, which is 30% polyester and 70% cotton (see Fig. 6), the conductivity was still too low under 0.05W/m K for the highest sample temperature to the extent that it can be isolatedly used in construction if we compare it with the order of thermal insulators conductivities developed from textile waste to those of other known insulators [13, 14] and [15] illustrated in Table 1.

It is also clear from Fig. 6 that the conductivities of brick 1 containing more cotton and less polyester than brick 2 is the lowest. The tows breaks are very good isolators as illustrated in comparison with the references illustrated in Table 1.

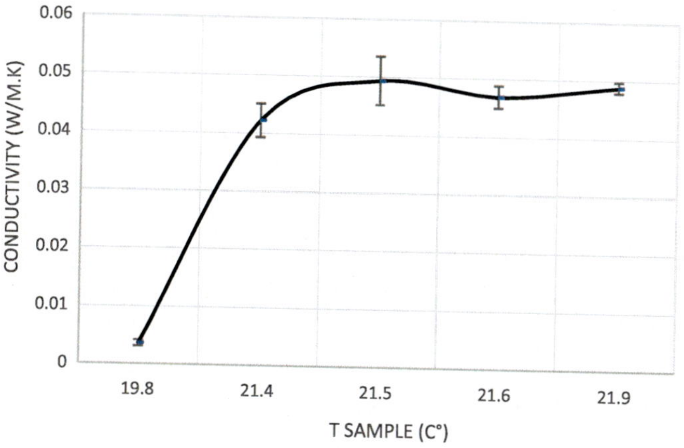


**Fig. 6** Variation of conductivity as a function of the sample temperature (brick 2)

**Table 1** Thermal properties of insulators developed from textile waste to those of other known insulators [13–17] and [10]

| Material | ρ(Kg/m$^3$) | λth (W/m K) | αth (m$^2$ /h)0.10$^{-3}$ Parker |
|---|---|---|---|
| Glass wool | 13–100 | 0.03–0.045 | 1.4–5.2 |
| Stone wool | 30–180 | 0.033–0.045 | 1.3–4.7 |
| Extruded polystyrene | 20–80 | 0.030–0.034 | - |
| Expanded polystyrene | 18–50 | 0.036–0.042 | 3.14–11.29 |
| Foam polyurethane | 27–60 | 0.030–0.032 | 1.89 |

## 5.3 *Diffusivity*

The thermo-physical properties of the studied samples were determined and compared to known civil construction insulators summarized in the table (Fig. 7).

The measurement of the diffusivity of a brick made from waste cuttings and polyester by the box method [11] with a result of 1.88.10$^{-6}$ m$^2$ /S is an important information in thermal insulation. Indeed, thermal diffusivity is a property that measures the material ability to conduct heat. A brick with a low thermal diffusivity will have a higher capacity to retain heat and therefore a better thermal insulation capacity. In this case, the low measured diffusivity of brick made from scrap cuttings and polyester can be an advantage in thermal insulation as well as a good indicator of a brick's good thermal insulating capacity. However, other factors must be considered to assess the overall thermal performance of the brick and its suitability for a specific thermal insulation use.

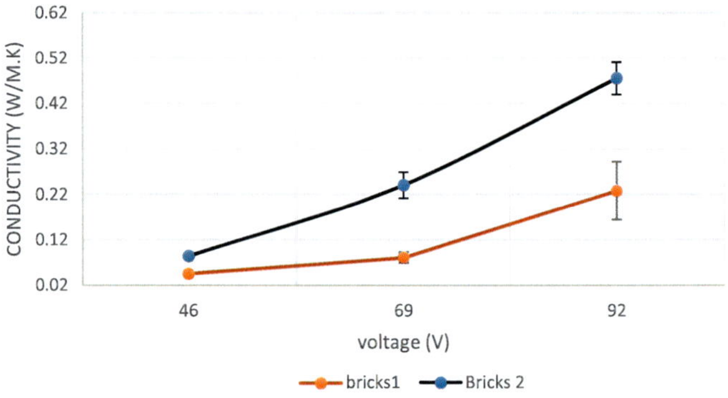


**Fig. 7** Variation of conductivity as a function of voltage

**Table 2** Summary of thermal analysis results

| Material | Thermal conductivity λth at 22°C (W/m. K) | Thermal diffusivity ($m^2$ /h)0.$10^{-3}$ | Thermal RESISTANCE $R_{th}$ ($m^2$ k/W) | specific heat Cp (J/Kg K) |
|---|---|---|---|---|
| Brick 1 (46V) | 0,046 | 6,768 | 0,42 | $5{,}72 \times 10^2$ |
| brick 2 (46V) | 0,058 | 8,496 | 0,31 | $5{,}72 \times 10^2$ |

In our case, more cotton leads to a better insulation (Table 2). For that, we must conduct a study for an optimization to have a brick with the lowest ratio of polyester that allows the cotton to stick on and the specimen to be durable.

## 6 Conclusion

In addition to being a promising solution for mitigating the environmental impact of the textile industry, the conversion of textile waste into thermal insulation products also aligns with the principles of eco-conception. This approach considers the environmental impact of a product or process throughout its entire life cycle, from raw materials extraction to end-of-life disposal. By incorporating eco-conception principles into the manufacturing process, the creation of sustainable and practical goods from textile waste as a raw material is achieved. The results of this study demonstrate the potential of energetic valorization of cutting textile waste to create high-quality thermal insulation for certain temperature that can compete with traditional insulation products, with an average thermal conductivity of 0.046 W/m K for the first sample (20% polyester-80% cotton) and with 0,058 W/m K for the second sample (30% polyester-70% cotton). Our results are very close to what is found in the literature. We have noticed that the average conductivity varies and increases by 0.2 W/m·K for bricks 1, and by 0.1 W/m·K for bricks 2. These changes in conductivity can have an impact on thermal insulation. Additionally, the thermal diffusivity of the first sample is measured at $6.768\times10^{-6}$ ($m^2/s$), while the second sample shows a higher value of $8.496\times10^{-6}$ ($m^2/s$). This indicates that the bricks with higher capacity have a better ability to retain heat. To ensure the accuracy of the results and optimize the manufacturing process, further testing is necessary. This includes measuring the product's insulation properties in different environments, optimizing the manufacturing process, and measuring the porosity, humidity as well as the effect of density. Overall, the energetic valorization of cutting textile waste into thermal insulation products has the potential to address the issue of textile waste and create a sustainable and useful product. Further research in this area is essential for its development.

**Acknowledgements** We would also like to express our gratitude to Mr. Ben Chiekh Maher, the head of the Study Laboratory Thermal and Energy Systems (LESTE), for providing us with the opportunity to conduct our research in the lab. His guidance, support, and encouragement throughout the project were invaluable and greatly contributed to its success.

We are immensely grateful to Mr. Mehouachi Mohamed Ameur for his assistance during the experimental phase of our research. His technical expertise and support were instrumental in carrying out the experiments effectively and obtaining accurate data.

Finally, we extend our gratitude to all other individuals, including laboratory technical staff of the LESTE laboratory, who provided support and assistance throughout the research process.

## References

1. Cartographie des déchets textiles en Tunisie et Maroc. Tunisie. [Performance]. Organisation des Nations Unies pour le développement industriel, (2021).
2. Preferred Fiber & Materials Market Report, october ,(2022)
3. Tasha L. Lewis, Huiju Park, Anil N. Netravali, Helen X. Trejo, «Closing the loop: a scalable zero-waste model for apparel reuse and recycling,» nternational Journal of Fashion Design, Technology and Education, vol. 10, p. Taylor and Francis Online homepage, (2017).
4. Shafiqul Islam, Gajanan Bhat, «Environmentally friendly thermal and acoustic insulation materials from recycled textiles, » Journal of Environmental Management, vol. 251, (2019).
5. Ankang Kan, Xuexiang Zhang, Zhaofeng Chen, Dan Cao, «Effective thermal conductivity of vacuum insulation panels prepared with recyclable fibrous cotton core. china, » International Journal of Thermal Sciences, vol. 9, (2023).
6. Institut National de la Statistique, » Statistique Exportation Déchets,(2021). [En ligne]. Available: http://www.ins.tn
7. Todorka Samardzioska, Milica Jovanoska, Vesna Grujoska, « Sustainable Thermal Insulation Derived from Recycled Textile Waste,» Cyril and Methodius University in Skopje, Faculty of Civil Engineering, vol. 4,(2023).
8. I. NABLI, Identification des matières textiles, Support de cours de Matériaux textiles, tunisia, 2006.
9. ISO-14062 Management environnemental —Intégration des aspects environnementaux dans la conception et le développement de produit, (2002).
10. A. HADDAD, recovery of textile waste: use in the field of thermal and acoustic insulation, Tunisie: Université de Monastir, Ecole Nationale d'Ingénieurs de Monastir, (2018).

11. W. J. Parker, R. J. Jenkins, c. P. Butler, G. L. Abbott, Flash Method of Detennining Thennal Diffusivity, Heat Capacity, and Thennal Conductivity,
12. Journal of Applied Physics, Volume 32, Issue 9, p.1679–1684, (1961).
13. GINETEX TABLE OF FIBERS, Belgium: European House of Textiles and Clothing.
14. Bruno HAY, Jean-Rémy FILTZ, Jean-Christophe BATSALE, «Mesure de la diffusivité thermique par la méthode flash - technique d'ingenieur,» [En ligne]. Available: https://matwo.fr/thibaud-marcel/wp-content/uploads/2021/09/matwo-r2955.pdf.
15. Ines Boulaoued, Guellouz Mohamed Sadok, Mhimid Abdallah & Jemni Abdelmajid, experimental determination of the thermo-physical properties of building insulating materials, vol. 3(2), Tunisia , (2012).
16. L.F. Cabeza, A. Castell, M. Medrano, I. Martorell, G. Pérez, I. Fernández, «Experimental study on the performance of insulation materials in Mediterranean construction,» Elsevier, vol. 42, pp. 630–636,( 2010).
17. Papadopoulos Agis, «State of the art in thermal insulation materials and aims for future developments, » Elsevier, vol. 37(1), pp. 77–86, (2005).
18. H. Lakrafli, S. Tahiri, A. Albizane, M. Bouhria, M.E. El Otmani, «Experimental study of thermal conductivity of leather and carpentry wastes,» elsevier, vol. 48, pp. 566–574, (2013).

# Functionalization of a Textile Surface by N2 Jet Technology to Improve Its Roughness for Medical Applications

**Maleke ZidI, Foued Khoffi, Yosri Khalsi, Abdel Tazibt, Frédéric Heim, and Slah Msahli**

**Abstract** Foreign Body Reaction (FBR) is a major problem when PolyEthylene Terephthalate (PET) textile implants are studied in the medical applications to treat conditions such as hernia development, arterial stenosis, or heart valve malfunction. Whatever the application, the natural porosity of textile materials tends to cause exaggerated tissue development after implantation, which may keep the implant from maintaining its flexibility. It has been shown in literature that one potential strategy to reduce the FBR process is to improve the surface roughness of the implanted material. The objective of this study is to adjust the topography of textile surfaces by inducing hairiness, generated by particle projection. For that purpose, a supercritical $N_2$ jet is used in order to supply sufficient velocity to sprayed particles, which result in local filament ruptures on the treated surface. In particular, the study examines the influence of jet projection parameters such as jet pressure, standoff distance SOD, particle size, and density on the roughness obtained on woven PET textile surfaces (plain weave constructions characterized by 34 and 28 weft yarn/cm). The results show that under certain projection conditions, particles can generate a layer of frayed fibers on the textile surface. It comes out that appropriate and regular hairiness is obtained on the sample surface with the following setup adjustment parameters: pressure of 900 bars, SOD of 400 mm, and a particle mass flow rate of 0.6 g/s.

**Keywords** PET · Medical application · Surface modification · N2 supercritical jet

M. ZidI (✉) · F. Khoffi · S. Msahli
University of Monastir, Textile Engineering Laboratory, LGTex, Monastir, Tunisia
e-mail: Maleke.zidi01@gmail.com

Y. Khalsi · A. Tazibt
Centre de Recherche d'innovation et de Transfert de Technologie, TJFU, Bar-Le-Duc, France

F. Heim
University of Haute-Alsace, LPMT, Mulhouse, France



S. B. Abdessalem et al. (eds.), *Proceedings of the Second International Conference of Innovative Textiles and Developed Materials-ITDM'2; 05-06 May 2023; Tunisia*,
https://doi.org/10.1007/978-981-99-7950-9_11

# 1 Introduction

Foreign body reaction (FBR) is a critical problem to address when polyethylene terephthalate (PET) textile implants are considered in the medical field to treat cardiac pathologies such as aneurysm or heart valve replacement [1]. The natural porosity of textile materials tends to address exaggerated tissue growth that may prevent implants from remaining flexible. FBR is marked by the development of fibrous cellular tissue (fibrosis) and the formation of calcifications that result in stiffening of the porous material, which inhibits the movement of the implant. Hence, it is interesting from a clinical perspective to monitor this process. [2, 3]. The FBR mechanism seems to be significantly influenced by the surface features of the implanted material (hydrophilicity, roughness, porosity, etc.…). In particular, it has been proven that the development of the fibroblast tissue relies on the properties of the textile pores of a surface and on the topography of the yarns used to manufacture the textile [4, 5, and 6]. One approach to restrict fibroblast proliferation and textile stiffening is to raise the roughness of the textile surface.

For this topic, a mechanical surface treatment was studied to modify the surface roughness by including plastic deformation with particle jetting which is the supercritical N2 jetting technique by studying the supercritical N2 jetting parameters such as static jetting pressure, standoff distance (SoD), particle size, and density on the roughness that can be obtained on medical woven textile surfaces [7, 8]. The results show that the particles generate a layer of frayed fibers on the textile surface, which increases in correlation with jet pressure and SoD.

# 2 Material and Methods

The surface treatment technology used in this work consists of spraying onto the surface to be treated solid particles embedded within a dense High Speed Supercritical Nitrogen Jet "HSNJ." This process is based on the use of liquid nitrogen at very low temperature (−180 °C) and very high pressure (up to 3500 bar). It helps to provide low mass particles. (In this work, the abrasive particle used is the corundum abrasive) with critical kinetic energy to superficially degrade in an irreversible way the material surface and locally modify the roughness of the polymer to be treated.

The material used in this work was based on two different densities of weft yarns of Polyester plain weave fabric: 34 and 28 yarn/cm, relatively fine for application in the medical application. The goal was to analyze in particular the performance of a less tight structure compared to a tighter structure in the performance of heart valve.

The samples were positioned on the experimental work table by adjusting the shooting distance and the static pressure of the jet. Once the nitrogen jet attained its supercritical level and the experimental setup was validated, the particles accelerated instantaneously to supersonic speeds (Table 1).

**Table 1** Experimental conditions

| Configuration | Pressure (bars) | SoD (mm) | Particle size(μm) |
|---|---|---|---|
| Beam of particles 0.6 g/s flow rate | 900 to 1300 | 400 | 212–300 |

**Table 2** Roughness profile

| | Control samples | P 900 | P 1100 | P 1300 |
|---|---|---|---|---|
| T34 | 3.05 | 3.35 | 3.75 | 4.18 |
| T28 | 2.58 | 2.87 | 3.75 | 5.61 |

# 3 Results and Discussion

## *3.1 Influence of the Pressure on the Hairiness Level*

Results show that the hairiness of the fabric increases with the pressure whatever the fabric that is considered. When the flow rate is 0.6 g/s and the SOD is 400 mm and when the pressure varies from 900 to 1300 bars, the hairiness level can be defined by three levels, as shown in Table 2.

In fact, when the value of the pressure rises, the velocity of the jet also rises and delivers more kinetic energy to the particle. This is converted into cutting energy as the sharp edges of a particle strike some of the individual filaments involved in the construction of the tissue substrate. The cut filaments are frayed and contribute to the obtained hairiness of the surface.

The table shows also that when the material is less tight ($P_{28}$), the roughness is more important as more filaments are cut and the fraying is increased. Actually, when the construction is tighter filaments are squeezed together in a more stable configuration, which makes them more difficult to fray especially at high pressure under increased stress.

A synthesis of the obtained average roughness values can be seen in Table 2. Several observations can be made. First, it comes out that the effect of pressure variation starts to become significant from 900 bars. The roughness increases for both the $P_{28}$ and the $P_{34}$ constructions compared to the control samples (from 2.58 to 2.87 μm and from 3.05 to 3.35 μm respectively). Moreover, the increase in roughness is proportional to the increase in pressure (from 2.87 to 5.61 μm for $P_{28}$ and from 3.3 to 4.18 μm for $P_{34}$ at pressure values of 900 bars and 1300 bars, respectively).

## *3.2 Bending Stiffness*

Figure 1 presents the results obtained regarding the bending stiffness of the tested materials. It comes out that the control samples are the stiffest (0.128 and 0.152

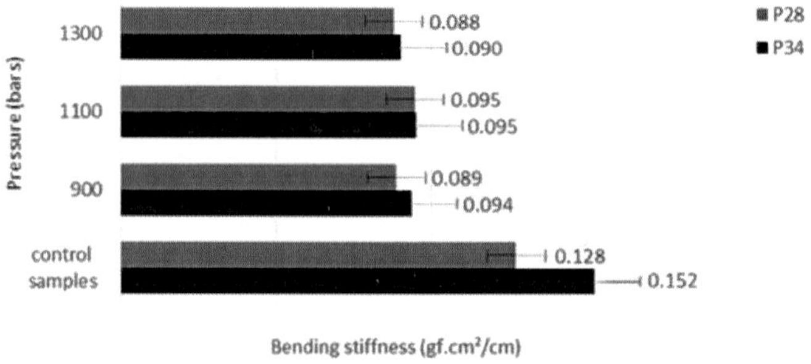


**Fig. 1** Influence of the jet pressure on the bending stiffness profile

respectively for $P_{28}$ and $P_{34}$). The $P_{34}$ sample being tighter is also less flexible as larger friction phenomena occur between neighbor filaments in the fibrous construction. After treatment at 900 bars, as soon as some filaments start to break under the impact of the particles and fraying occurs, the bending stiffness decreases down to 0.089 and 0.094, respectively for the $P_{28}$ and $P_{34}$ samples. As filaments are broken, less friction occurs in the material and the stiffness value goes down. At 1100 bars, more filaments undergo cutting, and the number of frayed yarns increases. This causes them to be entangled against each other, creating some mechanical bonds. These bonds add some bending stiffness to the sample. This hooking stiffness compensates for the drop in friction stiffness induced by the cutting, which makes that the stiffness doesn't evolve much (around 0.095 for both configurations). As the pressure increases up to 1300 bars, the surface frays more until partial degradation: this leads to decrease in flexural rigidity.

## 4 Conclusion

The objective of this study was to investigate the potential of a new mechanical surface treatment technique to modify the surface topography of a textile material used for medical implants by increasing its roughness. Solid corundum particles embedded in a high-velocity supercritical nitrogen (N2) jet were sprayed onto PET woven surfaces, varying the jet parameters (static pressure, SoD, particle size, and particle mass flow rate).

We found that increasing the static jet pressure from 900 to 1300 bar increases the fraying density of the yarns constituting the textile surface. Moreover, a SoD projection threshold of 400 mm and a pressure of 1100 bars are necessary to obtain a uniform roughness on the surface of the material (from 3.05 to 4.18 μm for P34 and from 2.58 to 5.61 μm for P28) are adapted to what is expected from a rigid woven structure, i.e., the plain fabric (P34 and P28). Below this value, the turbulence of the jet induces an irregular velocity in the jet, which does not allow the necessary control of the roughness obtained.

The parameters of the supercritical N2 sputtering technology could be adjusted to be tested with smaller particles, further emphasizing the suitability of this new process.

## References


1. .J. Wilson, R.E. Clegg, D.I. Leavesley, M.J. Pearcy, Mediation of biomaterial cell interactions by adsorbed proteins: a review, Tissue Eng. 11 (1–2), 1–18 (2005)
2. Vaesken A, P. A.-D. . Heart valves from polyester fibers: a preliminary 6-month in vivo study . *Journal of Biomedical Engineering* , 6(3):271–278 (2018)..
3. Zhao Q, T. N . Foreign-body giant cells and polyurethane biostability: in vivo correlation of cell adhesion and surface cracking. *Journal of Biomedicals Material Research*, 177–183 (1991)..
4. H. Cao, K. M. The topographical effect of electrospun nanofibrous scaffolds on the in vivo and in vitro foreign body reaction. Journal of Biomedical Materials Research, 1151–1159 (2010).
5. Ammar, Y., Swailes, D., Bridgens, B., & Chen, J. Influence of surface roughness on the initial formation of biofilm. Surface and Coatings Technology, 284, 410–416 (2015).
6. Cao, H., Mchugh, K., Chew, S. Y., & Anderson, J. M. The topographical effect of electrospun nanofibrous scaffolds on the in vivo and in vitro foreign body reaction. Journal of Biomedical Material Research - Part A, 93, 1151–1159 (201).
7. F.Khoffi, Y.Khalsi, J.Chevrier, H.kerdjoudj, A.Tazibt, F.Heim, Surface modification of polymer textile biomaterials by N2 supercritical jet: Preliminary mechanical and biological performance assessment, (2020).
8. Y.Khalsi, F Heim, J.T.Lee, A.Tazibt, N2 Supercritical Jet to Modify the Characteristics of Polymer Material Surfaces: Influence of the Process Parameters on the Surface Topography, (2018).

# The Living Design: The Soul of Regenerative Design and the Key to an Ecological Transition

**Mariem Fakhfakh and Amine Hadj Taieb**

**Abstract** Climate change, pollution, overexploitation, and the decline of biodiversity are alarming signals of the current state of our ecosystem. We must rethink our relationship with the built environment and natural systems, generating models aimed at regenerating rather than depleting life support systems and underlying resources. Regenerative design, originating in the United States, provides a framework for catalyzing this ecological transition and readapting biodesign to the urgent state of natural ecosystems. We must reposition scientific and biotechnological advances in relation to the functioning of natural systems, seeking to create a more sustainable future through living design. Living design seeks to increase the vitality of natural systems in a way that allows the planet to meet the needs of current and future generations. This research questions the relationship between scientists and designers in the context of climate change and ecosystem degradation, highlighting the key role of the biologist designer and living materials as an interface for a more sustainable future. In this context, we explore the recent work of Carole Collet, a designer-biologist in speculative biotextiles, entitled "Hybrid Futures," which addresses these questions. Collet examines alternative synthetic biomaterials for the textiles of the future, presenting synthetic life as the key to this ecological renaissance to preserve biodiversity and ecosystem health.

**Keywords** Living design · Regenerative design · Synthetic biotextiles · Sustainability

* Supported by organization LGTex.

M. Fakhfakh (✉)
Higher Institute of Arts and Crafts of Sfax, University of Sfax, Sfax, Tunisia
e-mail: mariemfakhfakh1@gmail.com

A. H. Taieb
ISAMS, University of Sfax, Sfax, Tunisia



S. B. Abdessalem et al. (eds.), *Proceedings of the Second International Conference of Innovative Textiles and Developed Materials-ITDM'2; 05-06 May 2023; Tunisia*,
https://doi.org/10.1007/978-981-99-7950-9_12

Faced with our alarming ecological state, designers are seeking to rethink our relationship with the built environment and natural systems, aiming to change the paradigm of sustainable design. Regenerative design offers a framework to catalyze this ecological transition by generating models that aim to regenerate rather than deplete life support systems and underlying resources. In this article, we have attempted to integrate ecological reflection into the practice of living design by examining the crucial problems of the twenty-first century and the distinction between biotechnology and nature. This research has examined the relationship between scientists and designers in the context of climate change and ecosystem degradation, highlighting the key role of the biologist designer and living materials as an interface for a more sustainable future. We have explored scenarios of future textile materials that combine science and design in favor of a restorative design, going beyond the neutral or negative performance of sustainable development. We have emphasized the repositioning of scientific and biotechnological advances in relation to the functioning of natural systems, seeking to create a more sustainable future through living design. In this context, we have posed several questions: How to rethink design modes for the future while considering ecological constraints and changes? How to reconcile ecology and the progress of biotechnologies? How can the progress of biotechnologies help us design new systems and materials in partnership with natural systems? In what way does living design create conditions conducive to the regeneration of healthy ecosystems?

In this perspective, we have explored recent research reflections on future textile materials that address these questions, particularly Carole Collet's project focused on the synergy between biology and design aimed at restoring ecosystems rather than simply causing less damage.

## 1 How Has Sustainable Design Served the Natural Ecosystem State?

Conventionally, sustainable design aims to slow the degradation of the planet's natural systems by creating products and systems that have minimal impact on the environment. Conventional ecological designs and materials are not sufficient to stop the destruction of biodiversity and the complete collapse of ecosystems. To address this, the Sustainable Development Goals (SDGs) adopted by the United Nations in 2015 represent a global call to action to protect, restore, and promote the sustainable use of terrestrial and marine ecosystems [1]. In this sense, ecological design aims to regenerate the life support systems and underlying resources rather than deplete them. To overcome these challenges, a paradigm shift is necessary to design ecologically and sustainably by adopting participatory thinking. The concept of "regenerative design," which originated in the United States and is relatively new in the field of design, provides a theoretical framework to catalyze this change. Regenerative design aims to design systems that are in harmony with natural ecosystems, creating processes that contribute to environmental restoration rather than simply minimizing damage.

## *1.1 Regenerative Design: A Solution for the Future*

Etymologically, "regenerative" is the adjective form of "regeneration," which refers to the process of renewing or reconstituting something. This ability is unique to living beings, allowing them to regenerate and be reborn after the destruction of a part of themselves. Regenerative design is a concept that takes inspiration from this inherent regenerative ability in living beings. It aims to create systems that restore, renew, and revitalize the environment and human well-being. Initially, the term "regenerative" was used as an adjective to describe a system's self-healing ability. However, over time, it has evolved into a broader term that encompasses a design approach focused on creating sustainable and regenerative systems. These systems are intended to restore and strengthen ecosystems instead of depleting them. As a result, regenerative design can now be applied at various levels, ranging from food systems to the design of buildings and construction materials. Its purpose is to have a positive impact on natural systems and foster beneficial synergies for both the environment and humans.

In this context, Bill Reed and Pamela Mang, founders of the Regenesis group, defined regenerative design as a system of technologies and strategies, based on an understanding of the inner working of ecosystems that generates designs to regenerate rather than deplete underlying life support systems and resources within socio-ecological wholes [2]. Reed and Mang envisioned regenerative design as an alternative solution for ecological design while engaging in a reversal process. Regenerative design is a radical approach to sustainable development that aims to catalyze positive impacts for the local ecosystem and regenerate it while developing a sustainable and environmentally respectful built environment. By adopting a regenerative approach, ecological design is rethought to achieve a transition toward long-term sustainable solutions while reconnecting with nature.

Michael Pawlyn, architect and founder of Exploration Architecture, presented regenerative design as an approach that aims to create designs that give more than they take [3]. This means that design concepts must be designed to maximize their positive impact on the environment and local communities while minimizing their resource consumption and negative impact. This approach focuses on creating positive feedback loops that enhance the health of ecosystems and communities. It involves all key actors and processes of the place: natural systems, human systems, cultural systems, and economic systems. They engage in a continuous and healthy relationship through coevolution. This approach considers humans as an integral part of the ecosystem, rather than as an observer. In general, the European research project *COST RESTORE* proposed a definition synthesizing several key elements of contemporary discourse on regenerative design: Regenerative: allowing social and ecological systems to maintain a healthy state and evolve [4].

In this sense, the main objective of regenerative design is to restore, preserve, and promote the health of all systems by favoring their function and evolution. Design has often been criticized for its use of non-renewable resources and toxic materials, which have significantly contributed to ecological collapse and economic degradation, both locally and globally. By positively observing and actively participating in the resolution of these problems, design, with its objects, materials, and systems, is seen as a solution for restoring and repairing ecosystems. By adopting a regenerative approach, design can contribute to the creation of a more sustainable and environmentally friendly economy for future generations. This approach has paved the way for a new generation of environmentally friendly practices based on the principles of circular design. The Ellen MacArthur Foundation, founded by British sailor Ellen MacArthur, played a key role in popularizing the concept of the circular economy with the publication of its report "Towards the Circular Economy" in 2012. This report defined the circular economy as a regenerative economic system by design, aiming to preserve the intrinsic value and quality of products, components, and materials at every stage of their use [5]. The circular economy goes beyond simple recycling by recovering end-of-life objects to create resources for new products.

In summary, regenerative design urges us to reconsider our relationship with nature and collaborate with it to build a more sustainable and resilient future. The resulting ecological innovations and materials offer alternative solutions for revitalizing the local economy, society, culture, and environment, with the potential for long-term positive impact on both the environment and local communities. By embracing the principles of the circular economy, regenerative design provides sustainable solutions to address current environmental and social challenges.

To achieve this harmonious coexistence with the natural environment, the ecological approach relies on new pillars that foster responsible design concepts promoting ecosystem and community regeneration. In the next section, we will explore these pillars that enable us to envision a future where design is seen as a solution rather than a problem.

## *1.2 The Pillars of Regenerative Design*

As previously mentioned, the aim of regenerative design is to reconcile the relationship between design, materials, and the surrounding ecosystems, taking into account the needs of the planet. In this context, biomimicry and bio-inspiration are gaining interest in this regenerative approach. According to Pawlyn, the best way for architects to create regenerative architecture is to use biomimicry, a functional discipline that translates the functions and processes found in biology into solutions that meet human needs [6]. To achieve this goal, biomimetic design must be considered a catalyst for the emergence of positive solutions that aim to design closed-loop systems rich in feedback, thereby regenerating degraded natural ecosystems. Thus, biomimetic creation must be based on a salutogenic intention that aims to strengthen the health of ecosystems and communities [7]. In practice, this can be translated into creating solutions that strengthen local ecosystems and contribute to the well-being of all living beings, using the principles of nature to design resilient and sustainable systems. This leads us to the concept of "neo-materialist biomimicry," which challenges the traditional conception of conventional biomimicry and Benyus' mainstream organization in biomimicry [8]. However, inspiration is not limited to the relationship between the original and the copy the model and the reproduction. It involves a deep understanding of natural systems and their functioning, as well as how form and matter emerge together in a process of growth. This approach makes it possible to design structures that are in harmony with nature by taking into account the characteristics and needs of living organisms.

Design biologists have proposed an alternative framework based on inspiration rather than simple imitation, working with living organisms to create new forms of nature, generate materials, and establish beneficial synergies for humans and the environment. This leads us to highlight three main pillars that allow for the adoption of a regenerative design that promotes coevolution between design and the natural and human systems that surround us.

- "Neo-materialist biomimicry" as a conceptual basis

Rather than simply reproducing the form (or function) of an organism in a material object, this approach involves understanding form and matter as co-emerging in a growth process, in order to design innovative solutions that integrate harmoniously into natural ecosystems.

- Coevolution as a conceptual strategy

The coevolution as a conceptual strategy is about designing in partnership with nature to ensure mutual positive impacts in the long term. This conceptual strategy focuses on the process rather than the result, as regeneration is a direction and a life process.

- Multi-species thinking as a conceptual principle

This approach involves designing for each living species, taking into account their behaviors, needs, and survival parameters [9].

These three pillars of regenerative design allow for a holistic approach to design, aimed at promoting plural coexistence of living beings and the preservation of natural capital. By combining human expertise with the wisdom of nature, design aims to provide innovative and sustainable solutions to contemporary environmental challenges.

To achieve this goal, designers have proposed an innovative alternative framework based on inspiration rather than simple imitation. This has revealed that the designer's task is to increase the potential of living beings rather than simulate and imitate them. Based on scientific progress and the synergy between design and biology, the designer has the ability to create new forms of nature, generate materials, and establish beneficial synergies for humans and the environment. This leads us to address the concept of living design at the heart of this ecological transition.

Indeed, this article has questioned the relationship between scientists and designers in the context of climate change and ecosystem degradation, highlighting the key role of the designer-biologist and living materials as an interface for a more sustainable future. Our research has explored the potential of living design in achieving the goals of regenerative design while examining the distinction between biotechnology and nature. Thus, it is essential to question the different potentialities of living design to create regenerative living designs and materials, and how they can contribute to creating a more sustainable future.

## 2 How Can Living Design Contribute to Creating a More Sustainable Future?

With the progress of biology and technology, designers have experimented with new ways of approaching living beings and their vital processes by imagining new forms of life and creating various designs that question our relationship with nature while determining scientific concepts and methods. In this sense, Thierry Marcou, author of the article "Towards a Design of Synthetic Life," mentions that biology and synthetic biology have given rise to a new branch of design: living design, in which science becomes the designer's toolbox [10]. As a sign of both an evolution of the relationship to living beings and the omnipresence of new technologies, designers have taken hold of scientific knowledge and the new biological tools at their disposal to creatively and scientifically transcribe the potentialities of the living in the viable design process. In this sense, the designer coordinates the actions carried out by living beings with conceptual objectives in order to produce living systems and devices applicable on a conceptual level. A new generation of designer-researchers is attempting to adopt a new axis of research that directs their work toward scientific reflection in order to experiment with new ways of approaching living beings by valuing nature rather than exploiting it to achieve a new age of symbiosis between our bodies, the microorganisms they host, our environment, and our objects. In this context, the designer-biologist Carole Collet has presented three different ways of integrating living beings into human design by imagining a new form of life created by the designer [11]. She proposed three axes by taking the living as a model, collaborator, and a hackable system allowing for redesigning while relying on the synergy between design and biological science. This synergy is gradually taking a fundamental place in this ecological transition to "naturalize" design [12]. Designers are increasingly interested in synthetic biology and biofabrication to produce living designs and materials that fit into a virtuous cycle of regeneration of natural, social, cultural, and economic systems. In this perspective, we have explored the different ways in which the design of the living can engage in this restoration process through living materials and designs.

## 2.1 *Biofabrication: A Process of Reversal*

Designers are playing an increasingly important role in this movement by using biofabrication and new "disruptive" living technologies that no longer focus on form, but on process [13]. Drawing on biology, they focus on using biological processes to design sustainable alternatives to conventional materials. This means they can work with nature to produce biomaterials made from biological organisms such as mycelium from mushrooms, kelp, bacteria, or yeast. These biomaterials can generate new biodegradable objects, sustainable buildings, and urbanism. Microorganisms are thus transformed into an architectural medium and construction material. By using biofabrication, designers can design solutions to reduce environmental impacts and promote the natural regeneration of ecosystems. Biomaterials are seen as living organisms that grow, evolve, and interact with their environment to create ephemeral architectural structures in harmony with natural systems. This approach has focused on creating closed resource systems, maximizing ecological productivity, and actively participating in biological systems.

## 2.2 *Synthetic Biology is the Soul of Biomaterials*

Synthetic biology is an emerging field of biomaterials science that aims to create artificial biological systems that can be programmed and controlled. This innovative approach uses principles from biology, engineering, and computer science to design semi-natural biological structures. This synergy allows us to explore how humans can act on living organisms and their growth processes by developing living biomaterials with properties suited to our needs and those of nature. In this context, Carole Collet has defined the synergy between design and synthetic biology as a response to the challenges of regenerative design. This regenerative approach is defined by Collet as an approach to design based on a living system that integrates the learning of natural sciences and humanity in service to the regeneration of biodiversity, climate, and communities [14]. By integrating environmental, social, and economic concerns into the design process, designers emphasize the use of natural systems and promote collaboration among designers, scientists, and communities. This collaborative approach aims to develop more sustainable solutions that are well-suited to the specific needs of local social and environmental systems. Synthetic biology allows us to design objects, buildings, and environments that are in harmony with living systems and that help restore environmental balance. Living design is seen in this sense as a key means of achieving restorative and integrated sustainability in our social and environmental systems.

### *2.3 The Hybrid Nature: Restoration Process*

Designers, architects, and artists explore the idea of merging the fields of biology, chemistry, and nanotechnology to create new hybrid organisms that combine living and non-living elements. This convergence of sciences creates links between previously distinct domains and makes living technologies increasingly interconnected with current scientific progress. The boundaries between the living and non-living are blurring and these creators are developing a new creative language to illustrate the future potential of these hybrid systems [13]. They have imagined a future where nature is enhanced and where new definitions of "life" are explored. These creators are like alchemists who develop scenarios to save the current ecological degradation while improving nature. They are strongly interested in ecology and have been introduced to the bio-digital world, where nature could regain its place with the help of digital programs easily manipulated by humans. Indeed, hybridization allows us to develop innovative ecological architectural protocols and new composites from living and synthetic materials using techniques such as bio-hacking, computational design, and digital prototyping [15].

In conclusion, biomimicry, biotechnology, and co-fabrication are considered key elements in creating an ecological design capable of adapting to future changes and ensuring sustainable living for future generations. The proposed framework allows for visualizing the various ways in which living design can be used for regenerative design. Living design is seen as an integral part of regenerative design, creating biomaterials and biotextiles that aim to regenerate life support systems and preserve local biodiversity (Fig. 1).

## 3 "Futures Hybrids" by Carole Collet: An Example of Using Living Design to Preserve Biodiversity

By blurring the boundaries between creative expression and scientific experimentation, living design and specifically biohackers, allow us to produce programmed synthetic materials while promoting benefits for the environment and humankind. Biological designers work in collaboration with synthetic biologists or employ advanced biotechnology to replace natural nature with synthetic nature through the programming of living organisms. This positions synthetic nature as an approach that aims to increase the vitality of natural systems in order to improve the conditions of the environment and society as a whole. In this context, Collet worked on exploring biomaterials in the textile industry, specifically by creating new fur fibers for a new generation that contributes to the protection of the planet. She raised the question of the ethics of fur production and questioned whether synthetic biology could allow us to grow fur more sustainably and responsibly. By programming nature, she seeks to create a new form of animated entity that can produce fur without causing harm to animals or the environment. In this speculative project entitled "Futures Hybrids,"

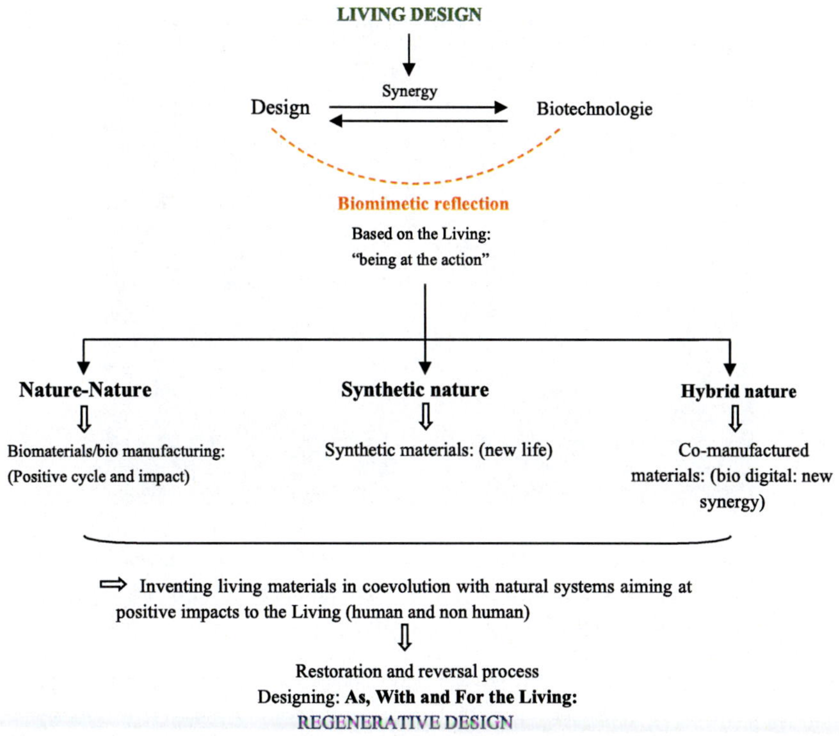


**Fig. 1** Summary diagram: living design for sustainable regenerative design

completed in 2015, C. Collet examined alternative synthetic biomaterials for future textiles by exploring a synthetic topology where animal and plant worlds converge to form a new hybrid animated entity [16]. The concept of recombinant DNA or "cutting and pasting" genetic materials, allows us to imagine selected traits from different species being edited into new living hybrid material factories. In this scenario, a fungus and a plant are presented as reprogrammed to produce the fur of an endangered raccoon. The designer emphasizes the need for innovative solutions to address the ethical and ecological challenges of fur production. Futures Hybrids targeted the luxury market and suggest possibilities for bio-fur that could be designed to integrate brand values into seasonal fur (Fig. 2).

Indeed, luxury brands, who are the main users of fur, are looking for alternatives to animal fur. They have questioned the ethics of fur production and the possibility of producing fur without exploiting farm animals or threatening endangered species. In this context, LVMH's environmental strategy, LIFE 360, includes a research and innovation program dedicated to sustainable luxury. Hélène Valade, LVMH's Environmental Development Director, stated that the invention of new materials, regeneration practices, and technologies will provide their Houses with the opportunity to

"Raccoon Fungi": new fast-growing mushroom species reprogrammed to cultivate synthetic fur.

"Raccoon Alocasia Zebrina": Plant leaves programmed to produce fur

**Fig. 2** "Futures hybrids" series by Carole collet: an exploration of biomateriality in 2015 [16]

achieve their climate and biodiversity ambitions [17]. Under the keyword "sustainable luxury," Carole Collet and Tom Ellis, a professor of synthetic biology, worked on the potential of academic research in design and science to prototype new biotextiles for the LVMH/Fendi luxury industry [17]. By 2022, this new programmable bio-luxury aims to integrate intelligent properties derived from natural organisms and develop new characteristics to generate a type of laboratory-grown fur that does not exist in nature. Carole Collet emphasized for this project that Biodesign research will allow us to bring different perspectives to life and move toward new production models inspired by living things and their functioning. Learning from nature and working at the intersection of design and biotechnology is essential in exploring such futuristic innovations [18]. This project is not about continuing this metaphorical approach, but rather about developing keratin grown in a laboratory using synthetic yeast, which can then be harvested and transformed into a fur-like textile using textile manufacturing technologies. In this context, Professor Tom Ellis explained that in recent years, our understanding of biological materials and their synthesis by nature has progressed considerably. The time is now ideal for embarking on a project to create fibers for the fashion industry made from microbes, in an environmentally friendly way [18]. This approach was put in place following the announcement by several luxury brands to stop using animal fur in their collections, including Dolce and Gabbana and all brands under the Kering group.

Fashion, design, and science converge to form a crucial alliance in favor of protecting and restoring the environment. In this interrelation, the biological world is the starting point and the object of study to achieve a sustainable and regenerative world of luxury. Regarding synthetic biology, speculative and critical design can highlight different perspectives and help test the purpose and future relevance of this technology. As Balsamo emphasized in a quote by Adams, designers work within the context of technological emergence by hacking the present to create conditions for the future [19]. However, to realize these conditions, it is also essential for designers, who understand how to specify the use and performance of a material, to proactively work with scientists in the laboratory. By going beyond purely scientific advances to enable valid design applications, we can help shape the future promises of this technology. It is crucial for designers, whether they employ speculative and critical design or directly engage in modifying living cells, to actively participate in the field of synthetic biology. This involvement is essential for understanding and addressing the ethical and social implications that arise from exerting control over nature through such practices. To achieve this, the designer-biologist must possess a mastery of both the language and techniques of biology. Additionally, they must have a deep understanding of the ethical implications associated with manipulating living matter.

Designing our future biomateriality has brought us back to the traditions of horticulture and animal management, while stemming from collaboration with advances in synthetic biology and tissue engineering. However, ethics must be at the center of the design thinking. In order to preserve and regenerate the environment, it is essential to draw inspiration from the living world and explore new working methods such as biohackers which combine microorganisms and biological systems. As an active catalyst in our design, programmed living must be effectively and positively integrated into local systems, which involves collaboration between natural and conceptual systems.

## 4 Conclusion: Living Design: A Promising Path for Regenerative Design

Living design is gaining growing attention globally, leading to frequent discussions within scientific and design communities. These discussions revolve around the future of design and the integral role of ecology within that future. By adopting this approach, we can work in harmony with nature by developing solutions that contribute to the preservation of biodiversity and the creation of natural habitats for wildlife and flora. Although living design concepts are still in their early stages, their use in our daily lives is generating a great deal of interest and projecting a future that abolishes the distinction between nature and human technology. We are talking about a future strongly inspired by existing mechanisms and reality. We imagine a future where systems designed by and for nature become the only solutions to the climate challenges created by humans themselves in order to preserve our environment and planet.

## References


1. UNPD, https://sdgintegration.undp.org/, last accessed 2023/02/10.
2. Mang, P., Reed, B.: Regenerative Development and Design. Encyclopedia Sustainability Science & Technolog, 115–141 (2020).
3. Fairs, M.: Vers l'architecture zéro carbone et le design régénératif. In: Le french design 2059, 1–3 (2020).
4. Brown, M., Haselsteiner, E., Apró, D., Kopeva, D., Luca, E., Pulkkinen, K., Rizvanolli, B.: Sustainability, Restorative to Regenerative. COST Action CA16114 RESTORE, Italy (2018).
5. Ellen Macarthur Foundation, https://ellenmacarthurfoundation.org, last accessed 2023/02/02.
6. Pawlyn, M.: Biomimétisme et architecture. Rue De L'echiquier (2019).
7. Wahl, D C.: Interview. In: Eggermont, M., Hoeller, N., Mckeag, T.: Zygote Quarterly, V3 (17), 66–79 (2016).
8. Kamili, L.: Biomimétisme et bio-inspiration: nouvelles techniques, nouvelles éthiques ?. Techniques & Culture Varia (2019).
9. WGSN, https://www.wgsn.com/en/blogs/regenerative-design-repairing-world-carole-collet-0, last accessed 2023/03/02.
10. Thierry, M.: Vers un design de la vie synthétique. Internetactu Net (2014).
11. Collet, C.: En Vie Aux frontières du design. Fondation EDF, Paris (2013).
12. Chiambaretta, P.: Naturaliser l'architecture. PCA-Stream 03,161–175 (2014).
13. Benayer, A.: La Fabrique Du Vivant: Mutations/Créations 3. Ircam Centre Pompidou, Paris (2019).
14. Central Saint Martins, The Forest School Talks: Regenerative design, https://www.youtube.com/watch?v=NIEXF3q8ItI, last accessed 2020/12/21.
15. Ecologicstudio,https://www.ecologicstudio.com/projects/h-o-r-t-u-s-xl-astaxanthin-g, last accessed 2023/01/20.
16. Collet, C.: Designing Our Future Bio-Materiality. AI & SOCIETY (36), Springer, 1331–1342, London (2020).
17. LVMH, https://www.lvmh.fr/actualites-documents/actualites/lvmh-sassocie-a-des-laboratoires-de-recherche-novateurs-pour-developper-de-nouveaux-biomateriaux/, last accessed 2023/02/20.
18. Fashion Net Work, https://fr.fashionnetwork.com/news/Lvmh-et-fendi-travaillent-sur-une-fourrure-synthetique-creee-en-laboratoire,1400056.html, last accessed 2023/01/21.
19. Yelavich, S., Adams, B.: Design as Future Making. Bloomsbury (2014).

# Using Fabric Warping in Architecture: A Contemporary Conception for Technical Use

**Amine Hadj Taieb and Wouroud Turki**

**Abstract** The textile envelope is now one of the components of the building system which aims to meet the new needs of contemporary construction: environmental protection, well-being, and energy saving. In fact, for several years now, the building envelope has been the subject of numerous studies centered on the challenges of energy saving and efficiency. From this point of view, climate and environmental issues are fundamental. Also Introducing textiles in architectural envelope is closely linked to the sustainability requirements of construction, they can affect the innovative textile market, by studying the possibilities of increasing industrial production in the commercial sector of ecological textiles, which have a very low impact on the environment, and producing textiles for technical uses, which are very efficient for particular uses.

**Keywords** Textile warping · Architecture · Environment

## 1 The Textile Envelope in Architecture

### *1.1 The Building Envelope*

The building envelope often represents the boundary between the interior and exterior of a building. It includes three main systems: the roof, the wall (including doors and windows), and below-grade waterproofing elements (foundation) (Fig. 1).

In the first place, the envelope should not only be an aesthetic element, but also an element of evaluation of the architectural quality with the objective of ensuring the comfort of the user through its efficiency.

A. H. Taieb (✉) · W. Turki (✉)
Higher Institute of Arts and Crafts of Sfax, University of Sfax, Sfax, Tunisia
e-mail: falseamineht@yahoo.fr

W. Turki
e-mail: falsewouroudturki@gmail.com



S. B. Abdessalem et al. (eds.), *Proceedings of the Second International Conference of Innovative Textiles and Developed Materials-ITDM'2; 05-06 May 2023; Tunisia*,
https://doi.org/10.1007/978-981-99-7950-9_13

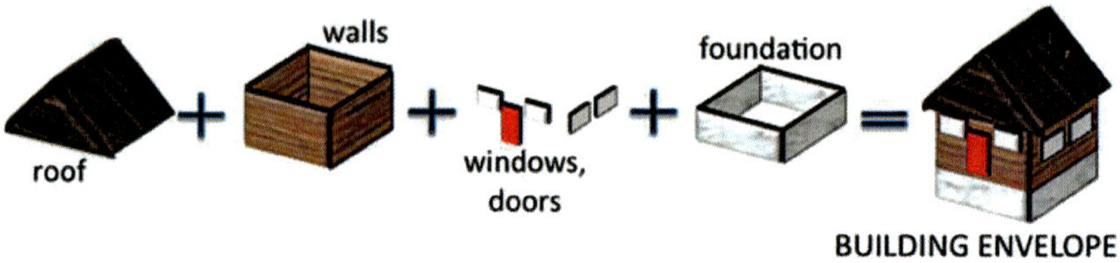


**Fig. 1** Components of the building envelope [1]

In addition, it must ensure airtightness and insulation and also allow for good regulation of air and light circulation. With the rise of environmental concerns, the building envelope is also required to participate in the regulation of the energy consumption of the building.

## *1.2 The Tent, the Origin of Textile Envelope*

The textile envelope originated in Moldavia 40,000 years ago, the oldest tent was found in Moldavia and was circular in shape. The need to provide a house in a short time is linked to the concept of the tent: a functional house that represents an advanced form of shelter, with a high degree of adaptation to the environment [2].

For example, when traveling along the paths south of the Atlas Mountains, the traveler will see black dots leaning against a hill or on the edge of a ravine. These are the tents of the nomads. When studying the tents of nomadic peoples, they were made of camel wool yarn mixed with goat hair. Thus, by analyzing the relationships between the constituent elements and the methods of construction, it becomes clear that all forms of nomadic life are based on the criteria of lightness, reversibility, and adaptability. It is possible also to find a convergence with present-day forms of architecture. Let's also take the example of the free-standing Indian teepee and the Mongolian yurt, which have typical elements of the support structure of circus tents. They retain the simple structural approach with point support elements (poles) that support the loads of the overlying flexible material (fabric).

Another origin of today's tensioned structures must be sought in Roman times, when the ability to steer the sails of ships and the use of canvas surfaces, called "velarium," to cover the auditoriums of theaters and amphitheaters were put to use.

Today, the tent has become a residue. Abandoned for a long time, given away, sold occasionally, or left in a corner of the sedentary person's house, it has been recovered by the tourist industry which has assigned it new functions. Tents today are no longer just for shelter or living! It is also used for leisure activities such as camping, hiking, and traveling by bicycle [3].

## 2 The Properties of the Textile Envelope for Contemporary Design

In recent years, the building envelope has been the subject of much scientific research, above all for its function as a separating filter between the interior and the exterior, with specific performance requirements in terms of environmental comfort.

In fact, this issue is increasingly pressing and present in current technological research, in order to satisfy new contemporary needs closely linked to sustainable design, in particular the need to create interior environments with a good level of thermal, acoustic, and visual comfort, guaranteed not only by mechanical air-conditioning systems, but also by the passive and active use of atmospheric agents and renewable energy resources [4].

Analyzing the history of architecture, three main "environmental control models" can be identified, each of which can be associated with a type of building envelope with specific characteristics. The first model is the conservative model, characterized by a type of environmental control that uses large masses of masonry with few openings to reduce heat loss in cold climates and to mitigate the heating effects of direct solar radiation in warm climates. The second model, defined as selective, is characterized by similar general principles to the first, but "uses the structure not only to preserve the desired environmental conditions, but to let those conditions in from the outside." To this end, large transparent walls can be provided for lighting and passive heating. The third model, the regenerative one, characterizes the buildings typical of the American building tradition in which the control of the environment is totally entrusted to the plant systems, precisely defined as "regenerative installations," while the envelope is understood only as a barrier capable of limiting the interaction between the interior and the exterior. For our research, the performance of the textile envelope for the contemporary project could fall under the definition of "selective envelope" due to the connoting textile characteristics and technology applied to the facade [5].

In this perspective, we ask the following question:

*How can the use of textile warping optimize the building structure (indoor and outdoor)?*

The textile material, which is lightweight and flexible, is used in the construction of the building envelope using innovative textiles, which can improve qualities of durability and resistance to atmospheric agents. Textiles differ according to the way the material is used: Textiles, which are strong, lightweight, and stretchable at the same time, find an interesting application in tensile architecture, in particular for roofing structures for stadiums, or airport or railway station halls.

For example, "Morphosis" has created the façade of a textile research center from fibers [6]. Morphosis founder Thom Mayne has collaborated with Korean textile manufacturer Kolon to create a new research facility that incorporates the company's high-tech materials into its façade:

**Fig. 2** Morphosis façade for a textile research center in Seoul [7]

- The façade is made from a high-tech fiber used to manufacture bulletproof vests.
- Aramid, the material chosen by Morphosis for the interconnected sun protection system, is a reinforced fiber manufactured by Kolon that has five times the tensile strength of iron, used for everything from bulletproof vests to aircraft parts.
- Morphosis has used recycled materials where appropriate and a bubble deck technology that uses 30% less concrete than standard slabs by adding plastic balls to the material (Fig. 2).

## 3 Cases of Study

If at the beginning of their diffusion textile envelopes were imposed thanks to their structural properties (covering of large spans, reduction of the weight of the structures), these last years they are often used for their characteristics of transparency, translucency, or for the advantages of performance obtained for thermal insulation.

Any type of building can be covered with a textile envelope, from office buildings to public buildings such as schools. The textile building is essentially composed of three elements: membrane, supporting system, and tensioning system [6].

The case study proposed and analyzed in this article refers to buildings whose envelope is made of textile materials, giving the building functional, environmental, and typological properties. The exanimated examples are:

**The Inflatable Tea House**

Based in Frankfurt since 2007: is a joint between sculpture and temporary room for ceremonies designed by Kengo Kuma Materials: Gore Tenara 3T40 (630 gr/$m^2$ PTFE-fabric with laminated fluoric foil) and 38% translucency (Figs. 3 and 4).

**Fig. 3** Inflatable tea house exterior [8]

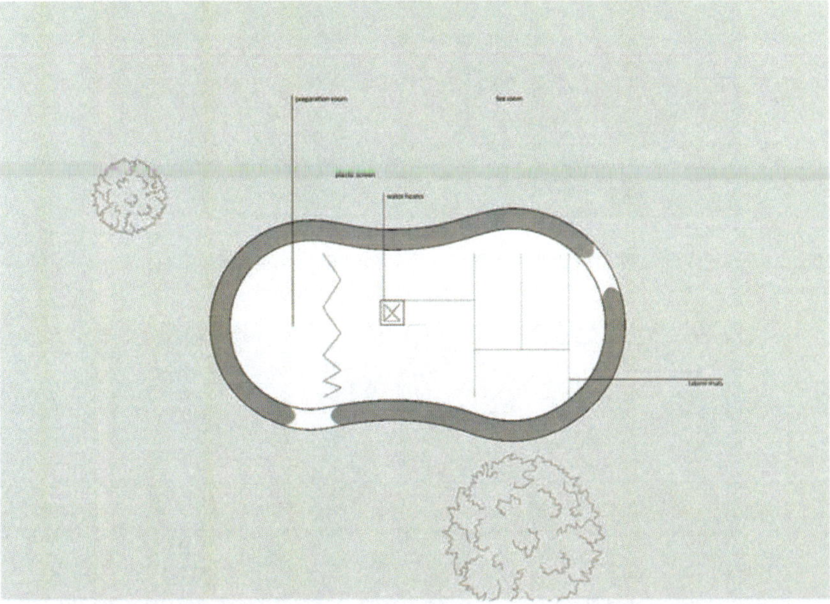

**Fig. 4** Inflatable tea house plan [9]

## The New Experimental Audiovisual Production Center "Magical"

Built in 2012 in Spain, the 7448 $m^2$ project is designed by Pitcharchitects. The material used for the building envelope is Polytetrafluoroethylene (PTFE). In the form of fibers. The textile structures, with their fluid and airy forms, are the signature of the building. Their translucency allows the play of light and shadows to enhance the covered spaces [10] (Figs. 5 and 6).

**Prada Transformer, Seoul 2009**; was conceived by OMA/Rem Koolhaas. The pavilion shape is a tetrahedron; hexagon, cross, rectangle, and circle into one pavilion. The material used, called "Cocoon Membrane" (produced by Cocoon Holland BV) is a synthetic material which is normally used to wrap and protect large pieces of machinery: PVC fabric, which was apparently developed to protect mothballed US aircraft in the aftermath of the Second World War (Figs. 7 and 8).

**Fig. 5** Experimental audiovisual production center "magical" façade detail [11]

**Fig. 6** Experimental audiovisual production center "magical" façade [12]

**Fig. 7** Prada transformer exterior

**Fig. 8** Prada transformer interior

The Prada Transformer played host to four events, being lifted and repositioned on to a different face (Fig. 9).

From the analysis of the above examples, the following table summarizes the advantages and performance of using textile materials in the field of architecture (Table 1) [13].

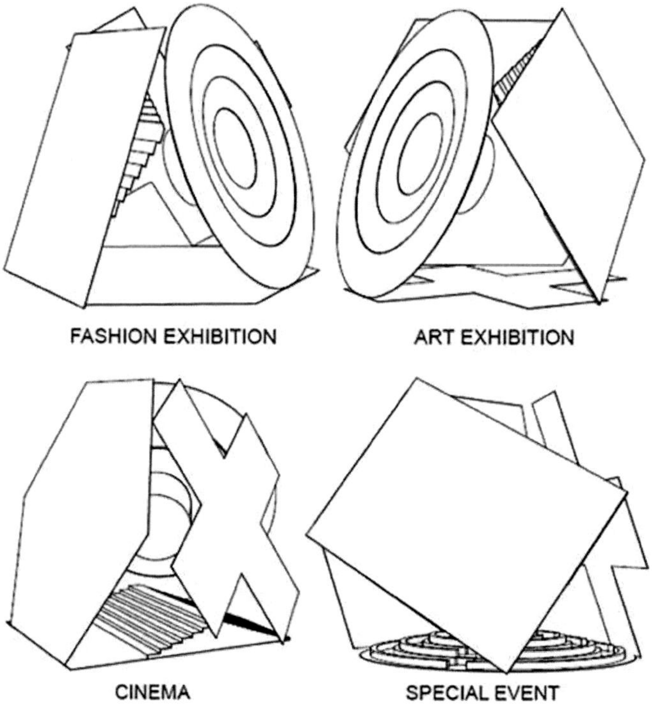


**Fig. 9** Prada transformer different positions

**Table 1** Benefits of textile materials used in architecture summarizing

| | |
|---|---|
| Aesthetically | Textile structures and materials offer large, aesthetic closures and a wide variety of shapes. Unlike traditional building materials, textile structures can take a variety of complex forms |
| Physical properties | Stretched fabric structures are known to withstand large deformations caused by wind or rain |
| | Low floor load (lightness): The lightness of the fabric makes it possible to design lightweight structures. This is particularly important for textile facades and stretch ceiling structures |
| | Lighter load-bearing structure: The material density is optimized to allow building structures to withstand constraints |
| | Strong waterproofing: Thanks to the surface treatment, the architectural textile materials offer exceptional resistance to UV, atmospheric agents, soiling, and ageing |

(continued)

**Table 1** (continued)

| | |
|---|---|
| Aesthetically | Textile structures and materials offer large, aesthetic closures and a wide variety of shapes. Unlike traditional building materials, textile structures can take a variety of complex forms |
| | Easy to maintain |
| | Translucency: preserves natural light thanks to the translucency of the fabric |
| | Movement: Because of the lightness of the building structure, it can be turned or carried on to other place (ephemeral building) |
| | Sustainability: the use of a textile facade makes the building more sustainable, lowering the environmental impact |
| Material saving | The total investment cost of a textile membrane is more economical than a steel roof and a traditional tile roof |

# 4 Conclusion

Due to its high mechanical properties, textile is considered the fifth most important material in construction, after wood, glass, concrete, and steel. They are a mark of contemporary architecture [14]. The finesse of the frameworks combined with the elegance of the assembly and connection details allows for a wide variety of structural and architectural expressions. Textile fibers in architecture enable significant architectural innovations due to weight savings. Today, major industrialists are studying new emerging uses of textiles in the building industry [15]. Current trends include composite materials, filtration and indoor air quality, anti-ageing treatments for surfaces, and nanotechnology in coating.

# References

1. https://energyeducation.ca/encyclopedia/Building_envelope#cite_note-3
2. Bechmann, R. Le Langage Textile Dans La Construction. *Ethnologie française* **1989**, *19* (1), 83–89.
3. Reyner Bahnam, The Architecutre of the Well-Tempered Envirnoment, Architectural press, London 1969.
4. Farrukh, A. H. *Textile in House Building & Civil Applications*; 2015. https://doi.org/10.13140/RG.2.1.1212.3927.
5. Bertino, Stefano, Andrea Giovanni Mainini, and Tiziana Poli, '4. Textile Façades', in *4. Textile Façades* (Birkhäuser, 2013), pp. 56–69 <https://doi.org/10.1515/9783034613507.56>
6. 'Morphosis Weaves Textile Research Facility Facade from Reinforced Fibre' <https://www.dezeen.com/2018/08/30/morphosis-textile-research-facility-facade-from-reinforced-fibre-seoul-south-korea-architecture/> [accessed 9 March 2023].
7. www.dezeen.com, https://www.dezeen.com/2018/08/30/morphosis-textile-research-facility-facade-from-reinforced-fibre-seoul-south-korea-architecture/ acceded 09/03/2023.
8. Priniotakis, Georgios, Laetitia Marrot, Urszula Stachewicz, Aleksandra Krstic-Furundzic, Enrico Venturini, and Vaida Jonaitiene, 'Smart Textile for Building and Living', *Autex Research Journal*, 22 (2021), 000010247820210041 <https://doi.org/10.2478/aut-2021-0041>
9. https://www.archilovers.com/projects/123223/modern-tea-house.html 09/03/2023.

10. archilovers.com https://www.tensinet.com/index.php/about/members-of-tensinet?view=project&id=4399 09/03/2023.
11. *Experimental Center for the Audiovisual Production "Magical" picharchitects.* Catalan-Architects. https://www.catalan-architects.com/ca/picharchitects-barcelona/project/experimental-center-for-the-audiovisual-production-magical (accessed 2023-02-25).
12. archdaily.com https://www.archdaily.com/413545/magical-science-and-technology-park-pich-aguilera-architects 09/03/2023.
13. archilovers.com https://www.archilovers.com/projects/113729/experimental-center-for-the-audiovisual-production-magical.html 09/03/2023.
14. archdaily.com https://www.archdaily.com/500362/5-years-later-a-look-back-on-oma-s-prada-transformer 09/03/2023.
15. I.ABDELSABOUR. FABRICS AS AN INVENTIVE BUILDING MATERIAL AND THEIR COMPATIBILITY WITH FLEXIBLE ARCHITECTURE. *Journal of Engineering Sciences Assiut University Faculty of Engineering Vol. 48 No. 1.* Assiut University January 2020, p PP. 121–135.

# Use of Date Palm Waste to Produce Bio-Composite Material

**Ferhat Maroua, Labed Adnane, and Djemai Hocine**

**Abstract** Bio-sourced composite materials have become a potential alternative to the currently used synthetic fiber composite materials and promising materials for the future. In Algeria, there is a growing interest in palm fibers as agricultural residues due to their high availability and mechanical properties. This study is an experimental investigation of the mechanical behavior of different composite materials based on palm date waste. Samples are made by thoroughly mixing natural and saturated fibers (leaflets fibers) in a polyvinyl acetate (PVA) matrix with dimensions (160, 15, and 10 mm). Three-point bending tests are performed to determine the modulus of elasticity; and good mechanical properties are obtained in comparison with values obtained from the literature.

**Keywords** Bio-composite material · Leaflets fibers · Three-point bending · Flexural modulus

## 1 Introduction

Recently, wide applications of natural fiber-reinforced composites in polymeric matrix have been spread into construction, automobile, marine, etc. [1–4]. Thus, natural fibers emerged as a viable alternative to conventional synthetic fibers in structural and semi-structural applications. Thus, recent studies have focused on enhancing the mechanical, physical, and chemical properties of natural fiber-reinforced composites and explore into their performance under different loads [5–9].

F. Maroua · L. Adnane (✉)
Laboratory of Mechanical Engineering (LGM), University of Biskra, B.P. 145 R.P. 07000, Biskra, Algeria
e-mail: a.labed@univ-biskra.dz

D. Hocine
Laboratory of Energy Engineering and Materials (LGEM), University of Biskra, B.P. 145 R.P. 07000, Biskra, Algeria



S. B. Abdessalem et al. (eds.), *Proceedings of the Second International Conference of Innovative Textiles and Developed Materials-ITDM'2; 05-06 May 2023; Tunisia*,
https://doi.org/10.1007/978-981-99-7950-9_14

Algeria (in northern Africa) has a large resource collected from the agricultural waste of date palm (*Phoenix Dactylifera L.*) [10, 11]. On average, fourteen new leaves and eight clusters per palm can be retained each year [12]. Only, some farmers use partially this palm wood, but most of the material is discarded as waste. Indeed, the development of reinforced composite materials from this type of waste can be an excellent solution from an environmental point of view. It has been reported that the use of date palm fibers with different types of polymers can improve the characteristics of the composites such as flexural strength, Young's modulus, bending strength, and flexural modulus.

This work is a comparative study of the mechanical behavior of new bio-composite materials for thermal insulation applications by using leaf palm residues (*Phoenix dactylifera L.*). These mechanical properties are compared with other material properties from the literature [12].

## 2 Material and Method

The biomaterial (leaflets fibers-PVA) is obtained from the leaflets fibers as reinforcements of PVA as a matrix.

### *2.1 Reinforcement (Leaflets Fibers)*

To prepare the fibers of the palm, we follow these steps: collect the palm leaflets, clean, grind, and sieve by several cuts and choose the size [0.315 mm].

### *2.2 Matrix (White Glue-PVA)*

White glue is special glue for wood, this provides a strong, and as far as a durable joint.

### *2.3 Specimens Preparation*

To prepare (leaflets fibers-PVA) material, the mass fractions between the reinforcement and the matrix are taken into account (Table 1). The armature is mixed into the matrix. Then, the mixture (reinforcement/matrix) is poured into a metal mold to obtain a plate of dimensions (240 × 120 × 100) mm$^3$. As shown in (Fig. 3), the plates obtained are air-dried under ambient conditions for at least 10 days. After drying, the

**Table 1** Mass fractions of composite material

| Composite material | Matrix | Reinforcements |
|---|---|---|
| (Leaflets fibers–PVA) | 80% | 20% |

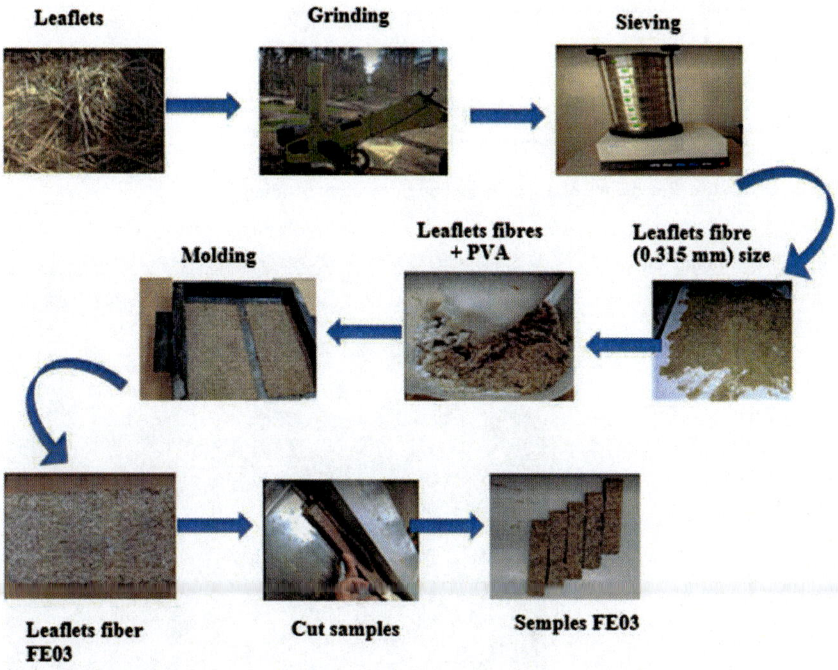


**Fig. 1** Experimental procedure to obtain the bio-composite material samples

plates are cut to obtain specimens of dimensions (200 × 15 × 10) mm, according to EN SO14125 (Fig. 2).

The pictures of Fig. 1 present the experimental protocol to obtain the desired samples.

## 3 Results

Flexibility modulus ($E_f$) and maximum stress ($\sigma_{max}$) are the mechanical behavior that is determined in this work by three-point bending tests. These tests are carried out in a mechanical test laboratory on a universal machine (TEST), with a speed of 01 mm/min and a force sensor of 05 kN. Four specimens were tested in these studies.

We consider a composite material (leaflets fibers–PVA)) with ($B$) in width, ($L$) in length, and ($tf$) in thickness. Flexibility modulus is calculated from the linear part of a curve load ($P$)–deflection ($f$), we can calculate by this formula:

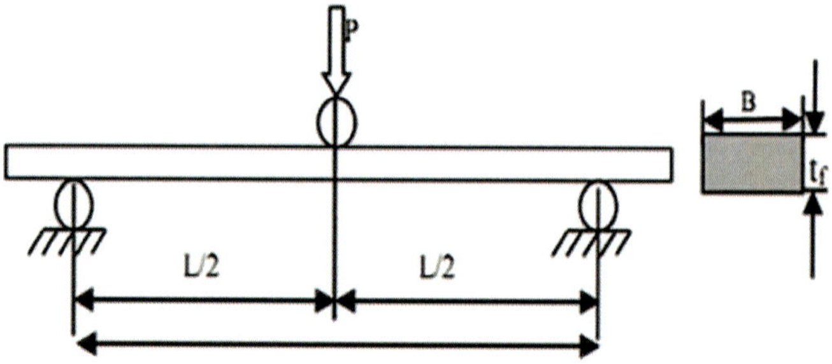


**Fig. 2** Geometric dimensions of a specimen n by three-point bending

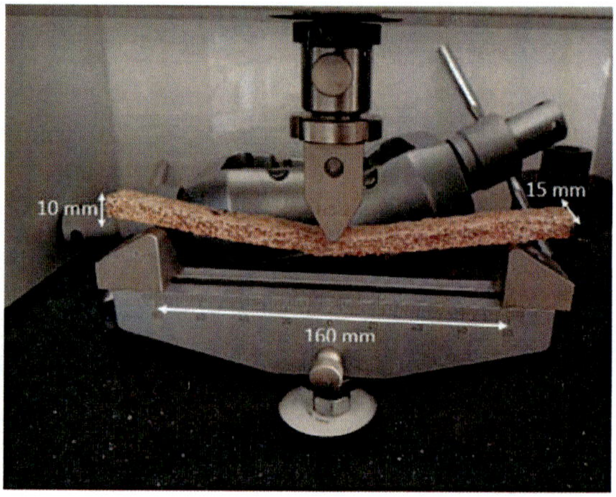


**Fig. 3** Three-point bending test of MC specimen

$$E_f = \frac{\Delta P}{\Delta f} \cdot \frac{L^3}{4B.t_f^3} \quad (1)$$

With: $\Delta P/\Delta f$ is the slope of the linear part of curve $P = f(f)$ (Figs. 2 and 3).

Figure 4 presents load–deflection curves in a three-point bending test of composite material (leaflets fibers–PVA). They are started with a linear increase, then a nonlinear increase, and finally they are reduced in load. We can calculate the modulus of flexibility in the linear part by formula (1). Table 2 presents the values of this parameter in composite material (leaflets fibers–PVA) compare with the results of another material (leaflet fibers-expanded polystyrene) which is studied by Masri [12].

This material composite is isotropic because its properties are the same in all directions. The modulus of the flexibility of our composite is 97.99 MPa, while, it is 250.3 MPa in (leaflet fibers-expanded polystyrene). It is clear that, the composite material (leaflets fibers–PVA) is more fragile than the composite material of Masri [12], and the same remark for the maximum stress (Table 2).

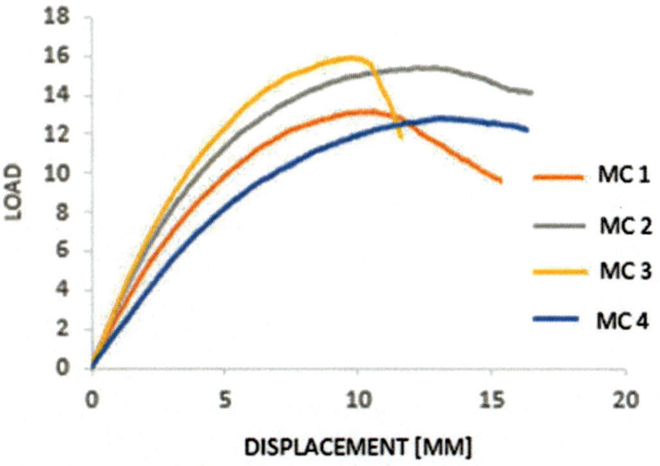


**Fig. 4** Load–displacement curves of composite materials stressed by three-point bending

**Table 2** Mechanical behavior of two composite materials differentiated by the matrix

| Specimens | $E_f$ (MPa) | $\sigma_{Max}$ (MPa) |
|---|---|---|
| (Leaflets fibers–PVA) | 97.99 ± 8.64 | 19.32 ± 2.23 |
| (Leaflet fibers-expanded polystyrene) | 250.3 ± 0.03 | 1.04 ± 0.08 |

## 4 Conclusion

This work presents an experimental investigation of the mechanical behavior of a composite material obtained from different parts of palm date waste. After experiment and comparison, we found that the bio-composite based on fiber leaflets and PVA matrix presents good mechanical properties in comparison with mechanical proprieties values obtained from the literature. The modulus of the flexibility of our composite is 97.99 MPa, while, it is 250.3 MPa in (leaflet fibers-expanded polystyrene).

However, materials based on date palm waste can present a promising alternative for the production of lightweight building materials and especially for thermal insulation applications.

Although, other parts of the date palm waste (petiole, trunk, leaves…) will be investigated in future studies for thermal insulation applications.

# References


1. Mohanty, A. K., Misra, M. A., & Hinrichsen, G. I.: Biofibres, biodegradable polymers and biocomposites: An overview. Macromolecular Materials and Engineering, 276(1), 1–24 (2000).
2. John, M. J., & Thomas, S. Biofibres and biocomposites. Carbohydrate polymers, 71(3), 343–364 (2008).
3. Krishnasamy, S., Thiagamani, S. M. K., Kumar, C. M., Nagarajan, R., Shahroze, R. M., Siengchin, S., ... & MP, I. D.: Recent advances in thermal properties of hybrid cellulosic fiber reinforced polymer composites. International journal of biological macromolecules, 141, 1–13 (2019).
4. Senthilkumar, K., Siva, I., Sultan, M. T. H., Rajini, N., Siengchin, S., Jawaid, M., & Hamdan, A.: Static and dynamic properties of sisal fiber polyester composites–effect of interlaminar fiber orientation. BioResources, 12(4), 7819–7833. (2017).
5. Da Costa Garcia Filho, F., da Luz, F. S., Oliveira, M. S., Pereira, A. C., Costa, U. O., & Monteiro, S. N.: Thermal behavior of graphene oxide-coated piassava fiber and their epoxy composites. Journal of Materials Research and Technology, 9(3), 5343–5351 (2020).
6. Garcia Filho, F. D. C., Oliveira, M. S., Pereira, A. C., Nascimento, L. F. C., Matheus, J. R. G., & Monteiro, S. N.: Ballistic behavior of epoxy matrix composites reinforced with piassava fiber against high energy ammunition. Journal of Materials Research and Technology, 9(2), 1734–1741 (2020).
7. Reis, R. H. M., Nunes, L. F., da Luz, F. S., Candido, V. S., da Silva, A. C. R., & Monteiro, S. N.: Ballistic performance of guaruman fiber composites in multilayered armor system and as single target. Polymers, 13(8), (2021). p1203.
8. Luz, F. S. D., Garcia Filho, F. D. C., Oliveira, M. S., Nascimento, L. F. C., & Monteiro, S. N.: Composites with natural fibers and conventional materials applied in a hard armor: A comparison. Polymers, 12(9), (2020). p1920.
9. Touloum, F., Younsi, A., Kaci, A., & Benchabane, A. Formulation of a composite of date palm wood-cement. Journal of Applied Engineering Science & Technology, 2(2), 57–63 (2016).
10. Almi, K., Benchabane, A., Lakel, S., & Kriker, A.: Potential utilization of date palm wood as composite reinforcement. Journal of Reinforced Plastics and Composites, 34(15), 1231–1240 (2015).
11. Agoudjil, B., Benchabane, A., Boudenne, A., Ibos, L., & Fois, M.: Renewable materials to reduce building heat loss: Characterization of date palm wood. Energy and buildings, 43(2–3), 491–497 (2011).
12. Masri, T., Contribution au développement des matériaux de construction à base des sous-produits du palmier dattier. Doctoral thesis, Universite Mohamed Khider Biskra (2018).

# Application of Biopolymers in Medical Textiles: Myriad of Opportunities

**Adel Elamri, Khmais Zdiri, and Mohamed Hamdaoui**

**Abstract** Nowadays, textile products are being used in several sectors for different purposes beyond imagination. Among these, a very important and emerging domain is medical and healthcare domain. Biopolymers, such as alginate, chitosan, cellulose, gelatin, and collagen, are polymers of natural origin produced by living organisms. They are known to be highly biocompatible and biodegradable. Current studies reported that biopolymers exhibit various health beneficial properties such as antimicrobial, anti-inflammatory, hemostatic, cell proliferative, and antioxidant activities. So biopolymers could be used in the production of medical and paramedical textile devices, with drug delivery property, in the form of fibers, nanofibers, nonwoven, fabrics, sponges, nanoparticles and thin films. In addition, thank to its wound healing properties, biopolymers could be employed in the treatment of soft tissue infections. Biomaterials are able to slow down, then stop the bleeding of recent wounds by activating platelet agglutination. They could also promote the regeneration of tissue cells like muscles, skin, and nerves. Moreover, biopolymers-based nonwoven textile wound dressings can be used drug delivery carriers for wound healing. This paper highlights the structure and characteristics of natural biopolymers. The recent research advances and trends of biomass-derived materials and their derivatives in healthcare and medical textile devices for promotion of wound healing, tissue regeneration, and drug delivery in soft tissue diseases are investigated.

**Keywords** Biopolymers · Medical · Health · Wound · Textile · Drug

A. Elamri (✉) · M. Hamdaoui
MPTex UR17ES33, ENIM, University of Monastir, 5000 Monastir, Tunisia
e-mail: amri.adel2201@gmail.com

K. Zdiri
LPMT UR 4365, ENSISA, UHA, 68093 Mulhouse Cedex, France

 
S. B. Abdessalem et al. (eds.), *Proceedings of the Second International Conference of Innovative Textiles and Developed Materials-ITDM'2; 05-06 May 2023; Tunisia*,
https://doi.org/10.1007/978-981-99-7950-9_15

# 1 Introduction

Textile products and devices for medical and healthcare purposes could be situated at the middle between life sciences and technology disciplines. It presents two different features, one concerning the technical aspect, includes textile engineering and chemistry, and the other enclosing the aspects of life sciences like pharmacy, medicine, and biology [1]. Textiles are omnipresent when it comes to the field of human health and medical practice. Advanced medical textiles are being rapidly developed due to their growth and improvement in the fields of bandaging and pressure garments, wound healing, controlled drug release, and implant devices [2].

The employment of textile materials and products for medical and paramedical sectors goes from simple dressing materials to biologically active scaffolds for tissue regeneration with drug delivery capacity and various permanent body implantable devices. Their use undergoes a very fast expansion due to interesting textile materials characteristics, such as flexibility, absorption, filtering, lightness, and softness [3].

Researchers and scientists in medical and pharmaceutical areas are being interested in biomass-based textile materials because of their advanced performances, low cost, and environmental friendliness. Due to their important use in biomedical applications, biopolymers are now very used in antimicrobial wound healing, tissue regeneration, drug release, etc. [4].

Biopolymers are naturally widespread and easily available. They include mainly polysaccharides, like chitin, chitosan cellulose, alginate, hyaluronic acid, starch, dextran, etc., and protein materials, like collagen, silk, keratin, and others. The use of biomaterials in medical textile sector is developing from simple antimicrobial wound dressings to advanced bioengineering applications, such as implanted blood vessels, cells regeneration, drug delivery, and so others (Fig. 1), thanks to their versatile, biodegradable, non-toxic, and chemical reactive properties [5].

This paper deals with three main research topics in the medical textile applications of biomaterials: (i) wound healing, (ii) tissue engineering, and (iii) delivery of therapeutic molecules and discusses recent progress and prospects.

# 2 Biopolymers Used in Medical Textile Applications

Polysaccharide and protein large molecules could be used to fulfill a physical function (dressing), or to carrier a substrate (drug release), or to fulfill a biological activity by enhancing the bioactivity of the final medical textile product (scaffold).

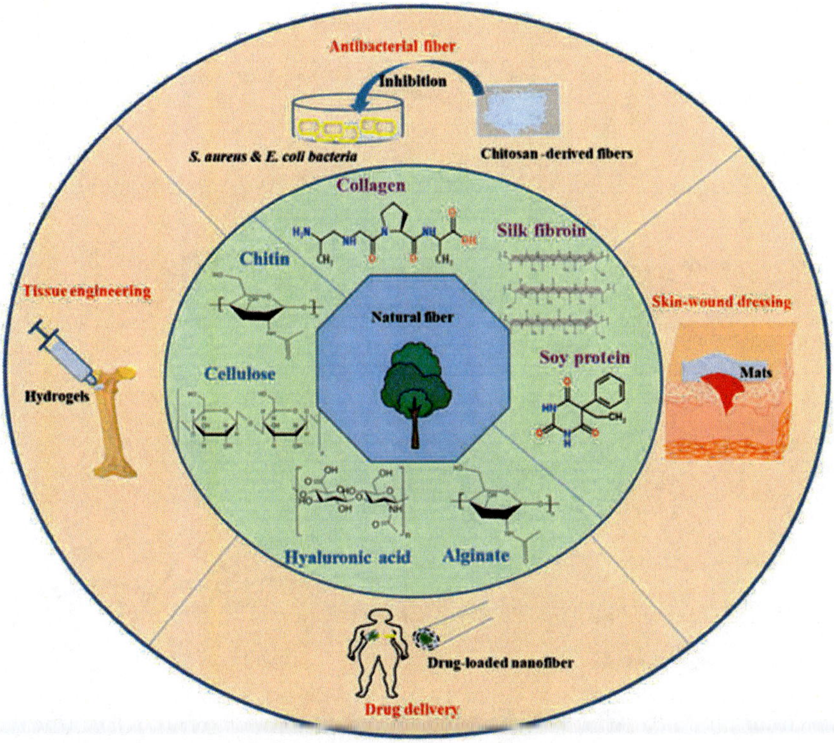


**Fig. 1** Biomass-derived polymers for biomedical applications

## *2.1 Polysaccharide Materials*

Generally, polysaccharides come from animals, plants–, and microorganisms. These biomass materials were used in many sectors, such as environment, packaging, energy, cosmetics, textile, and food. They are becoming widely applied in biomedical and paramedical textile area [6].

### 2.1.1 Cellulose and Cellulose Derivatives

Cellulose (Cell), $C_6H_{10}O_5$, is the most abundant biomass material in nature, and has been widely used in textile industry. It remains the most widely used for the preparation of various products for wound management [7]. Cell could originate from plants (cotton), fungi, algae, and bacteria with versatile and excellent properties. Despite their excellent features, cellulosic materials have limitations in their finishing processes coming from their poor solubility in most solvents due to the presence of large number of intra- and inter-molecular hydrogen bonds (Fig. 2) and high

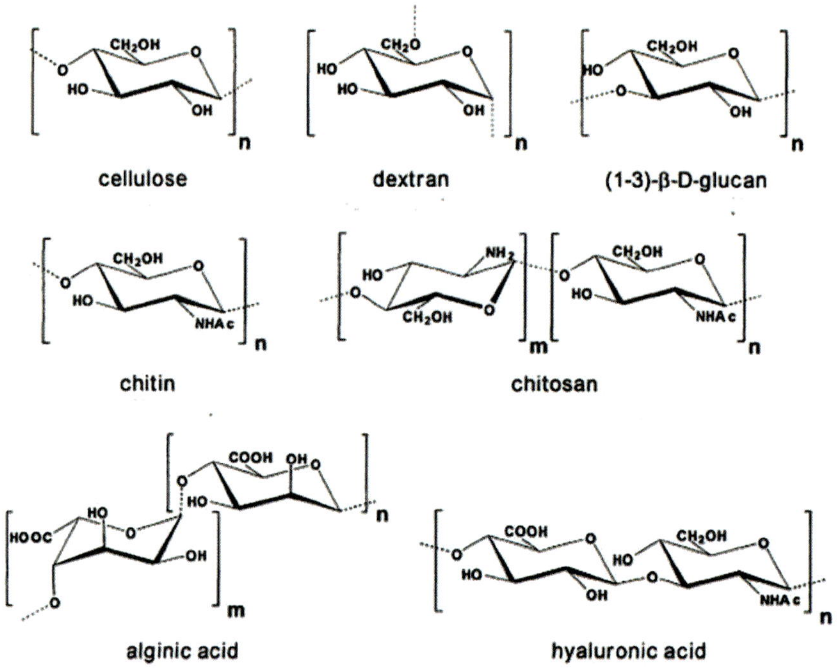


**Fig. 2** Polysaccharide biopolymers used in medical textiles

crystallinity. Hence, researchers has been focused on the use of cellulose derivatives, such us cellulose acetate, that present high solubility in various solvents allowing spinning and finishing operations [8].

Regenerated cellulose materials can be obtained by viscose process. The lyocell process is a likely eco-friendly alternative to the viscose process. Cellulose acetate, produced by cellulose acetylation, could also give acetate textile fibers. The acetate fiber is a semi-artificial material with excellent processing performances that could be used to produce textile products [9]. Another cell derivative is cellulose triacetate is also produced. Cell ethers are also biopolymers deriving from cell etherification in alkali baths.

### 2.1.2 Chitin and Chitosan

Chitin, poly-(1,4)-2 acetomido-2-deoxyß-D-glucose, occurs in nature as ordered macrofibrils in fungal cell walls and in the exoskeletons of mollusks and crustaceans, as well as in insect cuticles, but also in some algae and micro-algae [10]. With technological development and a deep investigation of its physiological and biological behavior, chitin is becoming transformed into chitosan (CSN), a water-soluble biopolymer [11]. Due to their excellent biological activities, and metal ion chelation capacity, nanomaterials based on chitin and chitosan are extensively applied in biomedical and pharmaceutical focuses, such as in wound treatment, hemostasis, antibacterial agents, tissue engineering, and cosmetics [12].

### 2.1.3 Alginate

Alginate is a naturally occurring polysaccharide obtained from seaweed, mainly algae. Due to its high biocompatibility, biodegradability, and mucosal adhesion, with low toxicity, this biopolymer is used in wound healing, drug delivery, and pharmaceutical packaging. Its structure is composed of irregular blocks of guluronic acid and mannuronic acid units linked by glycosidic linkages [13]. In a recent work, Hecht et al. [14] reported the development and characterization of sodium alginate and calcium alginate with focus on their structure. The characteristics of alginates could be modified by altering the number and the position of their and carboxyl and hydroxyl groups. These chemical modifications improve significantly the hydrophobicity, the biological activity and the solubility of alginate biopolymer. Alginate hydrogels are widely used in biomedical textile area, they are generally designed by crosslinking method [15].

### 2.1.4 Hyaluronic Acid

Hyaluronic acid (HA) is a negatively charged amino-glycan existing in human body mainly in eyes, skin, and connection tissues. HA exists in vivo as a disaccharide polymer of repeating d-glucuronic acid and n-acetyl glucosamine. HA is an important element of the extracellular matrix (ECM), which has a major role in skin wound healing, cells regeneration, and human body inflammatory response, etc. HA plays a crucial role in different steps of wound healing, such us proliferation and migration of keratinocytes and fibroblasts, improving inflammatory responses, increasing angiogenesis and collagen deposition at the wound location and inhibiting scar formation [16].

Hyaluronic acid biopolymer is composed of various molecules with different sizes, where each molecule is specified for each wound healing stage. In fact, large hyaluronic acid molecules (2000–4000 kDa saccharides) serve as space-filling molecules with regulation functions, but small size HA molecules are active in angiogeneis, inflammation, and immune stimulation phases [17]. These water-attracting properties give HA its characteristic elastic and cushioning properties illustrated by its concentration in skin and cartilage. When HA exists in high concentrations in this form, it creates a porous scaffolding network, allowing for select diffusion of cells and proteins, creating pathways for cell migration. This is a key feature of HA, as cell migration and proliferation are essential to successful wound healing. Smaller molecular size HA fragments also encourage cell migration, with very short, low-weight HA fragments inducing chemotaxis, and medium-sized HA chains stimulating inflammatory cytokines expression [18].

#### 2.1.5 Dextran

Dextran is a carbohydrate produced b bacteria, it is composed of α-1,6 linked D-glucopyranose with a weak content of α-1,2, α-1,3, or α-1,4-linked side chains. Dextran biomaterial is becoming applied in biomedical field due to its interesting properties like low cost, non-toxicity, and easy modification. This polysaccharide has been widely used as a plasma expander, peripheral flow enhancer, anti-thrombolytic agent, and for the rheological improvement of artificial tears. Dextran is highly water-soluble and very stable under mild acidic and basic conditions. It contains a large amount of hydroxylic groups which turns it suitable for derivatization and subsequent chemical and physical crosslinking. Moreover, dextran could be used as a drug delivery carrier, mainly to improve the systemic circulation of therapeutic molecules [19]. Dextran scaffolds were proved to be advantageous for soft tissue regeneration, as they present high resistance against protein adsorption and cell adhesion [20].

### 2.2 *Proteins*

Proteins used in the formulation/production of medical textiles include collagen, keratin, silk fibroin, and elastin.

#### 2.2.1 Collagen

Collagen is the most widespread mammalian protein characterized by a high mechanical stability. The collagen family englobes 28 types, the predominant ones are type I (cornea, ligament, dermis, tendon, and bone), then type II (vitreous and cartilage), and type III (intestine, blood, and skin). All collagens are structured as triple helices with each helices composed of near 1000 amino acids [21].

Soluble collagen type is obtained from young animals and is used in the design of collagen-based biopolymers [22]. Collagen threads were prepared: short fibers were generated by discontinuous methods, i.e., by infusing collagen into tubes or baths filled with a coagulant, followed by a coagulation time of up to several hours. Endless monofilaments were produced by means of continuous wet spinning methods using syringes and cannulas, a relatively flat microfluidic device, or a 3D-printed spinneret [23].

### 2.2.2 Silk Fibroin

Silk fibroin (SF) is an animal natural fibrillar protein derived from Bombyx mori cocoons and spiders. SF fiber is composed of 17–18 amino acids wrapped with sericin layer. Silk fibroin is processed from mulberry silk after removal of the outer silk sericin which may potentially elicit an immunological response when it is associated with fibroin [24]. SF has been considered a famous polymer matrix in tissue engineering and regenerative medicine, due to its great biocompatibility, tunable mechanical properties, and controllable degradation activity [25]. Silk fibroin has been also used for skin regeneration due to its antioxidant, hemostatic, and low inflammatory biological activities. Gil el al. [26] proved that SF nanomaterials allowed human fibroblasts and keratinocytes adhesion and improved in vitro collagen deposition mechanism. In addition, they stated that SF proteins are safe under high dermal toxicity, acute dermal irritation, and skin sensitization. Nevertheless, the safety and efficacy of silk fibroin material on skin damage repair have been rarely evaluated in clinically relevant large animal models or controlled human clinical trials. SF-based membranes could facilitate cutaneous wound healing and reduce inflammation compared with commercial dressings [27].

### 2.2.3 Keratin

Keratin is extracted from epithelial tissues and could be obtained from hair, feathers, and wool. It is able to be self-assembled as fibrous structures and is more cyto-compatible and non-immunogenic than collagen and elastin [28]. Keratin protein molecules are not soluble in aqueous medium and have strong mechanical properties due to the existence of cell binding motifs such as the arginine-glycine-aspartic acids [29]. Electrospinning of keratin is more or less difficult due to its limited viscoelasticity and rheological behavior. So, keratins are generally electrospun with incorporation of other polymers such as polyvinyl alcohol (PVA), poly ethylene oxide (PEO), and chitosan (CSN). The higher cyto-compatibility is obtained at increased keratin content into fibers [30]. Because antimicrobial activity of keratin is not remarkable, agents like irgasan, mupirocin, and silver are incorporated into keratin fibrous structures [31] (Fig. 3).

### 2.2.4 Elastin

Elastin is a protein of the extracellular matrix (ECM) of vertebrates. It is responsible for the mechanical resilience and viscoelasticity of some parts like skin, cartilage, ligaments, and vessels. It is mainly constituted of hyper cross-linked units tropoelastin. Elastin possesses unique properties, such as stretchability (until 220%), a high half-life of over 70 years, with excellent chemical and physical non-degradability [32]. Elastin insolubility limits its application, so some hydrolyzed soluble elastin forms are performed. These soluble elastins were easily electrospun into textile

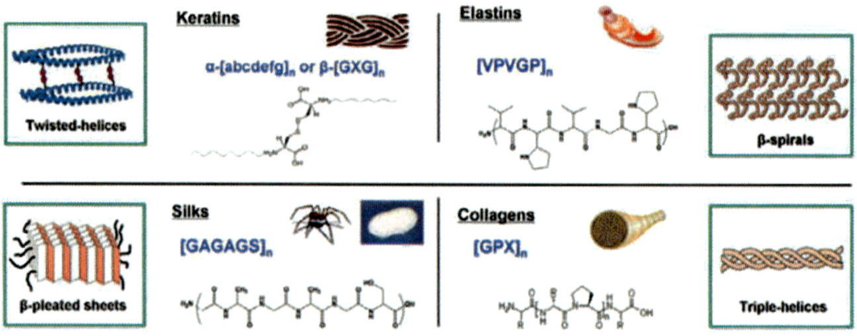


**Fig. 3** Molecular architecture and repetitive sequences of main fibrous proteins

nanofibers [33]. In a previous study, an elastin/collagen/PCL tripolymeric scaffold was tailored. The obtained scaffold possesses high mechanical behavior as well as biological characteristics such as proliferation, tissue regeneration, and cell adhesion [34]. Electrospinning of elastin and collagen blends have been applied to design different types of scaffolds. The incorporation of elastin has decreased scaffold contraction, degradation, and mechanical rigidity.

## 3 Electrospinning of Biomaterials for Medical Purposes

Electrospinning (ESP) is a relatively new technology applied to spin nonwoven membranes and fine nanofibers. It is a simple technique and low cost due to availability of biomass materials. The ESP apparel (Fig. 4) is composed of polymer propulsion system, a high voltage source, and a final collector [35]. The morphology of electrospun nanofibers depends on several factors, such as solution (viscosity, conductivity, molecular weight, and surface tension of the solution), process settings (voltage, spinning distance, and the flow rate), and processing environment (temperature and humidity) [36].

The properties of electrospun biomaterials that allow its use to promote wound healing are their high flexibility, high surface area, and volume ratio with important porosity. These characteristics improve permeability and cell respiration capacity, moisture absorption, good adaptability to wound skin surface, and the potential to leave no scars. The main biological activities of electrospun nanomaterials, such as skin cells' growth rate, biocompatibility with blood and tissues, and anti-inflammatory effects, vary with the nature of applied biopolymer.

Electrospun textile materials based on biopolymers, mainly polysaccharides and proteins, offer a myriad of opportunities in biomedical area due to their high biocompatibility and their easy accessibility in biomass sources. Low insoluble polysaccharides, like cellulose, are chemically modified to upgrade their spinnability [37]. Mixing biomaterials with other polymers is another strategy that facilitates the ESP

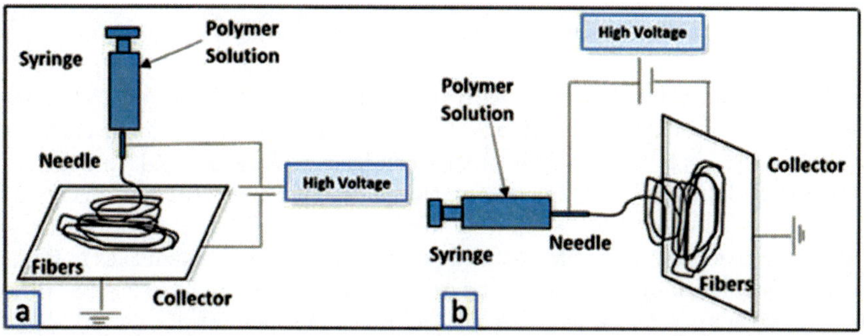


**Fig. 4** Schematic diagram of typical electrospinning apparatus: **a** vertical setup and **b** horizontal setup

process of polysaccharides and proteins. For chitosan biopolymer, when production of continuous filaments, suitable solvents should be used so that there is no interruption. Electrospinning of silk has also been reported [38]. The high mechanical strength of silk fibers allows its application in broad ranges of humidity and temperature.

## 4 Wound Healing Application

Wound healing is a multiphase process, where the cascade of healing is divided into four overlapping stages: Bleeding hemostasis, Inflammatory, Proliferative, and Remodeling or maturation (Fig. 5).

In modern life sciences, rapid and efficient wound treatment, including chronic wounds, is being of great interest. Protective wound dressings allowing suitable

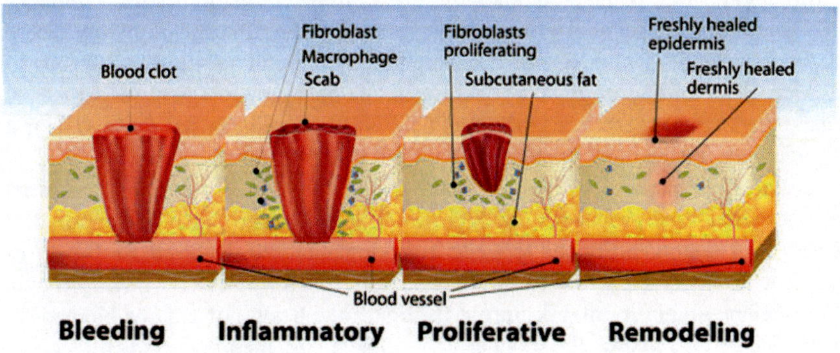


**Fig. 5** Wound healing stages

oxygen flow, humidity, and pH could inhibit bacterial invasion and accelerate wound healing stages.

Conventional wound dressings, like bandages, lints, and cotton gauzes, dry rapidly and glue to skin, leading to pain and discomfort in wound site. Moreover, these traditional dressings are not antimicrobial and luck of elasticity [39]. In dead, an ideal wound dressing should: (1) promote an equilibrating moisture environment, (2) absorb the wound extrudate, (3) allow sufficient air permeability, and (4) have antimicrobial activity. Natural biomass-derived polymers are potential candidates to fulfill the aforementioned functions as wound dressing, due to their high biodegradability, adsorption capacity, mechanical viscoelasticity, and moisture retention.

To improve functionalities of cellulose dressings, other biomass-derived polymers like chitosan, HA, and collagen and hormones were incorporated. The multifunctionalities of cellulose modified dressings at different phases of wound healing were investigated in further studies. Bakhsheshi-Rad et al. prepared cellulose/PVA/silk dressings possessing excellent hygroscopic behavior, as well as in vitro and in vivo antimicrobial and biocompatibility properties [40].

Hyaluronic acid biopolymer is also considered as an interesting candidate for biomedical exploitation, particularly for wound healing dressings. The structure of HA possesses numbered hydroxyl and carboxyl end groups leading to high hydrophobicity researched in adhesion and wound healing applications. In addition, hyaluronic acid inhibits bleeding, promotes collagen deposition and fibrosis, regulates inflammation via the enzymatic reaction of hyaluronidase and chemical interactions.

In the first steps of wound treatment, CSN could facilitate polymorphonuclear neutrophils migration into the wound site, allowing beginning of tissues regeneration. In the last stage of wound healing, CSN induces epithelialization of skin layers, as well as collagen formation and fibrosis synthesis [41]. Hydroxyl and amino groups of CSN allow its interaction with other polymers, such as PEO, polyvinyl alcohol, and collagen. These composites are useful to manufacture biomedial textiles in form of dressings, scaffolds, and sponges.

Biomass-derived collagen biomolymer has high biocompatibility that facilitates body keratinocyte combined with epidermal cells' proliferation and migration. These properties promote wounded skin regeneration. But, the major disadvantageous of collagen is its animal origin that could promote an immunological response and transfer undesirable pathogens to human tissues. Also, collagen stability and its fast degradability are the main drawbacks that limit its in vivo application.

Last years, clinicians and scientist have great interest to silk fibroin biomaterial thank to researched properties like high water uptake and oxygen content as well as very slow degradability and low rate immunogenicity. In addition, SF possesses excellent elasticity (superior to 30%), which are useful as elastic fibers in vivo. The excellent solubility of silk fibroin facilitates the design of various biomedical structures, such as hydrogels, sponges, (nano)fibers, and membranes [42].

Alginate nanofibers and nonwoven textiles have gained great attention in biomedical area, particularly in wound healing dressings. The main advantages of alginate-based textile biomaterials are easy and low cost production, good versatility, and flexibility. The web is fabricated by layers deposition, over one another, until obtaining the desired nonwoven structure. Recent years, other components have been added to alginate biopolymers to upgrade the final product characteristics. As an example, silver antimicrobial particles were incorporated to the alginate fiber. Moreover, calcium/alginate dressings were proven to decrease the wound pain [43].

## 5 Tissue Engineering and Regeneration

Any tissue damage beyond this crucial dimension requires external assistance approaches such as tissue engineering and regeneration (TER), which the external hollows are termed yardsticks. These tissues offer a platform for cellular activity and new tissue creation. TER has four interrelated prerequisites: suitable cells to colonize the scaffold and help the body to recover, a suitable environment/scaffold for cell adhesion and proliferation, appropriate biomolecules, and the right physical and mechanical forces for cell development In TER, the scaffolds have a crucial role. The development of artificial cellular scaffolds that mimic ECM is of great interest. During the last decade, there is a great interest to use biomass-derived substrates in TER area in order to guarantee high biodegradability and of scaffolds. Moreover, these biopolymers have intrinsic characteristics increasing their biological activities and therefore therapeutic function.

Electrospinning technology allows the production of textile nanomats (Fig. 6) with excellent properties mimicking those of native tissue ECM cells, such as mechanical elasticity, low density, and good porosity. In addition, ESP permits the integration of biological substances into fiber mats, which promotes tissue cells healing of biomedical textile structures.

In the previous work of Shalumon et al. [44], human stem cells were easily regenerated and spread into chitin/polyvinyl alcohol textile scaffolds. Moreover, these nanofiber composites induced hydroxyapatite in vitro deposition during mineralization phase. In another work, Kang et al. [45] deigned electrospun CSN-based nonwoven mats able to promote the regeneration of rat muscle derived stem cells.

Recently, Zhang et al. [46] synthetized biologically compatible cellulose/nanocrystals composite scaffolds charged with bone morphogenic protein BMP-2. The human bone stem cells have developed in vitro aligned with nanofibrous scaffolds together with calcium rate and alkaline phosphatase activity. Similarly, these scaffolds promoted collagen cells assembly and in vivo bone tissue formation in a rabbit bone defected model. In a similar work, Zhao et al. [47] used a water-soluble chitosan derivative carboxymethyl chitosan (CMCSN) in composite with polyethylene oxide. The CMCSN has interesting high properties of bioactivity, biodegradability, Ca2 + cation chelating capacity. Electrospun CMCSN/PEO nanomaterials promoted the

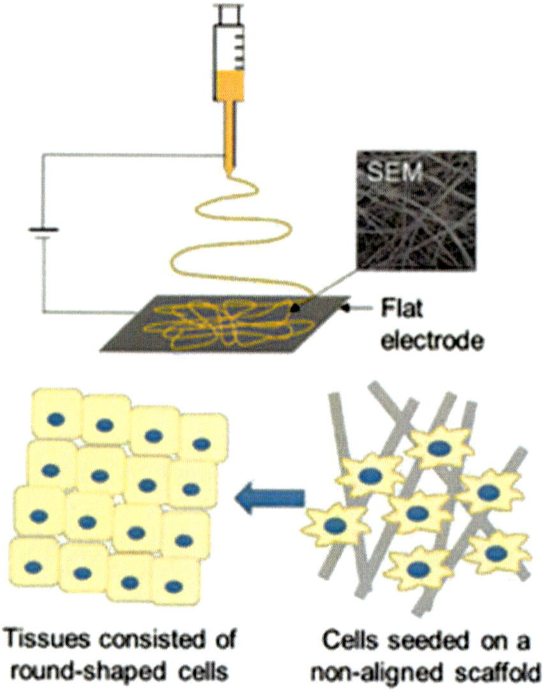


**Fig. 6** Electrospinning of textile mats for tissue engineering scaffolds

alkaline phosphatase activity and increased osteogenic differentiation of mouse bone in vitro cells.

In the area of nerve tissue engineering using biomedical textile, conductive materials were incorporated or coated on polysaccharide scaffolds during ESP process. Zha et al. [48] coated cellulose nanofibers with poly(N-vinylpyrrole) and poly(3-hexylthiophene). The obtained scaffolds possesses high hydrophilicity, conductivity, porosity, and thickness. Tests of electrical conductivity stimulation showed that poly(3-hexylthiophene)-coated textiles promoted well nerve cell proliferation.

In the field of vascular tissue regeneration and engineering, tubular-shaped textile nanofibres were developed by ESP method. In the work of Joy et al. [49], tubular-shaped scaffolds were electrospun from cross-linked hydrogel based on gelatin and carboxymethyl cellulose CMC. Later, Zhang et al. [50] fabricated nonwoven fabrics from SF/HA ribbon-shaped nanofibers. Cell viability experiments showed that the developed scaffolds improved well the proliferation of pig iliac endothelial cells.

## 6 Drug Delivery

Drug delivery carriers should carry and release therapeutic molecules to a targeted position under specified microenvironment, particularly for insoluble drug molecules. An ideal drug delivery carrier must present intrinsic properties, such as biocompatibility, inert character, and mechanical strength. It should be also able to deliver high load, simple to administer to patient, safe and easily sterilized. Bio-based textile materials are promising candidates to overcome the majority of the negative features of conventional drug delivery systems. Textiles could be loaded with therapeutics by various chemical (covalent conjugation and grafting) or physical (absorption or adsorption, coating, and encapsulation) methods [51]. Drug delivery biomedical systems based on textiles could be classified into three structures: filament, woven, and nonwoven fabrics. Depending on the based biopolymer, these textile materials could be degradable or non-degradable. Among the three aforementioned textiles categories, continuous filaments showed a great interest. Filaments could contain the active biomolecules inside them or chemically bond to their surface [52]. Such continuous filaments are generally fabricated using wet spinning technique (Fig. 7).

Cellulose biopolymer has interesting inherent properties, such as high biocompatibility, water stability and solubility, as well as mechanical strength, that facilitate its use in biomedical drug delivery. Cell fibers were coated with four drug types (ibuprofen, sulindac, indomethacin, and naproxen) [53]. In order to avoid the risk of burst release, a modified strategy of tri-axial electrospinning has been employed, this strategy allowed the improvement of the porosity and the hydrophilicity of the drug

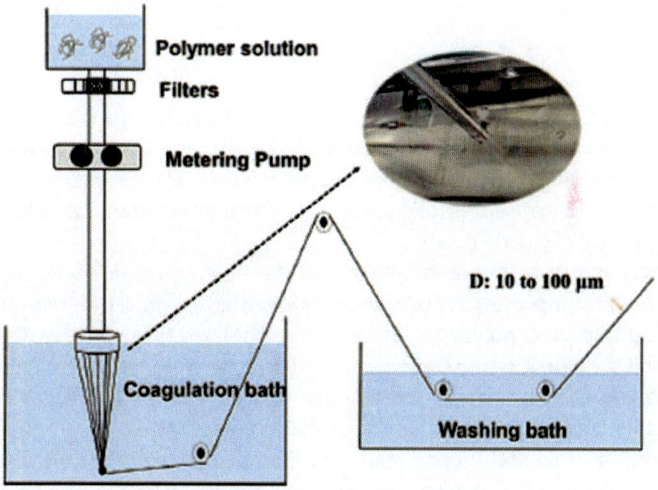


**Fig. 7** Wet spinning of continuous biopolymers' filaments

carrier surface [54]. Cellulose sulfate is a sulfated (or ester) cellulose derivative characterized by an increased solubility, antimicrobial activity and film forming ability compared to raw cellulose. These properties make it an efficient therapeutic (risedronate, rifampicin, and cytokines) encapsulating agent. Other cellulose derivatives, such as carboxymethyl cellulose, ethyl and methyl cellulose, nitrate, and acetate cellulose, have been proven to be efficient drug delivery carriers [55].

Textile applications of gelatin-based hydrogels have begun in drug delivery systems, where transdermal drug transport has been achieved, and thermo-responsive gelatin-based hydrogels are successfully building drug transport systems for textile applications. Textile-based transdermal treatment using drug-loaded gelatin-based hydrogel has become more successful in human skincare [56].

Composite electrospun nanofibers for biomedical drug delivery purposes consist the use of natural substances and biomaterials combined with drugs (sulfisoxazole, naproxen) and essential oils with antimicrobial activity (eugenol and tocopherol). This technique allows to fabricate poly (lactide-co-glycolide) scaffolds, allowing the permanent delivery of the therapeutic molecules from biomaterials [57].

Drug delivery using silk fibroin is rpvd to be sustained and pH-responsive systems. SF nanofibers-based hydrogels are successfully applied to insulin dependent type I diabetes. The delivery of doxorubicin, which is a breast cancer drug, has been performed by silk-based injectable hydrogels. Also, silk has been applied to design thermo-sensitive hydrogels used to deliver anti-vascular endothelial growing factor. Desferrioxamine silk nanofiber hydrogels were also developed and promoted rapid development of blood vessels in wound site. Ciprofloxacin silk fibroin/β-cyclodextrin citrate nanofibers were also developed for controlled drugs release [58].

## 7 Conclusions

Increasing attention toward sustainable development and environmental issues has resulted in many attempts at using renewables, i.e., biopolymers, in medical textile industry. Using biopolymers for textile practices reduces the dependency of the textile industry on petroleum resources and substantially promotes the environmental profile and sustainability of textile goods.

This study reviewed the developments in the field of medical and health care related to bio-based textiles. This review provided an appraisal on the innovative, intelligent and smart biopolymers' textile products related to medical textiles, particularly wound dressings, tissue engineering and drug delivery fields. This review also highlighted the current state of the technologies as well as future advances and enhancements potential of new medical textile biomaterials.

The addition of natural biopolymers modified the properties of the electrospun nanofibres. To name a few, the main biomaterials applied in biomedical textile applications are cellulose, silk, chitosan, hyaluronic acid, collagen, and gelatin. Mixing biopolymers with synthetic ones has been also applied to develop new biomaterials with inherent characteristics, like mechanical strength and thermal stability, beneficial for textile biomedical applications.

## References

1. King, M. W., Gupta, B. S., Guidoin, R.: Biotextiles as Medical Implants, Elsevier, Amsterdam, Netherlands, 2017.
2. Petros, S., Tesfaye, T., Ayele, M.: A Review on Gelatin Based Hydrogels for Medical Textile Applications. Journal of Engineering (2020). ID 8866582. https://doi.org/10.1155/2020/8866582.
3. Sumithra, M.: Development of medical textile product using chitosan incorporated herbal extract (Aristolochia bracteolate). Int. J. Pharm. Life Sci. 9(3), 5748–5754 (2018).
4. Akshay Kumar, K. P., Zare, E. N., Torres-Mendieta, R., Waclawek, S., Makvandi, P., Cernik, M.: Electrospun fibers based on botanical, seaweed, microbial, and animal sourced biomacromolecules and their multidimensional applications. Int. J. Biol. Macromol. 171, 130–149 (2021).
5. Li, D., Wang, Y., Huang, W., Gong, H.: Biomass-derived fiber materials for biomedical applications. Front. Mater. 10, 1058050 (2023).
6. Zhao, W., Liu, W., Li, J., Lin, X., Wang, Y.: Preparation of animal polysaccharides nanofibers by electrospinning and their potential biomedical applications. J. Biomed. Mater. Res. Part A 103, 807–818 (2015).
7. Alminderej, F.M., Ammar, C., El-Ghoul, Y.: Functionalization, characterization and microbiological performance of new biocompatible cellulosic dressing grafted chitosan and Suaeda fruticose polysaccharide extract. Cellulose 28, 9821–9835 (2021).
8. Jahandideh, A., Ashkani, M., Moini, N.: Biopolymers in textile industries. In Sabu Thomas, Sreeraj Gopi, Augustine Amalraj Editor(s), Biopolymers and their Industrial Applications, Chapter 8, pp. 193–218, Elsevier (2021).
9. Joyce, K., Fabra, G.T., Bozkurt, Y. et al.: Bioactive potential of natural biomaterials: identification, retention and assessment of biological properties. Sig Transduct Target Ther. 6, 122 (2021).
10. Elamri, A., Zdiri, K., Hamdaoui, M., Harzallah, O.: Chitosan: A biopolymer for textile processes and products. Textile Research Journal. 93(5–6), 1456–1484 (2023).
11. Elamri, A., Zdiri, K., Bouzir, D., Hamdaoui M.: Use of Chitosan as Antimicrobial, Antiviral and Antipollution Agent in Textile Finishing. Vlakna a Textil 29(3), 51–70 (2022).
12. Ahmad, S. I., Ahmad, R., Khan, M. S., Kant, R., Shahid, S., Gautam, L., et al.: Chitin and its derivatives: Structural properties and biomedical applications. Int. J. Biol. Macromol. 164, 526–539 (2020).
13. Saltz, A., Kandalam, U.: Mesenchymal stem cells and alginate microcarriers for craniofacial bone tissue engineering: A review. J. Biomed. Mater. Res. A 104, 1276–1284 (2016).
14. Hecht, H.; Srebnik, S.: Structural Characterization of Sodium Alginate and Calcium Alginate. Biomacromolecules 17, 2160–2167 (2016).
15. Wróblewska-Krepsztul J., Rydzkowski T., Michalska-Pożoga I., Thakur V.K.: Biopolymers for Biomedical and Pharmaceutical Applications: Recent Advances and Overview of Alginate Electrospinning. Nanomaterials (Basel) 9(3), 404 (2019).
16. Tan, G., Wang, L., Pan, W., Chen, K.: Polysaccharide Electrospun Nanofibers for Wound Healing Applications. Int J Nanomedicine 17, 3913–3931 (2022).
17. Castro, K.C., Campos, M.G.N., Mei, L.H.I.: Hyaluronic acid electrospinning: challenges, applications in wound dressings and new perspectives. Int J Biol Macromol. 173, 251–266 (2021).

18. Frenkel, J.S.: The role of hyaluronan in wound healing. Int Wound J 11, 159–163 (2014).
19. Maia, J., Evangelista, M.B., Gil, H., Ferreira, L.: Dextran-based materials for biomedical applications. Res. Signpost 37661, 31–53 (2014).
20. Samrot, A.V., Sathiyasree, M., Rahim, S.B.A., et al.: Scaffold Using Chitosan, Agarose, Cellulose, Dextran and Protein for Tissue Engineering-A Review. Polymers (Basel) 15(6), 1525 (2023).
21. Chattopadhyay, S., Raines, R.T.: Review collagen-based biomaterials for wound healing: Collagen-Based Biomaterials. Biopolymers 101, 821–833 (2014).
22. Ricard-Blum, S.: The Collagen Family. Cold Spring Harb Perspect Biol 3, a004978–a004978 (2011).
23. Tonndorf, R., Aibibu, D., Cherif, C.: Collagen multifilament spinning. Mater Sci Eng C Mater Biol Appl. 106, 110105 (2020).
24. Wang, Z., Zhang, Y., Zhang, J., Huang, L., Liu, J., Li, Y., Zhang, G., Kundu, S.C., Wang, L.: Exploring natural silk protein sericin for regenerative medicine: an injectable, photoluminescent, cell-adhesive 3D hydrogel. Sci. Rep. 4, 7064 (2014).
25. Kasoju, N. Bora, U. Silk fibroin in tissue engineering. Adv. Healthcare Mater. 1, 319 (2012).
26. Gil, E.S., Panilaitis, B., Bellas, E., Kaplan, D. L. Functionalized silk biomaterials for wound healing Adv. Healthcare Mater. 2, 206 (2013).
27. Zhang, W., Chen, L., Chen, et al. Silk Fibroin Biomaterial Shows Safe and Effective Wound Healing in Animal Models and a Randomized Controlled Clinical Trial. Adv. Healthcare Mater. 6, 1700121 (2017).
28. Fearing, B.V., Van Dyke, M.E.: In vitro response of macrophage polarization to a keratin biomaterial. Acta Biomater. 10, 3136–3144 (2014).
29. Yao, C.-H., Lee, C.Y., Huang, C.H., Chen, Y.S., Chen, K.Y.: Novel bilayer wound dressing based on electrospun gelatin/keratin nanofibrous mats for skin wound repair. Mater. Sci. Eng. C 79, 533–540 (2017).
30. Zarei, M., Tanideh, N., Zare, S., Aslani, F.S., Koohi-Hosseinabadi, O., Rowshanghias, A., Pourjavaheri, F., Mehryar, P., Muthuraj, R.: Electrospun poly(3-hydroxybutyrate)/chicken feather-derived keratin scaffolds: Fabrication, in vitro and in vivo biocompatibility evaluation. J. Biomater. Appl. 34, 741–752 (2020).
31. Akhmetova, A., Heinz, A.: Electrospinning Proteins for Wound Healing Purposes: Opportunities and Challenges. Pharmaceutics 13, 4 (2021).
32. Parham, S., Kharazi, A.Z., Bakhsheshi-Rad, H.R., Kharaziha, M., Ismail, A.F., Sharif, S., Razzaghi, M., RamaKrishna, S. and Berto, F.: Antimicrobial Synthetic and Natural Polymeric Nanofibers as Wound Dressing: A Review. Adv. Eng. Mater. 24, 2101460 (2022).
33. Hong, Y., Zhu, X., Wang, P., Fu, H., Deng, C., Cui, L., Wang, Q., Fan, X.: Tyrosinase-Mediated Construction of a Silk Fibroin/Elastin Nanofiber Bioscaffold. Appl Biochem Biotechnol. 178(7), 1363–1376 (2016).
34. Chong, C., Wang, Y., Fathi, A., Parungao, R., Maitz, P.K., Li, Z.: Skin wound repair: Results of a pre-clinical study to evaluate electropsun collagen-elastin-PCL scaffolds as dermal substitutes. Burns. 45(7), 1639–1648 (2019).
35. Mendes, A.C., Stephansen, K., Chronakis, I.S.: Electrospinning of food proteins and polysaccharides. Food Hydrocoll. 68, 53–68 (2017).
36. Wsoo, M.A., Shahir, S., Mohd Bohari, S.P., Nayan, N.H.M., Razak, S.I.A.: A review on the properties of electrospun cellulose acetate and its application in drug delivery systems: A new perspective. Carbohydr. Res. 491, 107978 (2020).
37. Poshina, D., Otsuka, I.: Electrospun Polysaccharidic Textiles for Biomedical Applications. Textiles 1, 152–169 (2021).
38. Chee, B.S., Nugent, M.: Electrospun natural polysaccharide for biomedical application. In Natural Polysaccharides in Drug Delivery and Biomedical Applications, Hasnain, M.S., Nayak, A.K., Eds., Chapter 26-pp. 589–615, Academic Press: Cambridge, MA, USA (2019).
39. Mostafalu, P., Kiaee, G., Giatsidis, et al.: A textile dressing for temporal and dosage controlled drug delivery. Adv. Funct. Mater. 27(41), 1702399 (2017).

40. Bakhsheshi-Rad, H. R., Ismail, A. F., Aziz, M., Akbari, M., Hadisi, Z., Omidi, M., et al.: Development of the PVA/CS nanofibers containing silk protein sericin as a wound dressing: In vitro and in vivo assessment. Int. J. Biol. Macromol. 149, 513–521 (2020).
41. Simoes, D., Miguel, S. P., Ribeiro,M. P., Coutinho, P.,Mendonca, A. G., Correia, I. J.: Recent advances on antimicrobial wound dressing: A review. Eur. J. Pharm. Biopharm. 127, 130–141(2018).
42. Yuan, M., Dai, F., Li, D., Fan, Y., Xiang, W., Tao, F., et al.: Lysozyme/collagen multilayers layer-by-layer deposited nanofibers with enhanced biocompatibility and antibacterial activity. Mater Sci. Eng. C Mater Biol. Appl. 112, 110868 (2020).
43. Varaprasad, K., Jayaramudu, T., Kanikireddy, V., Toro, C., Sadiku, E.R.: Alginate-based composite materials for wound dressing application: A mini review. Carbohydr Polym. 15(236), 116025 (2020).
44. Shalumon, K.T., Anulekha, K.H., Girish, C.M., Prasanth, R., Nair, S.V., Jayakumar, R.: Single step electrospinning of chitosan/poly(caprolactone) nanofibers using formic acid/acetone solvent mixture. Carbohydr. Polym. 80, 413–419 (2010).
45. Kang, Y.M., Lee, B.N., Ko, J.H., Kim, G.H., Kang, K.N., Kim, D.Y., Kim, J.H., Park, Y.H., Chun, H.J., Kim, C.H.: In vivo biocompatibility study of electrospun chitosan microfiber for tissue engineering. Int. J. Mol. Sci. 11, 4140–4148 (2010).
46. Zhang, X., Wang, C., Liao, M., Dai, L., Tang, Y., Zhang, H., Coates, P., Sefat, F., Zheng, L., Song, J.: Aligned electrospun cellulose scaffolds coated with rhBMP for both in vitro and in vivo bone tissue engineering. Carbohydr. Polym. 213, 27–38 (2019).
47. Zhao, X., Zhou, L., Li, Q., Zou, Q., Du, C.: Biomimetic mineralization of carboxymethyl chitosan nanofibers with improved osteogenic activity in vitro and in vivo. Carbohydr. Polym. 195, 225–234 (2018).
48. Zha, F., Chen, W., Hao, L., Wu, C., Lu, M., Zhang, L., Yu, D.: Electrospun cellulose-based conductive polymer nanofibrous mats: Composite scaffolds and their influence on cell behavior with electrical stimulation for nerve tissue engineering. Soft Matter 16, 6591–6598 (2020).
49. Joy, J., Pereira, J., Aid-Launais, R., Pavon-Djavid, G., Ray, A.R., Letourneur, D., Meddahi Pelle, A., Gupta, B.: Gelatin-Oxidized carboxymethyl cellulose blend based tubular electrospun scaffold for vascular tissue engineering. Int. J. Biol. Macromol. 107, 1922–1935 (2018).
50. Zhang, K., Fan, L., Yan, Z., Yu, Q., Mo, X.: Electrospun biomimic nanofibrous scaffolds of silk fibroin/hyaluronic acid for tissue engineering. J. Biomater. Sci. Polym. Ed. 23, 1185–1198 (2012).
51. Soni, B.: Cellulose-Based Graft Copolymers, Thakur, V.K. Eds, Taylor and Francis, London (2015).
52. Rostamitabar, M., Abdelgawad, A.M., Jockenhoevel, S. and Ghazanfari, S.: Drug-Eluting Medical Textiles: From Fiber Production and Textile Fabrication to Drug Loading and Delivery. Macromol. Biosci. 21, 2100021 (2021).
53. Tungprapa, S., Jangchud, I., and Supaphol, P.: Release characteristics of four model drugs from drug-loaded electrospun cellulose acetate fiber mats. Polymer 48 (17), 5030–5041 (2007).
54. Chen, Y., Qiu, Y., Chen, W., and Wei, Q.: Electrospun thymol-loaded porous cellulose acetate fibers with potential biomedical applications. Mater Sci. Eng. C Mater Biol. Appl. 109, 110536 (2020).
55. Oprea, M., and Voicu, S. I.: Recent advances in composites based on cellulose derivatives for biomedical applications. Carbohydr. Polym. 247, 116683 (2020).
56. Hassabo, A., Zayed, M., Bakr, M., Othman, H.: The Utilisation of Gelatin Biopolymer in Textile Wet Processing. Journal of Textiles, Coloration and Polymer Science 19(2), 125–136 (2022).
57. Soares, R.M.D., Siqueira, N.M., Prabhakaram, M.P., Ramakrishna, S.: Electrospinning and electrospray of bio-based and natural polymers for biomaterials development. Mater. Sci. Eng. C-Mater. Biol. Appl. 92, 969–982 (2018).
58. Troy, E., Tilbury, M.A., Power, A.M., Wall, J.G.: Nature-Based Biomaterials and Their Application in Biomedicine. Polymers 13, 3321 (2021).

# Smart & Technical Textiles

# Study of the Weight Loss of Abrasive Non-wovens from Industrial Waste

**Elaissi Arwa and Ghith Adel**

**Abstract** In this study, abrasive non-wovens made from industrial waste were produced. The Responsive Surface Methodology (RSM) was employed to comprehend the system's general behavior and calculate the weight loss of the abrasives. Using the Box-Behnken design, significant control parameters affecting the weight loss of abrasives were found. The weight loss and wear of the abrasive were found to be influenced by the size of the abrasive grain, according to the results. Results were not significantly affected by the percentage of fibers, resin concentration, or chemical treatment (cationization) of the reinforcements.

**Keywords** Abrasive non-wovens · Box-Behnken design · Main effects of mass loss.

## 1 Introduction

To increase system reliability, a technology known as the abrasive wear of materials is constantly developing. How it is applied in the sector is based on profitability. Abrasive costs make up the majority of operational expenses and amount to more than 75% of them [1, 2]. Due to their effectiveness and flexibility, non-wovens were used as reinforcements by several researchers [3]. A non-woven abrasive, for instance, is provided by Bret W. Ludwig's innovation and includes a fibrous web of non-woven, abrasive particles, and a polyurethane binder coupling [4]. To enhance interfacial

* Supported by MPTEX.

E. Arwa (✉) · G. Adel
Textile Materials and Processes Research Unit, Tunisia National Engineering School of Monastir, University of Monastir, Monastir, Tunisia
e-mail: arwa.elaissi@enim.u-monastir.tn

G. Adel
e-mail: ADEL.GHITH@u-monastir.tn

 
S. B. Abdessalem et al. (eds.), *Proceedings of the Second International Conference of Innovative Textiles and Developed Materials-ITDM'2; 05-06 May 2023; Tunisia*,
https://doi.org/10.1007/978-981-99-7950-9_16

adhesion performance, cationization treatments have been applied to the reinforcements [5]. In this situation, Santos et al. used a cationization technique followed by the grafting of nanoparticles to enhance the interfacial characteristics of sisal fibers with cementitious matrices [6].

Focusing on the effects of variables such as resin content, the amount and treatment of fibers used as reinforcements, and grain size on abrasive wear behavior is the main goal of the current research. By utilizing experience-based optimization approaches, it is possible to gain a greater understanding of the procedure in terms of the influence of the independent factors on the dependency of the variables. Using the quadratic model's analysis of variance (ANOVA), the variable parameters were evaluated. A statistically significant regression of the coefficient between the variables was discovered after measuring a total of 30 sets of experiments. Moreover, measures both before and after the abrasion test were evaluated.

## 2 Experimental

### 2.1 *Fibers Preparation*

Before developing composite abrasive materials, cotton-based textile waste and cellulose gypsum tow fibers were first pre-treated to remove any impurities. To perform the scrubbing process, the fibers were boiled for 2 h in a flask containing a mixture of 1.5 mL/L NaOH, 3 g/L $Na_2CO_3$, and 4 g/L non-ionic detergents (1:40). The treated fibers were then given a one-minute rinse in both hot and cold water. Lastly, the fibrous material was dried in a lab oven. The fibers were further treated to a cationization procedure in an alkaline environment employing a co-polymer of diallylamin and dimethyl diallyl ammonium chloride as a cationic agent (3 and 6%).

### 2.2 *Producing Non-woven Components*

The fibers were initially opened using a Shirley machine to create non-woven composites. The non-wovens were then created by mixing these basic materials with various fiber percentages utilizing carding, overlapping, and needling techniques [7–10].

### 2.3 *Preparation of Abrasive Particles*

Iron flakes were gathered from nearby metallurgical businesses. The desired grain size was determined after washing, drying, and inspection. Iron grains are divided

**Fig. 1** Used materials for the manufacture of abrasives

into two sizes based on particle size analysis, by the ISO 3310-1 standard. Less than 0.2 mm is the size of the powder grains. Large grains are chosen to be between 0.2 and 0.6 mm in size.

## 2.4 Abrasives Manufacturing

A thermosetting polyurethane resin with the name "Politex" is employed in this study. Once the reinforcement is placed on the coating table, the resin solution and hardener are slowly poured onto the textile material to create abrasives. The second resin layer is then completely mixed with the abrasive grains. A third layer of scattered grains covered the surface of the abrasives. The coating method determines the layer's thickness [11]. All of the samples were then adequately dried at 140 °C for the required time [12]. All the materials used for the manufacture of the abrasives are illustrated in Fig. 1.

## 2.5 Abrasion Test

The ISO 12947 standard for evaluating cloth abrasion resistance includes the Martindale method [13]. Instead of utilizing the normal abrasive paper, the prepared abrasives were used during the abrasion cycles over a denim backing (3/4 twill). The abrasive wear tests were carried out in a dry state at room temperature with a load of 9 N and a constant speed of 200 mm/s. All developed composites underwent a 5000-cycle abrasion test to simulate the deterioration of denim fabric. To determine the precise wear rate, weight loss data before and after polishing can be compared [14]. Five measurements were taken for each situation, and an average value was calculated.

## 2.6 *Choice of Factors and Field of Study*

We created an experimental strategy to identify the ideal circumstances that enable us to obtain the best value of the selected response to analyze the effects of the factors. To quickly filter the components of a multivariate system for the optimization of a particular process, statistical techniques are regarded as an effective and powerful instrument. It is challenging to run tests on all factorial combinations of the test variables due to the significant number of required experiments. To explore the impact of many response-influencing components, Response Surface Modeling (RSM), which permits the simultaneous change of a constrained number of response-influencing elements, can be helpful [15]. Based on the results from the experimental testing, basedased on the Box-Behnken design, which is a subset of Response Surface Methodology, the experimental tests are carried out (RSM). It takes prior knowledge about the procedure to build a statistical model. This optimization process essentially entails three steps: conducting statistically planned experiments, figuring out the coefficients of a mathematical model, forecasting the outcome, and assessing the model's fit [16, 17]. The preliminary tests conducted on the numerous parameters influencing the abrasion of the composites led to the selection of the factors and the study's focus for the current investigation. The wear rates of composites have been reported to be affected by fiber percentage and reinforcement treatment, resin concentration, and grain size. In the Box-Behnken program, the first 3 factors are continuous factors and the grain size represents a two-level category factor. The level of experimental design considered for this study is presented in Table 1.

The Box-Behnken design requires 24 design points plus 6 center runs, and therefore the number of trials is 30. Each experiment was performed three times.

**Table 1** Factors and areas of study

| Factors | Unit | Code factor | Levels not coded | | |
|---|---|---|---|---|---|
| | | | Level 1 | Level 2 | Level 3 |
| Resin concentration | % | A | 20/80 | 50/50 | 80/20 |
| Pre-traitements | – | B | Desized | Cationized 3% | Cationized 6% |
| Percentage of cellulosic fibers (fiber/cotton) | % | C | 20/80 | 50/50 | 80/20 |
| Grain size (category factor) | mm | D | ≤ 0.2 | 0.2–0.6 | – |

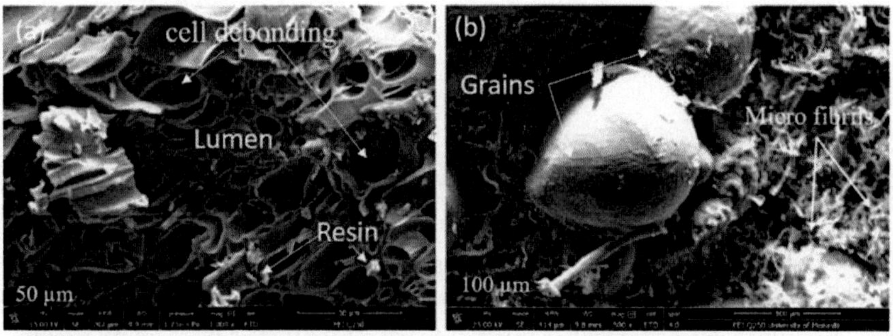


**Fig. 2** SEM photographs of **a** cross-section of the substrate (50 μm), and **b** abrasives surface (100 μm)

## 2.7 Scanning Electron Microscopy Characterization

Scanning electron microscopy (SEM) is a common tool for evaluating material shape and surface texture. To compare the different features of composites and examine their surface, the specimens were scanned with an FEI Q250 Thermofisher apparatus worked at 25 kV. To perform this research, the sample is metalized, which involves coating it with a gold film to turn it conductive.

# 3 Results and Discussion

## 3.1 Surface Morphology

As shown in Fig. 2, the surface structure of the produced composites was assessed by SEM at various magnifications. The appearance of the multilayered cell walls can be seen in the cross-section of the substrate made from cellulosic fibers of cotton and gypsum. Cell–cell dissociation and microfibrils were both appearing at the abrasive's surface.

## 3.2 Box-Behnken Analysis of Loss in Mass of Abrasives

To investigate the impact of the selected components on composite weight loss, RSM was performed using the design of experiments tool, Design-Expert version 8.0.4. Each experiment was repeated three times to estimate the random effect (residual or experimental variation). After carrying out the preliminary tests, the number of abrasion cycles leading to the deterioration of the specimen at approximately 5000 cycles was fixed. The abrasion cycles were carried out on a denim backing (¾ twill) using the Martindale machine.

### 3.3 Analysis of Main Effects Plots

We can examine the variation in the means of the levels of the components investigated using the main effects graph. When a factor's amount affects the response, the major effect happens. There are two cases. An effect of a factor is not substantial if the line linking the means of the responses from its various levels is horizontal. For a non-horizontal line, there is an effect that is more or less significant. In this instance, the levels of the component have various effects on the response. The relevance increases with an increasing slope (vertical line). An evaluation of the model's fit was done once the regression model had been constructed. The value of the correlation coefficient ($R^2$) is around 99.49%. According to Joglekar et alreport's, a decent model fit would have an $R^2$ value of at least 0.8 [18]. So, our model is entirely foreseeable. The information acquired from Fig. 3 demonstrates that the physical degradation (wear) of the abrasive composite is significantly influenced by the size of the abrasive grain, which results in a loss of mass. When the grain size goes from tiny (0.2 mm) to large (> 0.2 mm), the abrasive increasingly loses mass with a steep slope. The performance of the abrasive composite declines as grain size increases. This outcome was anticipated since large grains adhere less strongly than small grains.

After grain size, chemical pre-treatment has the greatest impact on abrasive composite wear. Compared to cationized fibrous supports, non-cationized fibrous supports exhibit weaker grain adhesion.

Certainly, we noticed constant abrasive degradation with Tow percentages ranging from 20 to 50%. With an increase in the amount of gypsum Tow fibers to 80%, this deterioration declines (see Fig. 3).

In general, the performance of the abrasive is enhanced to have the lowest mass loss by raising the resin content. The investigation of the outcomes of this element's fluctuation reveals that the variation in resin concentration is a factor that can affect the mass loss of the abrasives by up to 50%.

Increasing the amount of resin beyond 50% no longer has any effect on improving the wear resistance of the composite. This result is quite expected because the impregnation of the fibrous structure by the resin obeys the chemical reaction mechanisms. Namely the excess resin can no longer be absorbed by the structure because of the chemical affinity of the compounds (the phenomenon of saturation by the resin of the fibrous structure).

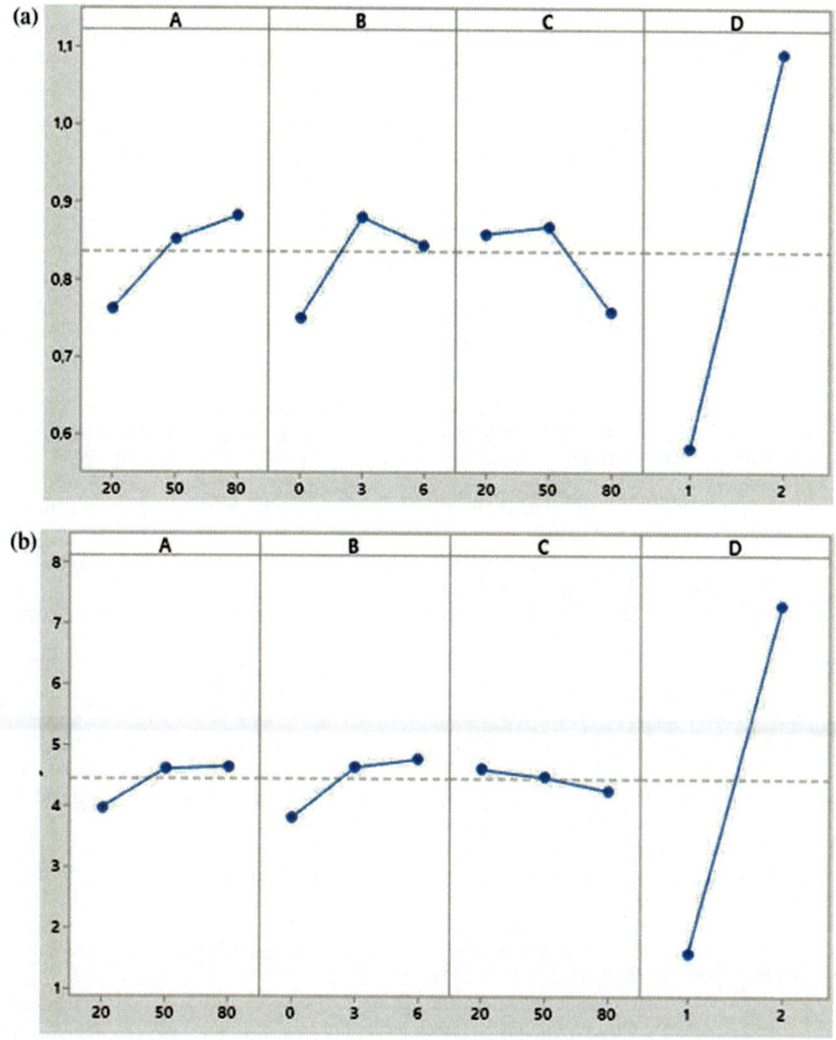


**Fig. 3** Graph of the main effects of mass loss of **a** abrasives and **b** denim

## 4 Conclusion

The effectiveness of abrasives derived from industrial waste on the wear performance of denim has been investigated and reported in this work. The SEM images showed the multilayered cell walls observed on the reinforcement. At the surface structure of manufactured abrasives, microfibrils, and abrasive grains have been observed. According to the Box-Behnken experimental design, the size of the abrasive grains present in the composite had a greater impact on the mass loss of the abrasive than resin concentration. By enhancing adhesion, abrasives' wear performance can be enhanced. The interfacial adhesion between the fibers and the polymer was improved by the cationization process. In comparison to abrasive materials based on non-cationized reinforcements, cationized reinforcement-based abrasives had a lower mass loss. The interfacial adhesion quality, which impacts the composite's resistance to abrasive wear, was significantly influenced by resin concentration. The increase in the percentage of gypsum tow in the fibrous structure makes it possible to reduce the wear of abrasives and their degradation.

## References


1. Singh, P., L. Singh, and S.J.J.o.m.p. Singh, Manufacturing and performance analysis of mechanically alloyed magnetic abrasives for magneto abrasive flow finishing. 2020. 50: p. 161–169.
2. Xie, W., et al., Green chemical mechanical polishing of sapphire wafers using a novel slurry. 2020. 12(44): p. 22518–22526.
3. LUKIĆ, S. and JOVANIĆ, P., Structural analysis of abrasive composite materials with nonwoven textile matrix. 2004. 58(3–4): p. 439–443.
4. LUDWIG, Bret W., Nonwoven abrasive articles and methods of making the same. 2011, Google Patents.
5. Elaissi, A., Alibi, H., Ghith, A., and Legrand, X., The Impact of Chemical Treatment of Cellulosic Fibers on Surface Properties and Matrix/reinforcement Interfacial Adhesion. Journal of Natural Fibers, 2022: p. 1–14.
6. Santos, R. D., Thomas, S., Ferreira, S. R., Silva, F. A., Combariza, M. Y., Blanco-Tirado, C., ... and Toledo Filho, R. D., Molecular grafting of nanoparticles onto sisal fibers-adhesion to cementitious matrices and novel functionalities. Journal of Molecular Structure, 2021. 1234: p. 130171.K. Elissa, "Title of paper if known," unpublished.
7. Merotte, J., Le Duigou, A., Bourmaud, A., Behlouli, K., and Baley, C., Mechanical and acoustic behaviour of porosity controlled randomly dispersed flax/PP biocomposite. Polymer Testing, 2016. 51: p. 174–180.
8. Elaissi, A., Ghith, A., Alibi, H., & Legrand, X. Study of the mechanical, chemical, and surface behaviors of non-woven abrasive made from waste chemically modified tow fibers. Journal of Industrial Textiles, 2022. 52, 15280837221126191.
9. Elaissi, A., Alibi, H., and Ghith, A., Surface Characterization of Textile Reinforcements. in International Conference of Applied Research on Textile and Materials. 2022. Springer.
10. Elaissi, A., Alibi, H., and Ghith, A., Development of Abrasives from Non-woven Based on Used Textiles. Journal of Natural Fibers 2020: p. 1–15.Y. Yorozu, M. Hirano, K. Oka, and Y. Tagawa, "Electron spectroscopy studies on magneto-optical media and plastic substrate interface," IEEE Transl. J. Magn. Japan, vol. 2, pp. 740–741, August 1987 [Digests 9th Annual Conf. Magnetics Japan, p. 301, 1982].
11. Elaissi, A., H. Alibi, and A. Ghith, Effect of pumice stone and pearlite abrasives characteristics on denim abrasion. Journal of Composite Materials 56.13 (2022): 2107–2116.
12. Mukhopadhyay, A. and V.K. Midha, A review on designing the waterproof breathable fabrics part II: Construction and suitability of breathable fabrics for different uses. Journal of Industrial Textiles, 2008. 38(1): p. 17–41.
13. Litim, N., Baffoun, A., Khoffi, F., Hamdaoui, M., Ben Abdessalem, S., and Durand, B., Effect of finishing resins on mechanical and surface properties of cotton Denim fabrics. The Journal of the Textile Institute, 2017. 108(11): p. 1863–1870.

14. Siddika, A., Uddin, M. N., Jalil, M. A., Akter, N. N., and Saha, K., Effects of Carded an d Combed Yarn on Pilling and Abrasion Resistance of Single Jersey Knit Fabric. 2017. 4(2): p. 39–43.
15. Škoc, M.S. and E. Pezelj, Abrasion resistance of high performance fabrics. Abrasion resistance of materials. Rijeka: InTech 2012: p. 35–52.
16. Adinarayana, K., and Ellaiah, P. Ellaiah, Response surface optimization of the critical medium components for the production of alkaline protease by a newly isolated Bacillus sp. 2002. **5**(3): p. 272–278.
17. Yadav, A., Bagotia, N., Yadav, S., Sharma, A. K., and Kumar, S., In-situ fabrication of surfactant modified CNT-based novel bio-composite and its performance evaluation for simultaneous removal of anionic dyes: Optimization by Box-Behnken design. 2022. **284**: p. 120262.
18. Elsayed, M. M., Aboelez, M. O., Elsadek, B. E., Sarhan, H. A., Khaled, K. A., Belal, A., ... and Elsadek, M. E. M, Tolmetin sodium fast dissolving tablets for rheumatoid arthritis treatment: preparation and optimization using Box-Behnken design and response surface methodology. 2022. **14**(4): p. 880.

# Numerical Investigations of the Tribological Behavior of G1151 Fabric Through Frictional Sensitivity Study During Drawing Using a Hybrid Elastic Discrete Model

**Walid Najjar, Mondher Nasri, Philippe Dal Santo, Xavier Legrand, and Damien Soulat**

**Abstract** The quality of a woven composite part is influenced by various factors, including the distribution and orientation of fibers and the occurrence of forming defects. The mentioned defects can have various causes, such as tool configuration, process settings, woven reinforcement structure, or ply orientation, among others. It is common to encounter issues with friction between the reinforcement and tooling material during the preforming process. The objective of this research is to investigate how friction impacts the distribution of shear angles during the dry reinforcement forming process, using a hybrid discrete elastic numerical model. This model, which is based on a unit cell composed of shell or membrane element reinforced by linear or nonlinear connectors, has previously been developed to simulate the behavior of woven reinforcement. The results demonstrate a significant impact of the friction coefficient on the local distribution of shear angles during the forming process.

**Keywords** Composite manufacturing process · Hybrid elastic model · Forming woven fabric · Friction coefficient sensitivity

W. Najjar (✉) · M. Nasri
LMPE, ENSIT, National High School of Engineering of Tunis, Tunis University, 5 Avenue Hussein, BP, 56, 1008 Bab Manara, Tunisia
e-mail: walid.najjar.ensit@outlook.com

P. Dal Santo
LAMPA, Arts et Métiers Institute of Technology-Angers, Angers, France

X. Legrand · D. Soulat
GEMTEX, Ecole Nationale des Arts et Industries Textiles, Lille University North of France, Lille, France



S. B. Abdessalem et al. (eds.), *Proceedings of the Second International Conference of Innovative Textiles and Developed Materials-ITDM'2; 05-06 May 2023; Tunisia*,
https://doi.org/10.1007/978-981-99-7950-9_17

# 1 Introduction

There are various methods available for shaping composite materials that incorporate woven reinforcements. One of these methods is liquid composite molding (LCM), which involves shaping composite parts using woven reinforcements and a polymer matrix [1, 2]. It is important to note that LCM processes differ from other commonly used methods like resin transfer molding (RTM) and vacuum infusion [3]. The RTM process involves three stages. Firstly, a preform is made to shape the reinforcement. Shaping can be done in a single step or multiple steps until the complete product architecture is formed [4]. The preform is then subjected to compaction pressure to obtain the desired volumetric ratio between the volume of reinforcement and the volume of the matrix. Side clamps can be used to prevent the formation of wrinkles during shaping by adding slight tension to the locks. Next, the resin is injected into the fibrous medium. Finally, after resin polymerization (crosslinking or hardening), the part is de-molded. Developing such processes requires the control of numerous design variables to improve their performance. These variables include mold temperature, gate pressure, reinforcement deformation, and injection point position. The preforming stage of the reinforcement is a critical phase involving complex physical mechanisms that depend on multiple parameters (shape of tools, preform characteristics, number and orientation of folds, applied loads, etc.). The literature has well-described the mechanical load such as tension, shear, compression, and bending to which the reinforcements are subjected during the preforming stage [5, 6]. However, defects' nucleation and growth are not yet fully comprehended [7]. Friction is one of the parameters observed on a macroscopic scale (preform scale) that can cause preform defects. Recently, Nasri et al. [8] have studied the effect of friction on the stamping of woven reinforcements. Numerous works in the literature have been published studying the frictional characterization of reinforcements and reinforcements with resin [9–12]. Najjar et al. [13] have experimentally studied the tribological behavior of the G1151 carbon fiber reinforcement. In order to verify the feasibility of the forming process and the fiber orientations in the final composite component, it is essential to simulate the initial preforming step. There are two main approaches used to simulate the preforming of fabrics: the geometric approach, which utilizes kinematic models and assumes that the threads are connected at the intersections of the weave, and the mechanical approach, which can be categorized into three types: continuous, discrete, and semi-discrete. The continuous approach considers the reinforcement as a continuous solid at the macroscopic level and accounts for the homogenized global behavior of the reinforcement. The discrete approach, on the other hand, represents the reinforcement as an elementary physical cell, utilizing structural finite elements such as trusses, beams, membranes, or shells, and is modeled at the mesoscopic scale [14]. The semi-discrete approach offers an intermediate solution between continuous and discrete methods. The current study uses a hybrid discrete modeling approach developed originally by Najjar et al. [15, 16] that phenomenologically describes the behavior of the fabric. This model has successfully simulated the stamping of woven fabrics and accurately predicted the yarn orientation of the deformed shape [16].

This paper aims to utilize this model to numerically investigate the impact of friction on the distribution of shear angles in the case of a hemispherical preforming process of a specific carbon woven reinforcement.

## 2 Material

The current paper focuses on the commercial G1151 woven carbon fabric manufactured by Hexcel, which is primarily used in aeronautical and transportation applications. The fabric is made up of "T300JB 6 K 40D" carbon fiber, with 51% of the yarns in the warp direction and 49% in the weft direction, and has linear densities of 396 tex. The yarns are made up of 6000 continuous filaments and are non-twisted and non-powdered. The interlock reinforcement weave style of the fabric is constructed with 7.5 warps/cm and 7.4 wefts/cm and has a nominal area density of 6300 g/m$^2$. The warp and weft yarns have widths of 1.9 and 2.2, respectively.

## 3 Presentation of Hybrid Elastic Discrete Model

The model utilized in this study was initially developed by Najjar [17] and has been thoroughly explained in previous works by the authors [15, 16]. It employs a hybrid discrete elastic model description of the fabric that employs a mesh consisting of 1D and 2D elements of the fabric's unit cell (see Fig. 1). The 1D elements are represented by axial connectors along the edges of each element to control the tensile stiffness in the fabric and predict anisotropy evolution during shaping. On the other hand, the 2D elements are represented by continuous shell or membrane elements to manage the contact phenomenon and control the in-plane shear rigidity. The 1D elements can possess either linear or nonlinear elastic behavior, depending on the fabric's behavior under tensile stress. Meanwhile, the 2D elements exhibit isotropic elastic behavior. The model incorporates the following parameters: $l_0$ represents the element dimension (the connector length) and $K_c$ (in the case of a linear approach) or $F_c$ and $U_c$ (in the case of a nonlinear approach) represent the connector stiffness. The elastic continuous elements of the woven fabric are characterized by $E1$, $E2$, $\nu$, and $e$, representing stiffness, Poisson's ratio, and thickness, respectively.

Table 1 gives an overview of a summary of the model parameters:

The identified parameters of the connector and the continuous element are presented in Table 2.

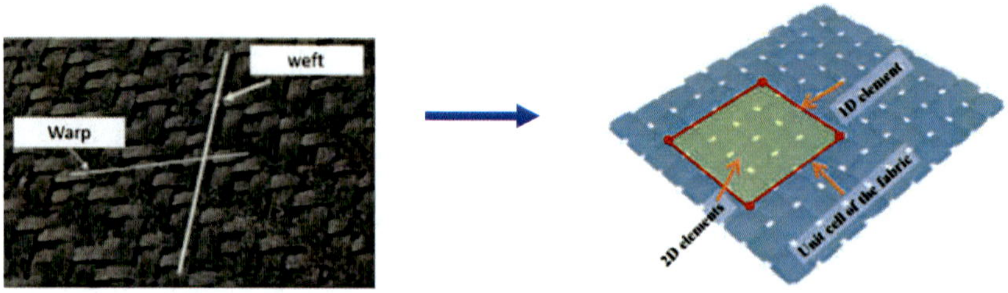


(a) (b)

**Fig. 1** Principle of the hybrid discrete mesoscopic modeling method utilizes a combination of 1D and 2D elements: **a** G1151 fabric, **b** numerical model

**Table 1** Hybrid elastic discrete parameters

| 2D elements | 1D element |
|---|---|
| $E$, Young's modulus of the shell material in the weft direction<br>$E2$, Young's modulus of the shell material in the warp direction<br>$e$, Shell thickness<br>$\upsilon$, Poisson's ratio<br>$l_0 = 1$, Size of the element | Stiffness for nonlinear connectors =<br>$F_c = K\,(u_c)$, where $K = f(u_c)$<br>$F_t = A.u_t^4 + B.u_t^3 + C.u_t^2 + D.u_t + c$<br>$A,B,C,D$ and $c$: coefficients to be identified<br>$(F_c, U_c)$: force and displacement of the connectors<br>The limits of $(F_c, U_c)$ are defined from the following formulations:<br>$u_c = \frac{1}{N}u_t,\ F_c = \frac{1}{n+1}F_t,\ N = \frac{X}{l_0},\ n = \frac{Y}{l_0}$<br>$N$ and $n$ refer to the quantities of connectors along the length and width of the specimen, respectively |

**Table 2** Parameters used for the simulation with linear and nonlinear connectors

| 2D elements | 1D element |
|---|---|
| $E1 = 12.5$ Mpa<br>$E2 = 12.5$ Mpa<br>$e = 0.005$ mm<br>$\upsilon = 0.16$<br>$l_0 = 1$ mm | Stiffness for linear connectors: $F_c = 22032$ N/mm<br>Stiffness for nonlinear connectors:<br>$F_t = -1.288E9u_t^4 + 4.32E7.u_t^3 + 3.16E5u_t^2 + 7.4E2u_t + 1.25$ |

## 4 Results and Discussion of the Study on the Sensitivity of Friction

The stamping set up utilized in this study aligns with those previously presented in [18].

The punch consists of a hemispherical head with a 50 mm radius, while the die cavities have individual spherical radii of 6 mm. The hole diameter is 55.4 mm, allowing for a 2.7 mm gap between the punch and die. The square composite blank

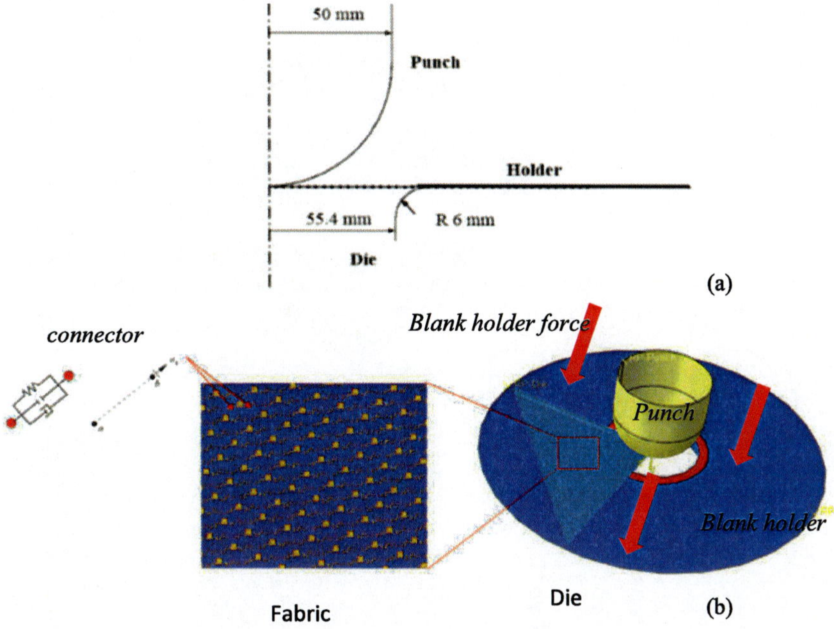


**Fig. 2** Hemispherical drawing simulation using hybrid discrete finite element of G1151 fabric: **a** tooling geometry; **b** numerical hemispherical model

used has dimensions of 260 mm × 260 mm. The binder, die, and punch are represented by analytical rigid bodies, each with their respective reference nodes defining boundary and loading conditions. The reinforcement is modeled using four-node shell elements S4R (refer to Fig. 2). This operation was simulated employing both the behavior model with linear and nonlinear connectors. A clamping force of 50N is adopted. The stamping velocity is set at 5 mm/s.

In order to computationally analyze the impact of friction on the G1151 fabric angles distribution during forming. A set of simulations of a reinforcement layer oriented at (0°/90°) is conducted using a hemispherical punch with a diameter of 100 mm, and the material parameters are obtained from [17] and presented in Tables 1 and 2. A range of friction coefficients (fabric/tooling) ranging from 0.1 to 0.7 are tested, with both linear and nonlinear models used for the connectors. In order to better understand the distribution of shear angles and apprehend the effect of friction, the shear angles along the path shown in Fig. 3 were analyzed for both the linear and nonlinear models. This analysis enables a better grasp of the shear angle distribution and evaluation of the influence of friction on these angles.

The evolution of shear angles along the path presented in Fig. 3 is plotted in Fig. 4 as a function of distance from the pole for various friction coefficient values in the linear model.

Fig. 3 The path where shear angles are measured and the three-area specific of a hemispherical preform

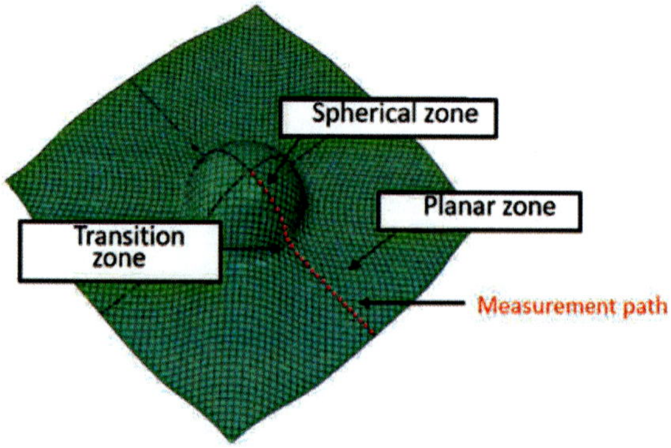


Fig. 4 Distribution of shear angles for the linear model

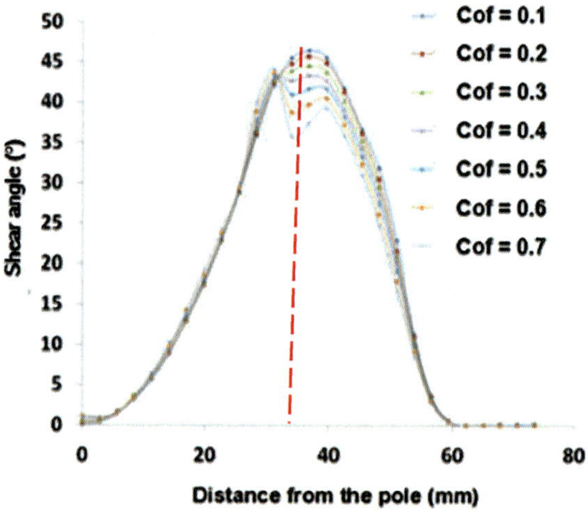


The results indicate that the shear angle is significantly influenced by changes in the friction coefficient, with lower friction coefficients leading to higher maximum shear angles. This observation is consistent with previous research [19], which suggests that higher tangential contact forces due to higher coefficients of friction can hinder yarns rotation. Moreover, the curves illustrate that the transient zone is particularly affected by variations in the friction coefficient. For instance, at a distance of 33.9 mm from the pole—which corresponds to the transitional zone—the shear angle is 45.5° when the friction coefficient is 0.1, while it is only 35.7° for a friction coefficient of 0.7.

It should be mentioned that the shape of the curves obtained changes progressively for high values of the coefficient of friction. This form passes from a Gaussian curve to a bimodal form. These two summits represent two maximums of the shear angles, one on the spherical zone and the other on the plane zone. This is also visible in Fig. 5b, which shows a quarter of the simulated draw with a high friction coefficient (tooling/fabric) of 0.6. Moreover, Fig. 5a shows that for the value of the friction coefficient 0.2, the most sheared zone is located in the transitional part between the spherical zone and the plane zone.

Figure 6 shows the evolution of the shear angles along the cited path, for simulations using the nonlinear connectors model, and for different values of friction coefficient.

The curves in Fig. 6 are in agreement with those obtained by the linear model in the spherical zone and the transient zone. The transition from Gaussian curves to bimodal curves is also recovered. However, it can be observed in this case that for the planar area, an increase in the friction coefficient leads to an increase in the shear angle.

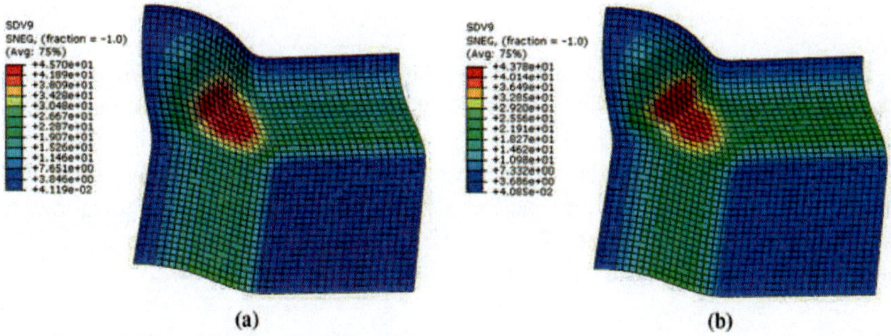


**Fig. 5** Shear angle distribution (in degrees °)after preforming simulation using a linear model: **a** with a friction coefficient value of 0.2; **b** with a friction coefficient value of 0.6

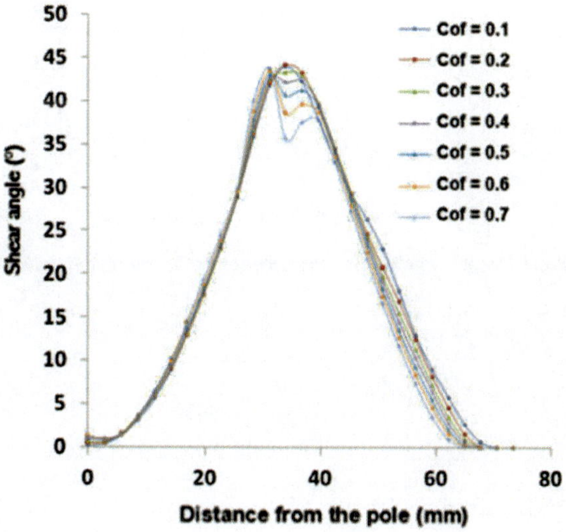


**Fig. 6** Distribution of shear angles for the nonlinear model

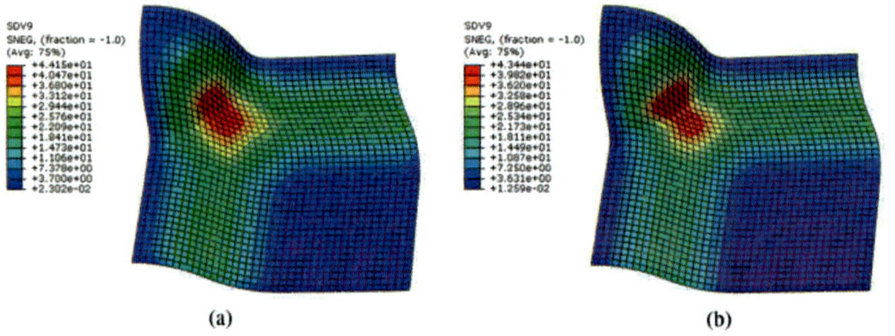


**Fig. 7** Shear angle distribution (in degrees°) after preforming simulation using a nonlinear model: **a** with a friction coefficient value is 0.2; **b** with a friction coefficient value is 0.6

It is also important to mention that the shear angle iso-contours derived from the nonlinear model (Fig. 7) exhibit consistency with those predicted by the linear model (Fig. 5).

The boundary curves of the drawn-in are also studied, Figs. 8 and 9 present the results, respectively, for the linear and nonlinear model. These results show that the variation of the friction has an effect on the drawn-in, and that a lower coefficient slightly favors the sliding of the fabric.

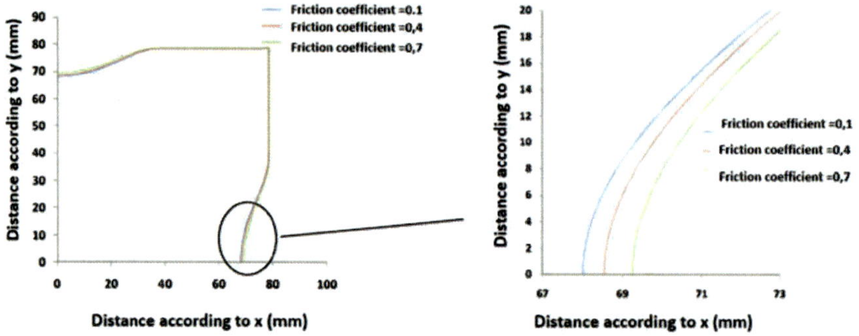


**Fig. 8** Impact of changing the friction coefficient on the drawn-in when using the linear model

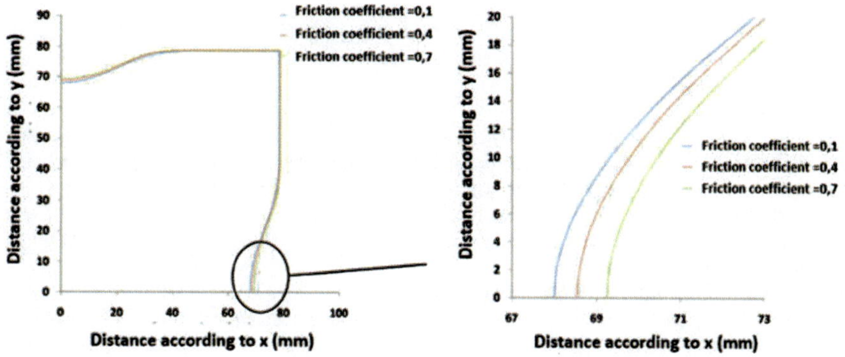


**Fig. 9** Impact of changing the friction coefficient on the drawn-in when using the nonlinear model

# 5 Conclusion

In this paper, a numerical hybrid discrete-continuum model that incorporated elastic shells and linear/nonlinear connectors has been used to investigate the impact of friction coefficient at the interface between the tools and G1151 preform on the shear angles distribution of the hemispherical obtained part. The results demonstrate that friction significantly affects the local distribution, location, and magnitude of the shear angles, especially in the planar-spherical transition zone. In the future research, it would be interesting to investigate the impact of shear angles and the friction coefficient in the case of multilayers forming. Furthermore, our approach enables the simulation of shaping processes by modifying the constitutive law of the continuous element. A promising suggestion would be to implement via subroutines a viscoelastic behavior and specific tribological laws that take into account temperature variations and the resin behavior, to investigate the influence of resin-induced friction during the shaping process of the polymerized reinforcement.

# References

1. Han SH, Cho EJ, Lee HC, Jeong K, Kim SS. Study on high-speed RTM to reduce the impregnation time of carbon/epoxy composites. Composite Structures 2015;119:50–8.
2. Boisse P. Mise en forme des renforts fibreux de composites: Ed. Techniques Ingénieur; 2004.
3. Parveez B, Kittur M, Badruddin IA, Kamangar S, Hussien M, Umarfarooq M. Scientific Advancements in Composite Materials for Aircraft Applications: A Review. Polymers 2022;14:5007.
4. Nasri M, Garnier C, Abbassi F, Labanieh A, Dalverny O, Zghal A. Hybrid approach for woven fabric modelling based on discrete hypoelastic behaviour and experimental validation. Composite Structures 2019;209:992–1004.
5. Boisse P, Hamila N, Guzman-Maldonado E, Madeo A, Hivet G, Dell'Isola F. The bias-extension test for the analysis of in-plane shear properties of textile composite reinforcements and prepregs: a review. International Journal of Material Forming 2017;10:473–92.

6. Gereke T, Döbrich O, Hübner M, Cherif C. Experimental and computational composite textile reinforcement forming: A review. Composites Part A: Applied Science and Manufacturing 2013;46:1–10.
7. Ouagne P, Soulat D, Hivet G, Allaoui S, Duriatti D. Analysis of defects during the preforming of a woven flax reinforcement. Advanced composites letters 2011;20:096369351102000403.
8. Nasri M, Abbassi F, Garnier C, Labanieh AR, Dalverny O, Zghal A. Analysis of woven fabrics forming through a comparison between discrete elastic and hypoelastic approaches based on mesoscale structure. Composite Structures 2022;284:115149.
9. Das A, Kothari V, Vandana N. A study on frictional characteristics of woven fabrics. AUTEX Research Journal 2005;5:133–40.
10. Fetfatsidis KA, Jauffrès D, Sherwood JA, Chen J. Characterization of the tool/fabric and fabric/fabric friction for woven-fabric composites during the thermostamping process. International journal of material forming 2013;6:209–21.
11. Fetfatsidis KA, Gamache LM, Gorczyca JL, Sherwood JA, Jauffrès D, Chen J. Design of an apparatus for measuring tool/fabric and fabric/fabric friction of woven-fabric composites during the thermostamping process. International journal of material forming 2013;6:1–11.
12. Akkerman R, Ten Thije R, Sachs U, De Rooij M. Friction in textile thermoplastic composites forming. Proceedings of the 10th international conference on textile composites-TEXCOMP2010. p. 271–9.
13. Najjar W, Pupin C, Legrand X, Boude S, Soulat D, Dal Santo P. Analysis of frictional behaviour of carbon dry woven reinforcement. Journal of Reinforced Plastics and Composites 2014;33:1037–47.
14. Boisse P, Aimène Y, Dogui A, Dridi S, Gatouillat S, Hamila N, et al. Hypoelastic, hyperelastic, discrete and semi-discrete approaches for textile composite reinforcement forming. International journal of material forming 2010;3:1229–40.
15. Najjar W, Legrand X, Pupin C, Dal Santo P, Boude S. A simple discrete method for the simulation of the preforming of woven fabric reinforcement. Key Engineering Materials: Trans Tech Publ; 2012. p. 213–8.
16. Najjar W, Legrand X, Soulat D, Dal Santo P. Experimental analysis and numerical modelling of dry carbon woven reinforcement preforming. Journal of Reinforced Plastics and Composites 2019;38:974–94.
17. Najjar W. Contribution à la simulation de l'emboutissage de préformes textiles pour applications composites: Paris, ENSAM; 2012.
18. Ding F, Peng X. Validation of a non-orthogonal constitutive model for woven composite fabrics via hemispherical stamping simulation. Fuhe Cailiao Xuebao(Acta Materiae Compositae Sinica) 2011;28:156–60.
19. Khan MA. Numerical and experimental forming analyses of textile composite reiforcements based on a hypoelastic behaviour: Lyon, INSA; 2009.

# Effect of the Twist and Composite Process on the Tensile Behavior of Carbon Multifilament Rovings. An Experimental Investigation Toward the Identification of Constituents' Behaviors

**Mohamed Medhat Salem, Haifa Mensi, Chayma Soltani, Adel Ghith, and Xavier Legrand**

**Abstract** The goal of this study was to determine the influence of the yarn twist on the tensile behavior of the carbon fiber rovings. One aim was to study less explored areas of the common stress strain graphs such as nonlinearity and plasticity and the effect of the twist on such areas. These findings should be coupled to other in-depth investigations in order to more faithfully model the behavior of the yarn as whole.

**Keywords** Carbon fiber · Tensile behavior · Composites

## 1 Introduction

In order to predict the mechanical behavior of textile structures, several studies on finite element analysis have been conducted on microscopic, mesoscopic, and macroscopic scales. On the macroscopic side, homogenization techniques take into account the global behavior of the reinforcement by replacing all the constituents by a single material that responds to homogeneous behavior laws that are easier to exploit [1, 2]. In order to predict the mechanical behavior of textile structures, several studies on finite element analysis have been conducted on microscopic, mesoscopic, and macroscopic scales. On the macroscopic side, homogenization techniques take into account the global behavior of the reinforcement by replacing all the constituents by a single material that responds to homogeneous behavior laws that are easier to

M. M. Salem · X. Legrand (✉)
École Nationale Supérieure Des Arts Et Industries Textiles - Université de Lille, 2 All. Louise et Victor Champier, 59100 Roubaix, France
e-mail: xavier.legrand@ensait.fr

H. Mensi · C. Soltani · A. Ghith
École Nationale d'ingénieurs de Monastir, Rue Ibn El Jazzar, 5000 Rue Ibn Jazzar, 5035 Monastir, Tunisia



S. B. Abdessalem et al. (eds.), *Proceedings of the Second International Conference of Innovative Textiles and Developed Materials-ITDM'2; 05-06 May 2023; Tunisia*,
https://doi.org/10.1007/978-981-99-7950-9_18

exploit [1, 2]. These techniques can predict the initiation of defects and failures as well as the in-plane and out-of-plane behavior of the reinforcement. However, these techniques are not capable of tracking the evolution of these defects [3]. To have a better simulation, it is necessary to take into account the interaction and inhomogeneity between the constituents of the reinforcement. Hence the interest of mesoscopic studies, where the components are modeled with their specific properties and behavior laws. Mesoscopic modeling gives results that are more faithful to the behavior of the overall reinforcement but requires higher computing power the larger the model [4]. Hybrid approaches between macro and meso allow to model the homogeneous reinforcement with the areas where defects are expected with the level of detail of a mesoscopic model. This reduces the computational cost considerably [5].

In order to have a model that is faithful to reality, mechanical tests on impregnated yarns are performed to determine the mechanical properties and use them in the constitutive laws [6–8]. In addition, to determine the Young's modulus of the yarns, tensile tests are performed. However, in most cases, these tests are only used to determine the modulus of elasticity and do not give any idea about the heterogeneity of the roving components (i.e., stress distribution, fiber alignment, level of twist per meter (tpm), variation of mechanical properties…) [9]. Hence, to identify and quickly summarize the behavior of carbon yarns, a tensile test is typically done, and the yield stress and tensile modulus is extracted from the linear slope of the stress/strain response. Yet this leaves plenty of valuable data in the curves [6]. A typical stress/strain graph for carbon yarns is not linear and will present different moduli depending on the aforementioned variables [10, 11].

The goal of this study is to primarily assess the effect that the roving twist coupled to the composite manufacturing process has on the tensile properties of the roving. This was done by comparing the behavior of rovings at different tpm either dry, resented then twisted or twisted then impregnated.

## 2 Material and Methods

For this study, the "Hextow IM7" 12 K filaments were used. These multifilaments are fairly widespread, and their characteristics are well documented (Young's modulus 276 GPa, yield stress 5.7 GPa, and an ultimate deformation of 1.8%). To consolidate the yarns, an epoxy "Sicomin SR8200" was used (Young's modulus 3 GPa, yield stress 67 MPa, and an ultimate deformation of 2.7%).

The yarns are to be twisted using a purposely made twisting device (see Fig. 1). The levels of twist chosen are 0, 10, 30, 50, 70, and 100 tpm. Three batches were prepared to be tested in the "INSTRON" tensile bench. The length of the samples was set to 200 mm.

The first batch is tested with dry fibers. The yarns were twisted in situ by clamping one end to the top grip of the bench, twisting the lower part and fixing it the lower grip. The second batch was tested with filaments that were soaked in the epoxy prior

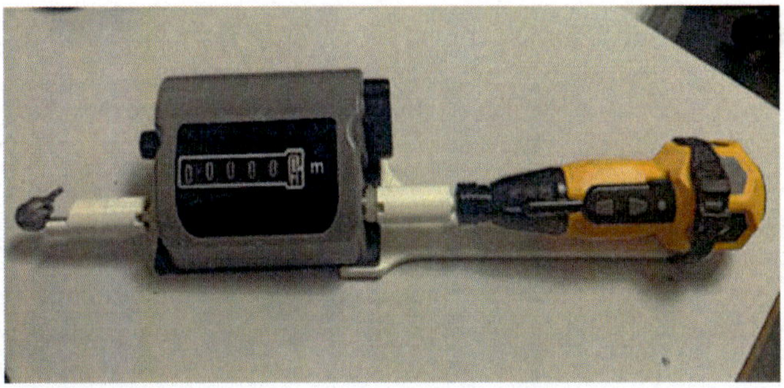


**Fig. 1** Yarn twisting device that was specifically designed for the study

to twisting. The matrix was allowed to set at the ambient temperature for 24 h. The third batch was with yarns that were twisted then impregnated and allowed to set in the same conditions. For each twist level, eight specimens were tested.

Once the composite yarns were tested, the rule of mixtures was used to determine the weight percentage of fibers and matrix. Assuming no void, the weight of the fibers part (mf) being already measured, the weight fraction of the fibers (Mf) is Mf = mf/mc with mc being the weight of the final composite. The weight fraction of the matrix (Mm) is then Mm = 1 − Mf.

# 3 Results and Discussions

The results of the weight fractions are displayed in Table1.

The twist level impacts significantly the impregnation of the matrix in the fibers. With no twist, the percentage of the matrix is up to 68. This fraction goes down to as low as 52%. It is also interesting to observe that twisting the yarn before impregnating it causes the yarn to take in less matrix. The tighter fibers block the

**Table 1** Weight fractions of the different samples

| Twist level | Fiber weight fraction (%) | | Matrix weight fraction (%) | |
|---|---|---|---|---|
| | Resin first | Twist first | Resin first | Twist first |
| 0 | 32 | | 68 | |
| 10 | 32.5 | 39 | 67.5 | 61 |
| 30 | 40 | 40 | 60 | 60 |
| 50 | 39 | 39 | 61 | 61 |
| 70 | 44 | 41 | 56 | 59 |
| 100 | 44 | 48 | 56 | 52 |

epoxy from penetrating to the core of the yarn. For the same final level of twist, the difference between twisting first and resin first is 4% less matrix.

The yarns then tested on the INSTRON tensile bench gave stress/strain graphs as presented in Fig. 2. In Fig. 2, we only represent the graph of 0, 50, and 100 tpm as a general idea. The graphs represent the mean values between eight tests for each configuration. The area around each mean curve is delimited between the maximum and minimum registered values for said configuration.

Figure 2c illustrates the different parameters that could be of interest to further identify the behavior of elementary fibers. E1, E2, and E3 are moduli of the graph. E1 can be referred to as a nonlinearity modulus that is indicator of fiber alignment/uncrimping as the whole yarn is put under stress. E2 is the modulus on the linear part of the graph and is commonly known as the elasticity modulus or Young's modulus. E3 can be referred to as a plasticity modulus where fibers start to break while others

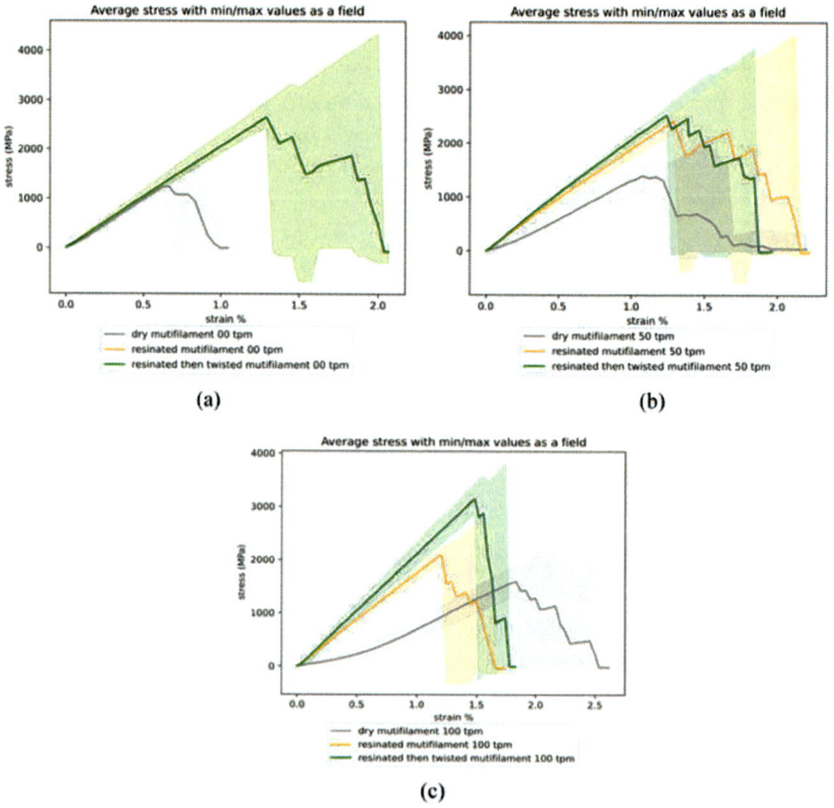


**Fig. 2** Examples of tensile tests on dry and impregnated carbon multifilament rovings at 0, 50, and 100 tpm. Average is represented as a solid line while the range between min and max tests is represented as a field

are still in their elastic domain. This confers a certain plasticity to the yarn as a whole. The yield stress (σ max) and strain (εmax) are also identified.

To sum up the results for all twist levels and processes, the values extracted are compiled in Fig. 3. E1 and E2 are represented as min and max of the field, and E3 is the line in the middle. Stress and strain are graphed as lines with whiskers that represent max and min values for a given configuration.

Results show that for dry multifilaments, the trend for up to 100 tpm is that the nonlinear part is increased as E1 goes down from 1756 to 410 MPa. The linear part E2 goes down from 2071 to 1061 MPa, and the plastic part E3 goes down from 1943 to 736 MPa.

The twist level makes it so the dry filaments go through a realignment phase during the test before they are fully constrained and under load. Since the increase in twist makes the filaments longer in the same roving sample, to get fully constrained, the yarn must deform further which explains the yield strain increasing 300% from 0.62 to 1.83. The yield stress slightly increases from 1293 to 1604 MPa due to the introduced inter-fiber friction that increases the roving cohesion and compensates for early breakages that might reduce its resistance.

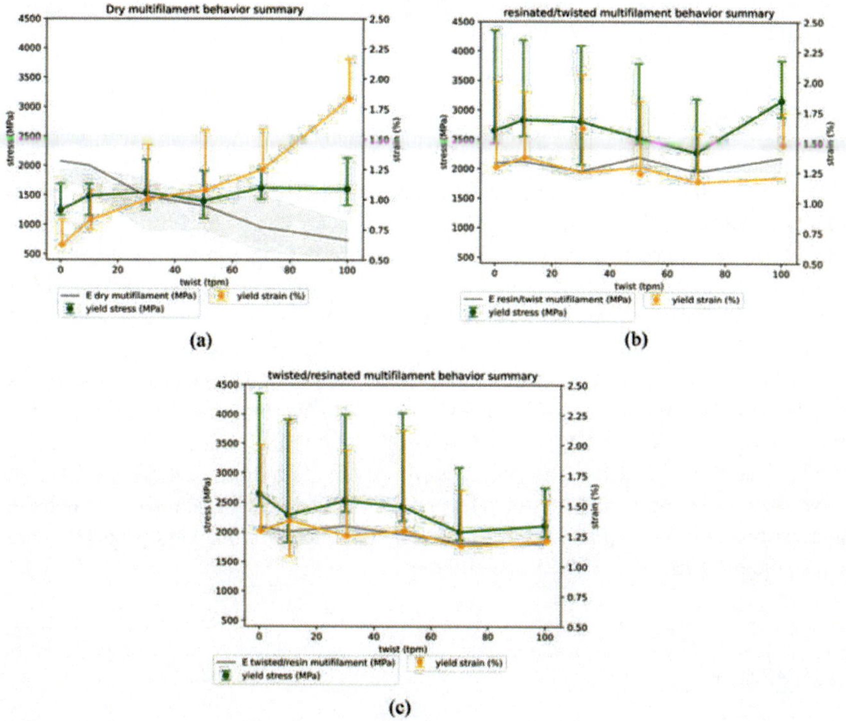


**Fig. 3** E1, E2, E3, and yield stress and strain at different twist levels represented for: **a** dry multifilaments, **b** resinated first multifilaments and **c** twisted first multifilaments

As for both composite rovings, the moduli are always fairly close and do not show a significant non-linearity or plasticity. For the characteristics of the twisted first rovings, they stay fairly similar till the 50 tpm mark where they start to get lower. The average moduli go down from 2035 MPa for 0 tpm to 1808 MPa for 100 tpm (a loss of 12%). The average yield stress goes down from 2651 MPa to 2113 Mpa (a loss of 22%). The average yield strain goes down from 1.28 to 1.2 (a loss of 8%). As for the characteristics of the resinated first rovings, they stay around the ± 5% with no significant modifications till the 100 tpm tests. The moduli go from 2035 MPa for 0 tpm to 2183 MPa for 100 tpm.

(a gain of 7%). The yield stress goes up from 2651 to 3155 MPa (a gain of 19%). The strain goes up from 1.28 to 1.47 (a gain of 14%).

Comparatively, as the twist level increases, either the twisted first or the resinated first behaves similarly up to 50 tpm. This is also visible in the weight fraction of the fibers/matrix where the rovings tend to take in the same amount of matrix. The matrix is able to prevent early slippages and loss of mechanical properties due to early fiber breakages by taking in the load where a filament might break but not debond from the matrix. The higher twist (from 70 to 100) makes it harder for the matrix to penetrate inside the core of the yarn and bond properly which explains the difference in behavior compared to rovings that had proper time to take in the matrix.

## 4 Conclusion

The effect of the twist was investigated experimentally. It was determined that up to a twist level of 50 tpm, no significant influence was observed. From a twist level of 70 tpm, the filament density becomes a hinder for a proper impregnation and this a loss of mechanical properties can be observed depending on the manufacturing process. The finding of this study will hopefully be used to determine the behavior of the roving's basic constituents (i.e., filaments) and their parameters using a genetic algorithm approach. In the literature, the identification of fiber properties in terms of peak tensile strength have been investigated using Weibull distribution have been investigated [12]. These have shown moderate degree of scatter, but cannot show the intricacies of identifying the distribution of tensile moduli and orientation that govern the behavior of the roving. The genetic algorithm approach would generate generations of random populations of carbon filaments with different properties and chose the most fitting to experimental data.

## References

1. Wang, J., Wang, P., Hamila, N., Boisse, P.: Mesoscopic analyses of the draping of 3D woven composite reinforcements based on macroscopic simulations, Compos Struct, (2020).

2. Hello, G., Schneider, J., Aboura, Z.: Numerical Simulations of Woven Composite Materials With Voxel-FE Models. In: 16th European conference on composite materials, p. 22–6 (2014).
3. El Said, B., Daghia, F., Ivanov, D., Hallett, S.R.: An iterative multiscale modelling approach for nonlinear analysis of 3D composites, Int J Solids Struct, 132–133, pp. 42–58 (2018).
4. Creech, G., Pickett, A.K.: Meso-modelling of non-crimp fabric composites for coupled drape and failure analysis, J Mater Sci, 41 (20) pp. 6725–6736 (2006).
5. Iwata, A., Inoue, T., Naouar,N., Boisse,P.,Lomov, S.V.: Coupled meso-macro simulation of woven fabric local deformation during draping, Compos Part A Appl Sci Manuf, 118, pp. 267–280 (2019).
6. ASTM, Standard test methods for properties of continuous filament carbon and graphite fiber tows, ASTM International, West Conshohocken, PA (2011).
7. Ji, X., Wang, C., Francis, B.A.P., Chia,E.S.M., Zheng, L., Yang, J., et al.: Mechanical and interfacial properties characterisation of single carbon fibres for composite applications, Exp Mech 55 (6), pp. 1057–1065 (2015).
8. Crabtree, J., Penumadu, D., Young, S.: Tensile properties of carbon fiber: single filament vs tow based testing, American society for composites: thirty-second technical conference. Purdue University, west Lafayette, Indiana, October 23–25, 2017. Lancaster, PA, USA: DEStech publications, inc, p. 11 (2017).
9. Matthew E. K., Joshua D. C., Stephen Y., Dayakar P.: Concept of limit stress for the tensile behavior of carbon fiber composite tows, Composites Part B: Engineering, Volume 201 15, 108384 (2020).
10. Zhifan, C., Tsu-Wei, C., Guoyi, S.: Determination of single fibre strength distribution from fibre bundle testings. Journal Of Materials Science 19, 3319–3324 (1984).
11. Yuanxin, Z., Yang, W., Yuanming, X., Shaik, J.: Tensile behavior of carbon fiber bundles at different strain rates. Materials Letters 64(3), 246–248 (2010).
12. Sweety, K., Nithya, S., Padmavathi, N., Eswara Prasad, N., Subrahmanyam, J.: Tensile properties and fracture behaviour of carbon fibre filament materials. J Mater Sci, 45:192–200 (2010).

# Investigation of the Properties of Developed Abrasives After Abrasion Test

**Elaissi Arwa and Ghith Adel**

**Abstract** This paper aims to compare the different abrasive nonwovens developed from textile waste with mechanical properties and surface states suitable for the washing treatment of jeans. The abrasives were analyzed using energy dispersive X-ray analysis (EDX). The abrasive wear behavior of the composites after application to the washing out of jeans (denim) was evaluated. Response surface methodology (RSM) was used to understand the overall behavior of the system and to determine the weight loss of the abrasives. Significant control factors and interactions influencing abrasives' weight loss were identified using the Box-Behnken design. Results indicated that the size of the abrasive grain affected the weight loss and the wear of the abrasive. The percentage of fibers, the concentration of the resin, and the chemical treatment of the reinforcements (cationization) had little significant influence on the results. The interactions appeared not very obvious compared to the influence of the granulometric parameter.

**Keywords** Abrasive nonwovens · Box–Behnken design · Mass loss interactions

## 1 Introduction

Abrasive wear of materials is a technology that is always evolving to improve system reliability. Profitability determines how it is used in the industry. Abrasive wear of materials is a technology that is always evolving to improve system reliability. Profitability determines how it is used in the industry.

E. Arwa (✉) · G. Adel
Textile Materials and Processes Research Unit, Tunisia National Engineering School of Monastir, University of Monastir, Monastir, Tunisia
e-mail: arwa.elaissi@enim.u-monastir.tn

G. Adel
e-mail: ADEL.GHITH@u-monastir.tn

 
S. B. Abdessalem et al. (eds.), *Proceedings of the Second International Conference of Innovative Textiles and Developed Materials-ITDM'2; 05-06 May 2023; Tunisia*,
https://doi.org/10.1007/978-981-99-7950-9_19

Recycling grains and fibers makes the product more economical, efficient, and environmentally friendly. Bret W. Ludwig's invention provides a nonwoven abrasive comprising a fibrous web of nonwovens, abrasive particles, and a polyurethane binder binding [1]. Nonwoven fiber webs are selected to be suitably compatible with adherent binders and abrasive particles while also being processable in combination with other article components. Cationization treatments have been carried out on the reinforcements to improve interfacial adhesion performance [2]. Scrap metal represents one of the materials discarded in large quantities as a by-product of the steel industry. It is the most commonly used metal in everyday life, usually in the form of various alloys. It is characterized by its solidity which gives it an abrasive character. A deeper understanding of the process in terms of the impact of the independent variables and their interactions on the dependency of the variables can be obtained by using experience-based optimization methodologies. The evaluation of the variable parameters was determined using the analysis of variance (ANOVA) of the quadratic model. A total of 30 sets of experiments were measured and significant regression of the coefficient between the variables was obtained. Also, measurements were tested before and after the abrasion test.

## 2 Experimental

### 2.1 Fibers Preparation

Cotton-based textile waste and cellulose gypsum tow fibers were first pre-treated to eliminate any impurities before being used to prepare composite abrasive materials. The fibers were boiled for 2 h in a flask containing a mixture of 1.5 mL/L NaOH, 3 g/L $Na_2CO_3$, and 4 g/L non-ionic detergents to carry out the scrubbing procedure (1:40). The treated fibers were then rinsed with hot and cold water for 1 min each time. Finally, a laboratory oven was used to dry the fibrous material. Under alkaline circumstances, the fibers were also subjected to a cationization process using a co-polymer of dimethyl diallyl ammonium chloride and diallylamin as cationic agents (3 and 6%).

### 2.2 Nonwoven Manufacturing

To prepare nonwoven composites, the fibers were first opened using a Shirley machine. Then, the nonwovens were blended from these raw materials using carding/overlapping/needling technology at different fiber percentages [3–6]. The tow fibers in the surface and the "Z" direction (across the thickness of the fabric) can be seen in the cross section of nonwoven fabrics after punching (see Fig. 1).

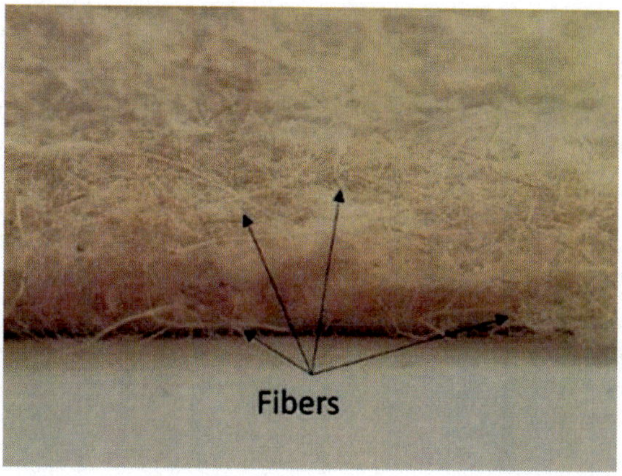


**Fig. 1** Cross section of nonwoven fabric after punching

## 2.3 Abrasive Grains Preparation

Iron grains were collected from local metallurgical industries. They were washed, dried, and examined to determine the desired grain size. According to the ISO 3310–1 standard [7], particle size analysis divides iron grains into two sizes. The powder grains have a size of less than 0.2 mm. Grains with a size between 0.2 and 0.6 mm are selected as large grains [8].

## 2.4 Abrasives Manufacturing

In this research, a thermosetting polyurethane resin named "Politex" is used. The resin solution and the hardener are poured step by step onto the textile material in the fabrication of abrasives once the reinforcement is set on the coating table. The abrasive grains are then thoroughly blended with the second resin layer. The abrasives' surface was covered with a third layer of dispersed granules. The thickness of the layer is determined by the coating technique [9]. Finally, all of the samples were dried for an adequate period at 140 °C [10].

## 2.5 Abrasion Test

The Martindale method is included in the ISO 12947 standard for determining fabric abrasion resistance [11]. The abrasion cycles were carried out on a denim backing (¾ twill) using the prepared abrasives instead of the standard abrasive paper. The abrasive wear tests were performed at room temperature in a dry state with a constant speed of 200 mm/s and a load of 9 N [12]. To deteriorate the denim fabric, all created composites were submitted to a 5000-cycle abrasion test. A comparison of weight loss data before and after polishing can be used to establish the specific wear rate [13]. For each condition, five measurements were obtained and an average value was determined.

## 2.6 Techniques of Characterization

Energy dispersive X-ray (EDX) analysis supplied by ThermoFisher was used to describe the chemical composition of the materials. The surface of the samples was coated with a very thin layer of gold by vapor deposition at an accelerating voltage of 25 kV to reduce charging and improve image quality. Energy dispersive X-ray analysis (EDX) is used to identify the elemental composition of the studied materials.

## 2.7 Choice of Factors and Field of Study

Based on the Box–Behnken design, which is a subset of response surface methodology, the experimental tests are carried out (RSM). It takes prior knowledge about the procedure to build a statistical model. This optimization process essentially consists of three steps: carrying out statistically planned tests, determining the coefficients in a mathematical model, forecasting the response, and assessing the model's fit [14, 15].

The preliminary tests conducted on the numerous parameters influencing the abrasion of the composites led to the selection of the factors and the study's focus for the current investigation. According to reports, the amount of fiber and how it is reinforced, the amount of resin present, and the grain size all have an impact on the wear rates of composite materials. The first three variables in the Box–Behnken program are continuous, while the grain size represents a two-level category variable. Table 1 lists the level of experimental design that was taken into consideration for this study.

The Box–Behnken design requires 24 design points plus 6 center runs, and therefore the number of trials is 30 (Table 2). Each experiment was performed three times.

**Table 1** Factors and areas of study

| Factors | Unit | Code factor | Levels not coded | | | Levels coded | | |
|---|---|---|---|---|---|---|---|---|
| | | | Level 1 | Level 2 | Level 3 | Level 1 | Level 2 | Level 3 |
| Resin concentration (resin/ catalyst) | % | A | 20/80 | 50/50 | 80/20 | 20 | 50 | 80 |
| Pre-treatments | – | B | Desized | Cationized 3% | Cationized 6% | 0 | 3 | 6 |
| Percentage of cellulosic fibers (fiber/cotton) | % | C | 20/80 | 50/50 | 80/20 | 20 | 50 | 80 |
| Grain size (category factor) | mm | D | ≤0.2 | 0.2–0.6 | – | 2 | 4 | |

**Table 2** Experimental trials derived from design-Expert software

| Number | 1 | 2 | 3 | 4 | 5 | 6 | 7 | 8 | 9 | 10 | 11 | 12 | 13[a] | 14[a] | 15[a] |
|---|---|---|---|---|---|---|---|---|---|---|---|---|---|---|---|
| A | 20 | 80 | 20 | 80 | 20 | 80 | 20 | 80 | 50 | 50 | 50 | 50 | 50 | 50 | 50 |
| B | 0 | 0 | 6 | 6 | 3 | 3 | 3 | 3 | 0 | 6 | 0 | 6 | 3 | 3 | 3 |
| C | 50 | 50 | 50 | 50 | 20 | 20 | 80 | 80 | 20 | 20 | 80 | 80 | 50 | 50 | 50 |
| D | 1 | 1 | 1 | 1 | 1 | 1 | 1 | 1 | 1 | 1 | 1 | 1 | 1 | 1 | 1 |
| Number | 16 | 17 | 18 | 19 | 20 | 21 | 22 | 23 | 24 | 25 | 26 | 27 | 28[a] | 29[a] | 30[a] |
| A | 20 | 80 | 20 | 80 | 20 | 80 | 20 | 80 | 50 | 50 | 50 | 50 | 50 | 50 | 50 |
| B | 0 | 0 | 6 | 6 | 3 | 3 | 3 | 3 | 0 | 6 | 0 | 6 | 3 | 3 | 3 |
| C | 50 | 50 | 50 | 50 | 20 | 20 | 80 | 80 | 20 | 20 | 80 | 80 | 50 | 50 | 50 |
| D | 2 | 2 | 2 | 2 | 2 | 2 | 2 | 2 | 2 | 2 | 2 | 2 | 2 | 2 | 2 |

[a]Indicates the focal point of the design

## 3 Results and Discussion

### *3.1 EDX Analysis*

The EDX scans were taken at four different locations in the lateral direction of the abrasives. The abrasive contains varying amounts of carbon, oxygen, potassium, calcium, and mainly iron. Figure 2a specifies the composition of the reinforcements alone, which is essentially composed of carbon and oxygen. In Fig. 2b and Fig. 2c, the element iron appears in the intermediate layer of the composites. Figure 2d indicates the composition of the surface of the abrasives, which is mainly constituted of iron.

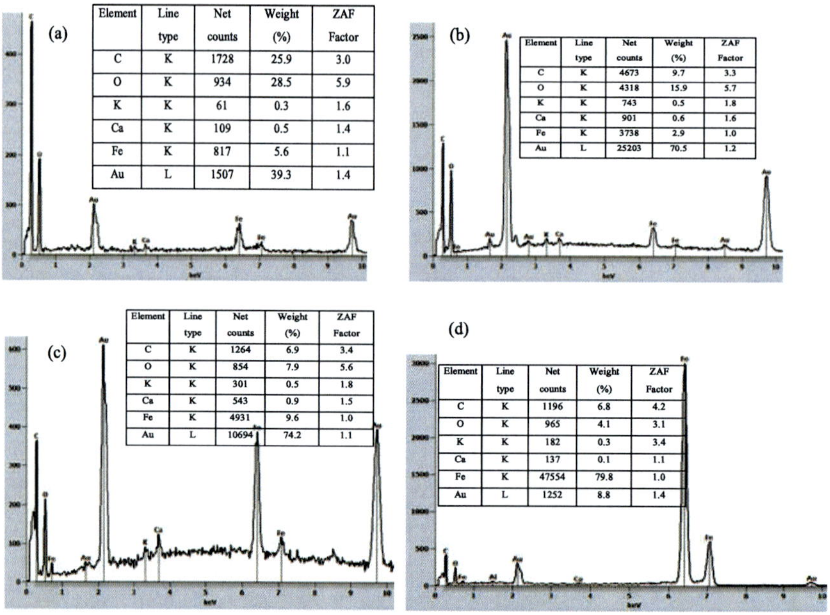


**Fig. 2** EDX analysis in the side view of abrasive manufactured by coating process before abrasion test: **a** reinforcement, **b** interface fibers-resin and **c** interface resin-grains, and **d** surface

## 3.2 Box–Behnken Analysis of Loss in Mass of Abrasives

To investigate the impact of the selected components on composite weight loss, RSM was performed using the design of experiments tool, Design-Expert version 8.0.4. Each experiment was repeated three times to estimate the random effect (residual or experimental variation). After carrying out the preliminary tests, the number of abrasion cycles leading to the deterioration of the specimen at approximately 5000 cycles was fixed. The abrasion cycles were carried out on a denim backing (¾ twill) using the Martindale machine.

## 3.3 Analysis of the Interaction Diagrams

Interaction graphs are a representation of the averages of response information for each level of factors. An interaction is present if the response for one level of parameters depends on the other levels. The interaction graphs are a representation of the information averages of responses for each factor level. In an interaction diagram, parallel lines indicate no interaction. In other words, the response to the one-factor level is not influenced by the level of the second factor. The more non-parallel the lines, the greater the degree of interaction. The graphical results of the influence of the interactions between the construction factors of the abrasive composite on its wear are given in Fig. 3.

From the shapes of the obtained curves, we note that the size of the grains prevails over all the factors. Indeed, the largest grain size causes the most significant degradation of the abrasive composite whatever the construction factor of the abrasive associated with it (treatment, resin concentration, percentage of fibers). This finding is indicated by the three lower-level diagrams. The interactions between the percentage of gypsum tow and the factors of the concentration of the resin and the pretreatment of the fibers have similar variation shapes with small amplitudes and deviations. Therefore, these interactions are weakly significant and have no important influence on the wear of the studied abrasive composites.

The interaction between the resin concentration and the pretreatment is more significant than all the other interactions since the variation lines intersect at the same point and have the same variation inclination. Based on the results of the obtained graphs, we can consider that the grain size is the most influential factor in holding the mass loss of the abrasive composite. The other factors have influences that are not highly significant. The interactions seem not very obvious because of the influence of the grain size parameter.

## 4 Conclusion

In this work, the optimization of the properties of abrasives obtained by iron grains by analyzing the effect of a certain number of manufacturing parameters has been reported. Based on the Box–Behnken experimental design, it has been observed that grain size is the most influential parameter on the appearance obtained. The wear performance of abrasives can be improved by improving adhesion. The cationization process improved the interfacial adhesion between the fibers and the polymer. The mass loss of abrasives composed of cationized reinforcements was lower than that of abrasive materials based on non-cationized reinforcements. The resin concentration played a major role in the quality of the interfacial adhesion, which affects the abrasive wear of the composite. The increase in the percentage of gypsum tow in the fibrous structure makes it possible to reduce the wear of abrasives and their degradation.

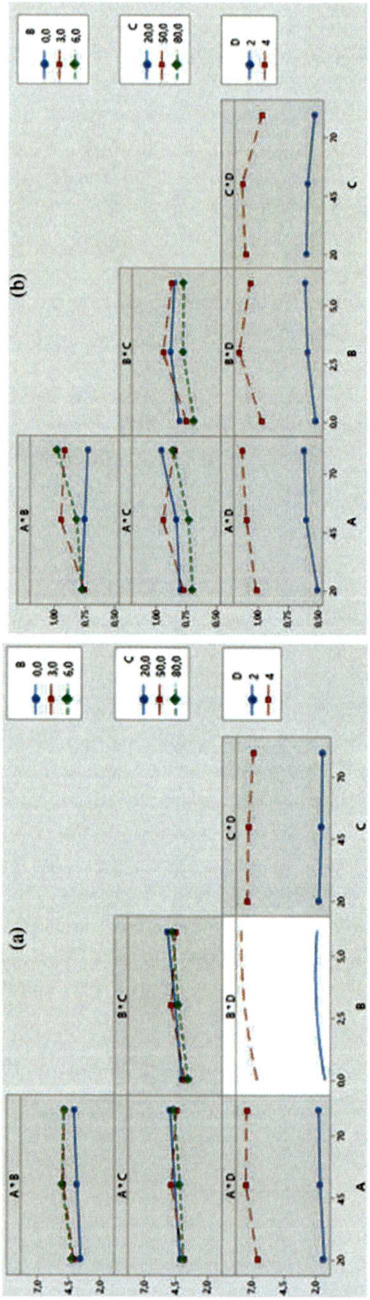


**Fig. 3** Graph of mass loss interactions of abrasives

**Acknowledgements** Supported by MPTEX.

## References


1. Ludwig, B. W., *Nonwoven abrasive articles and methods of making the same*. 2011, Google Patents.
2. Elaissi, A., et al., *The Impact of Chemical Treatment of Cellulosic Fibers on Surface Properties and Matrix/reinforcement Interfacial Adhesion*. Journal of Natural Fibers, 2022: p. 1–14.
3. Merotte, J., et al., *Mechanical and acoustic behaviour of porosity controlled randomly dispersed flax/PP biocomposite*. Polymer Testing, 2016. **51**: p. 174–180.
4. Das, D., et al., *Composite nonwovens*. Textile Progress, 2012. **44**(1): p. 1–84.
5. Arwa, E., A. Hamza, and G. Adel. *Surface Characterization of Textile Reinforcements*. in *International Conference of Applied Research on Textile and Materials*. 2022. Springer.
6. Elaissi, A., et al., *Development of Abrasives from Non-woven Based on Used Textiles*. Journal of Natural Fibers 2020: p. 1–15.
7. DSTU, I. J. T. R. and T. Part, *3310–1: 2017 (ISO 3310–1: 2016, IDT) Sieves*. **1**.
8. Elaissi, A., H. Alibi, and A. J. J. o. C. M. Ghith, *Effect of pumice stone and pearlite abrasives characteristics on denim abrasion*. 2022: p. 00219983221089712.
9. Mukhopadhyay, A. and V. K. Midha, A review on designing the waterproof breathable fabrics part II: Construction and suitability of breathable fabrics for different uses. Journal of Industrial Textiles, 2008. **38**(1): p. 17–41.
10. Litim, N., et al., *Effect of finishing resins on mechanical and surface properties of cotton Denim fabrics*. The Journal of the Textile Institute, 2017. **108**(11): p. 1863–1870.
11. Siddika, A., et al., *Effects of Carded an d Combed Yarn on Pilling and Abrasion Resistance of Single Jersey Knit Fabric*. 2017. **4**(2): p. 39–43.
12. Narayanaswamy, B., P. Hodgson, and H. Beladi, *Effect of particle characteristics on the two-body abrasive wear behaviour of a pearlitic steel*. Wear, 2016. **354**: p. 41–52.
13. Škoc, M. S. and E. Pezelj, *Abrasion resistance of high performance fabrics*. Abrasion resistance of materials. Rijeka: InTech 2012: p. 35–52.
14. Yadav, A., et al., *In-situ fabrication of surfactant modified CNT-based novel bio-composite and its performance evaluation for simultaneous removal of anionic dyes: Optimization by Box-Behnken design*. 2022. **284**: p. 120262
15. Elsayed, M. M., et al., *Tolmetin sodium fast dissolving tablets for rheumatoid arthritis treatment: preparation and optimization using Box-Behnken design and response surface methodology*. 2022. **14**(4): p. 880.

# Contribution to the Improvement of Certain Comfort Characteristics of Baby Diapers

**Wala Kallala, Adel Ghith, and Faten Fayala**

**Abstract** Baby diapers are products that offer optimal hygiene. In fact, at an age when baby cannot communicate yet, unforeseen events can happen. This means that there can be an emergency at any time, which is why they must use disposable diapers. Nowadays, these diapers have become a real concentrate of technology both in the choice of their components and in their implementation in the product in order to optimize their own functions and their combinations between them. Today, the industry has become highly competitive, and all manufacturers are trying to produce products that will satisfy their customers and they do this by providing attractive comfort qualities. In this context, and in order to contribute to the improvement of certain comfort characteristics of disposable baby diapers, we carried out a comparative study called benchmark between four of the most used brands in the Tunisian market, in order to determine the best composition. In this study, the analysis of baby diaper characteristics is divided into composition, fabrication and comfort characteristics. Diaper composition, mass, thickness, total absorption capacity, PLUTO absorption capacity, fluid runoff quantification and measurement of diaper rewets are all characteristics to be measured or identified. The results indicated that the most suitable diaper has a total absorption capacity of 1774.34% in saline solution, a balance of components (33.04% SAP/34.34% pulp/32.59 NAM), an absorbent core structure (homogeneous) containing 50% pulp and 50% SAP, a total mass of 38.32 g and average rewet, Pluto, runoff values.

**Keywords** Benchmark · Disposable baby diaper · Absorbent core · Superabsorbent polymer · Absorption capacity

---

W. Kallala (✉) · F. Fayala
Laboratory of Thermal and Energetic System Study (LESTE), National Engineering School of Monastir, Monastir University, Monastir, Tunisia
e-mail: wala.kallala@enim.u-monastir.tn

A. Ghith
Textile Materials and Processes Research Unit (UMPTEX), National Engineering School of Monastir, Monastir University, Monastir, Tunisia



S. B. Abdessalem et al. (eds.), *Proceedings of the Second International Conference of Innovative Textiles and Developed Materials-ITDM'2; 05-06 May 2023; Tunisia*,
https://doi.org/10.1007/978-981-99-7950-9_20

# 1 Introduction

Baby diapers are one of the most emerging markets for baby care products due to rising health awareness among consumers. In addition, the increase in the number of women in employment presents opportunities for countries to expand their workforces and generate additional economic growth. In addition, the greater the number of working mothers, the greater the market growth for baby diapers. In addition, disposable diapers are convenient, safe, and time-saving materials. Disposable diapers do not need to be cleaned and can be reused, which is the gospel of working women. As a result, household demand for diapers has increased, even among less affluent consumers [1].

The history of disposable diapers began more than a century ago, but they have been exploited commercially seventy years ago, in the late 1940s, in Europe and North America.

However, development was far from smooth, and the enormous success it achieved was unpredictable from the start. When looking at a product in the end, it is helpful to look at its development over four different times:

- Product invention (before 1930s), when many independent inventors sought better solutions to traditional diapers.
- Early commercialization (1930s–1950s), as private companies launched diapers and gradually expanded the market.
- Rapid adoption (1960s to 1980s), a surge in demand as disposable diapers became affordable.
- Continuous improvement (1990s to present), where leading manufacturers compete globally to provide consumers with continuously innovative and improved products [2].

Disposable baby diapers have become a veritable concentrate of technology, it is one of the most important consumer products of the twentieth century, continues to evolve rapidly in the beginning of the twenty-first century in terms of shape and components in order to improve their absorption properties and to solve problems encountered by the user such as leaks, irritation and is becoming thinner, simpler to use and more eco-friendly.

Disposable baby diapers have a complex structure as shown in Fig. 1.

The top sheet is the first element in direct contact with the baby's skin. It has a softness and a great capacity to keep the baby dry by retaining the fluid away from the skin, which avoids the problems of irritations [3]. The surface nonwoven is made of a melt formed polypropylene fiber (spunbond), thermo bonded (melt formed nonwoven) or a carded web. On the front side, the nonwoven is hydrophilic, allowing rapid absorption of liquid. On the reverse side, it is hydrophobic, which prevents the fluid from rising up. The second element is the high loft. The role of this part is to acquire and then distribute and/or transport the fluid from top sheet to the absorbent core in various directions in order to make the use of the super absorbent polymer (SAP) more efficient.

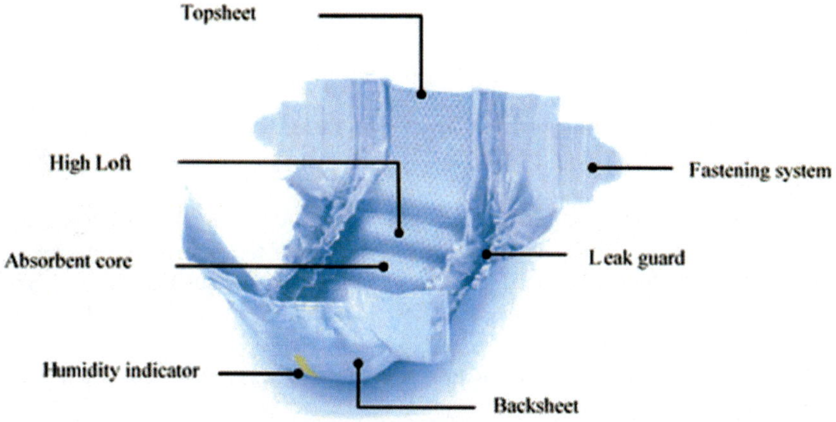


**Fig. 1** Representative scheme of the baby diaper composition

The main requirements of this acquisition layer is that it must provide a void to capture the liquid. It must also generate a capillary force gradient with the absorbent core to ensure rapid transfer of the fluid leaving these pores empty and ready to receive further liquid flow. The third element is the absorbent core. This is the most functional part of the diaper, and it is responsible for the retention of fluids. The quality of a diaper depends on the absorbency and retention capacity of this part. It is constructed of a cellulose structure (pulp) and variable levels of super absorbent polymer (SAP).

Cellulose pulp is a major component of absorbent hygiene products. It can approximately absorb up to 10 times its weight in fluid.

The super absorbent polymer (SAP) is a hydrophilic material called sodium polyacrylate. This polymer, in its raw state, has a powder form and retains at least 20 times its weight in urine. It turns into a gel that traps it when it comes into contact with urine, preventing it from coming into contact with the baby's skin.

In addition to the three principal components described before, the baby diaper contains other elements such as the fastening system, which is composed of a comfort band in front and fasteners on each side at the back of the diaper and the elastic between legs and the leakage barriers.

The back sheet is constituted of microporous polyethylene, and it retains the fluid but lets out the steam. It thus ensures a protection against the leakage.

According to EDANA [4], the main components in baby diapers are FLUFF and SAP:

- 24% Fluff pulp.
- 33% SAP.
- 21% Nonwovens.
- 13% Elastics and adhesive tape.

- 5% Polyethylene (PE) film
- 3% Adhesive.
- 1% others.

According to [5], the global baby diapers market has increased from 55.36 billion USD in 2022 to 59.62 billion USD in 2023, with a Compound Annual Growth Rate (CAGR) of 7.7%.

The baby diapers market is expected to grow steadily, driven by increasing demand for practical, eco-friendly products, as well as the rise of e-commerce channels.

Companies in the market need to focus on innovation and sustainability in their product offerings and marketing strategies, in order to remain competitive and respond to changing consumer demands [1].

In this context, a competitive benchmarking process was conducted to select the sample products for this comparative study.

As indicated in the rest of the work, four representative samples of commercial brands of baby diapers most sold on the Tunisian market are examined. It should be noted that for reasons of confidentiality and respect, we will maintain the anonymity of the companies, and their brands will not be mentioned in this study.

The four most used brands in the Tunisian baby diaper market are examined. These diapers were tested anonymously and were named A, B, C and D. The A and B diapers are made in Tunisia, the C diaper in Algeria and the D diaper in Egypt. All the diapers have the same size which is the size 5 in order to obtain comparable results.

## 2 Materials and Methods

### *2.1 Separation Methods of Diaper Components*

First, we have to separate the components of the diapers to test them. For this, we followed the steps below:

Remove the elastic and nonwoven from the diapers and weigh them.

- Weigh the mass of the absorbent core AC
- Separate the SAP polymer from the paste:
  - If the structure of the AC is non-homogeneous: the separation of SAP and paste is easy.
  - If the homogeneous structure of the AC where the SAP was bound to the cellulose pulp: place the AC in a container and then shake it to separate the SAP polymer from the pulp.

## 2.2 *Methods of Characterization of Disposable Baby Diapers*

### 2.2.1 Analysis of Diaper Components

*Dimensional Measurements of Disposable Baby Diapers*

We carry out the dimensional measurements of the baby diapers from the following sizes:

The width and length of the diapers.
The width and length of the absorbent core.

Measurements of the Widths, Lengths of Baby Diapers

The measurements of the width and length of the diaper are made with a metal ruler by fixing the ends of the diapers with magnets on a metal plate. The following figure shows the positions of the lengths and widths to be retained.

Measurements of the Widths and Lengths of the Absorbent Core

We first took off the absorbent core carefully by removing the nonwoven and plastic outer parts (non-absorbent material, NAM) of the diaper with scissors, and then measured the length and width with a metal ruler.

*Centrifuge Retention Capacity (CRC)*

The purpose of this method is to determine the amount of liquid trapped in the SAP after centrifugation. This method is called the "tea bag" method [6, 7]. For this purpose, tea bags of size 85*60 cm were filled with 0.2 g of SAP. The prepared bags were sealed with the sealing machine. We also closed an empty tea bag to use it as a tare. Subsequently, we immersed the samples either in distilled water or in a saline solution of concentration 9 g/l (NaCl + distilled water) for 30 min.

As soon as the time was up, the bags were placed in the centrifuge under a speed of 1250 rpm for 3 min. Once the centrifugation is finished, the balance is tared with the empty bag, the other bags filled with SAP are weighed and their final weights $W_f$ are noted.

$$\text{CRC(g/g)} = \frac{W_f - W_i}{W_i} \tag{1}$$

CRC Centrifuge retention capacity (g/g).
$W_f$ Final SAP weight (g).
$W_i$ Initial SAP weight (g).

*Free Swell Capacity (FSC)*

The free swell capacity (FSC) technique is the most practical and fastest way to evaluate the absorption capacity of SAP[6, 8].The purpose of this method is to determine how much the SAP can absorb when soaked for a defined period of time.

SAP samples are tested in the same manner as the (CRC) method but instead of using a centrifuge after the 30 min, the tea bags are left for 5 min to drain off excess solution, then the balance is tared with the empty bag and the bags filled with SAP are weighed and their final weights $W_f$ are noted.

$$\mathrm{FSC}(\mathrm{g/g}) = \frac{W_f - W_i}{W_i} \quad (2)$$

FSC Free Swell Capacity (g/g).
$W_f$ Final SAP weight (g).
$W_i$ Initial SAP weight (g).

*Determination of the Quantity of SAP*

This method allows us to calculate the amount of SAP contained in a diaper and to determine the centrifugal absorption capacity.

First, the dry weight of the diaper was recorded ($W_0$), then the diaper was immersed in a bath filled with a saline solution to approximate the urine for 30 min (PE on top and TNT in front of the bath). The saline solution is called synthetic urine. The solution is composed of 9 g of NaCl per liter of distilled water. Then, the sample is removed and put into the centrifuge at 1226 rpm (the nonwoven against the drum).

3 min later, the centrifuge stops and we note the weight $W_1$.

$$\text{Quantity of SAP} = \frac{[W_1 - (2 * W_0)] + (n * \mathrm{NAM})}{n * (\mathrm{RC_{SAP}} - \mathrm{RC_{fluff}})} * K \quad (3)$$

$W_0$ Diapers dry weight (g)
$W_1$ Diapers wet weight (g)
$n$ Number of diapers samples (n = 1)
NAM Non-absorbent materials (g)
$\mathrm{RC_{SAP}}$ Retention coefficient of SAP (g/g)
$\mathrm{RC_{fluff}}$ Retention coefficient of the fluff (g/g)
$K$ Correction coefficient.

### 2.2.2 Analysis of the Fabrication Characteristics of Baby Diapers

*Mass of Baby Diapers*

Using a scale, we measured the total mass as well as the mass of the absorbent bodies of the baby diapers. The operation is repeated on several diapers in order to have a representative average value for each brand of diaper analyzed.

*Thickness of Baby Diapers*

The thickness of the layer is measured according to ISO 9073-2:1995 using the thickness meter at three different points on the surface of the CA. Two measurements at points near the ends of the layer and one measurement in the center of the layer. The value considered is the average of the three values obtained.

*Density of Baby Diapers*

The density of a diaper is defined by the following Eq. 4. For each diaper, we have calculated this density.

$$\text{Density}\left(\text{g/cm}^3\right) = \frac{\text{Mass of the absorbent core (g)}}{L_{\text{AC}}(\text{cm}) * W_{\text{AC}}(cm) * \text{thickness (cm)}} \tag{4}$$

$L_{\text{AC}}$ Length of absorbent core (cm)
$W_{\text{AC}}$ Width of absorbent core (cm)

### 2.2.3 Analysis of the Comfort Characteristics of Baby Diapers

*Total Absorption Capacity*

The total absorption of a diaper is measured according to [9] ISO 11948-1 (1996).

First, the dry weight of the diaper is recorded ($W_1$). Then, the diaper immersed in a bath filled with a saline solution (9 g of NaCl per liter of distilled water) for 30 min. After this period, the wet layer is removed and suspended over a crossbar until no liquid falls for 5 min, then the suspended layer was removed and weighed ($W_2$). The total absorption capacity of the diaper was calculated according to Eq. 5.

$$\text{Total absorption capacity (\%)} = \frac{W_2}{W_1} * 100 \tag{5}$$

$W_1$ The weight of a wet diaper (g)
$W_2$ The weight of a dry diaper (g)

*PLUTO Absorption Capacity*

This method characterizes the diffusion of the liquid in a single direction, which is the plane of the baby's diaper as well as its rapid absorption. First, we cut 1 cm from the beginning of the absorbent core, weigh the sample and note $W_0$. Then, we put the sample in contact with a colored saline solution (9 g of NaCl per liter) in the PLUTO device (plane inclined by an angle of 30°) for 10 min. Once the time is up, we mark the level of the solution diffused into the sample with a marker and note the diffusion length. Finally, we weigh the exchange again and record ($W_1$).

$$\text{Ca}_{\text{PLUTO}} = W_1 - W_0 \tag{6}$$

$\text{Ca}_{\text{PLUTO}}$ PLUTO absorption capacity of a diaper (g)
$W_0$ Initial weight of the diaper
$W_1$ Final weight of the diaper

*Fluid Runoff Quantification*

This test is performed to quantitatively determine the surface effect of diapers according to ISO 9073-11:2002. First, the diaper was cut in half, then the ends opposite the center were clamped with a clamp on a 60° inclined table. The diaper sample was pulled to reduce wrinkles and twists on the surface. The travel distance was marked 10 cm above the edge of the layer. A glass separating funnel was clamped 1 cm above the travel distance and filled with 25 ml of saline solution mixed with a few drops of food coloring. At the bottom of the table, a balance was placed with a tissue folded over the weighing pan to catch any liquid that leaked out. A stopwatch was set for 10 min, and the tap on the separating funnel was opened.

The tissue absorbs any liquid that flows out. The runoff weight is called the primary run off value. During a 10 min interval, the process is repeated twice with new tissue, and the flow weights are named the secondary and tertiary runoff values [10].

*Measurement of Diaper Rewets*

The rewet was measured according to (ISO 9073-14, 2006) standard.

To do this, we place the rewet device on the diaper by positioning the tube opening on the micturition area (center of the absorbent body). Then, we weigh D(g) (1st micturition 100 g, 2nd and 3rd micturition 50 g) of the saline solution (9 g of Na Cl per liter). We pour it into the plastic tube and start the stopwatch immediately. When the diaper absorbs all the liquid, we stop the stopwatch, note the time A and leave the device at the same location on the diaper for 10 min. After these 10 min, we remove the rewet device and place the 1st set of filter paper (already weighed $W_0$(g)) on the micturition area of the diaper and then place a 2.5 kg mass on top for 2 min. When the time is over, we weigh again the set of filter papers $W$(g). For the 2nd and 3rd micturition, we repeat the same steps mentioned above.

The absorption rate and rewet value are calculated from the following formulas:

$$\mathrm{AR}_i = \frac{Di}{Ai} \quad (7)$$

$\mathrm{AR}_i$ Absorption rate.
$A_i$ Acquisition time in seconds relative to the dose $D_i$
$D_i$ Dose of the micturition

$$R_i = W_i - W_{0i} \quad (8)$$

$R_i$ Rewet value
$W_{0\mathrm{i}}$ Empty weight in grams relative to the sets of filter papers
1st Set 10 sheets of filter papers
2nd Set 20 sheets of filter papers
3rd Set 30 sheets of filter papers
$W_i$ Wet weight in gram relative to the dose Di

# 3 Results and Discussions

## 3.1 *Analysis of Characteristics of Composition*

### 3.1.1 Dimensional Measurements

Table 1 contains the dimensional measurements (width, length) of the baby diapers of the four brands studied.

The length values of baby diapers vary from 49.68 to 51 cm.
The width values of the baby diapers range from 23.08 to 24 cm.
Absorbent core length values range from 41.05 to 42 cm.
Absorbent core width values range from 10.05 to 12 cm.

The results indicate that the length and width values of the baby diapers as well as the absorbent core are approximately equal.

Generally, baby diapers with the same sizes have almost the same width and length measurements.

*Structure and Design of Baby Diapers*

Figures 2 and 3 show the structure and design of the components of the four baby diapers studied:

As shown in Figs. 2 and 3, all diapers contain a topsheet, a backsheet, an absorbent core and a backsheet.

**Table 1** Dimensional measurements of baby diapers

| | Baby diaper | | Absorbent core (AC) | |
|---|---|---|---|---|
| | Length (cm) | Width (cm) | Length(cm) | Width(cm) |
| A | 51 ± 0.2 | 24 ± 0.1 | 42 ± 0.5 | 10.05 ± 0.5 |
| B | 50.36 ± 0.3 | 23.5 ± 0.7 | 42 ± 0.5 | 12 ± 0.5 |
| C | 49.86 ± 0.4 | 23.58 ± 0.1 | 43 ± 0.5 | 11 ± 0.5 |
| D | 49.6 ± 0.1 | 23.08 ± 0.3 | 41.05 ± 0.5 | 10.05 ± 0.5 |

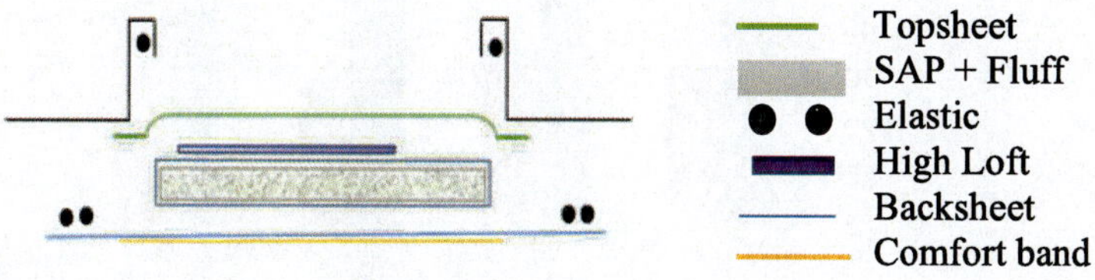


**Fig. 2** Structure and design of baby diapers A, B and C

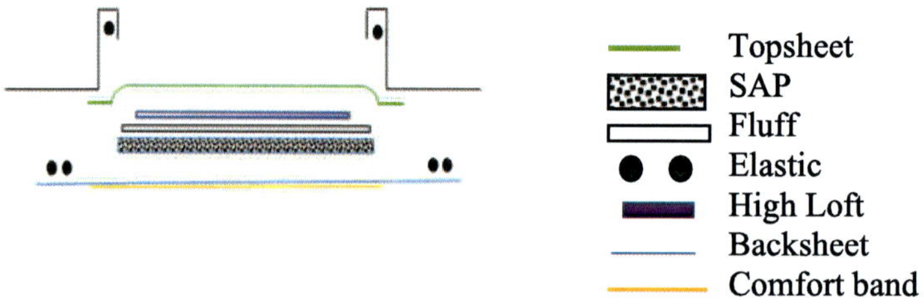


**Fig. 3** Structure and design of baby diapers D

We observe that the structure and design of the components of diaper D are different from the other diapers. The absorbent core of this diaper is non-homogeneous, i.e., the SAP is not bonded to the cellulose pulps, unlike the other structures in the rest of the diapers.

### 3.1.2 Centrifuge Retention Capacity (CRC)

The CRC method is used to determine the liquid retention from the weight of liquid lost from the swollen SAP upon centrifugation.

The CRC values are calculated from Eq. 1 and represented in the histogram shown in Fig. 4.

According to Fig. 4, we note that the retention capacity of SAP decreases in 0.9% saline solution.

We can deduce that the CRC of SAP varies with the concentration of the solution: as the amount of NaCl in the solution increases, the retention capacity of SAP decreases.

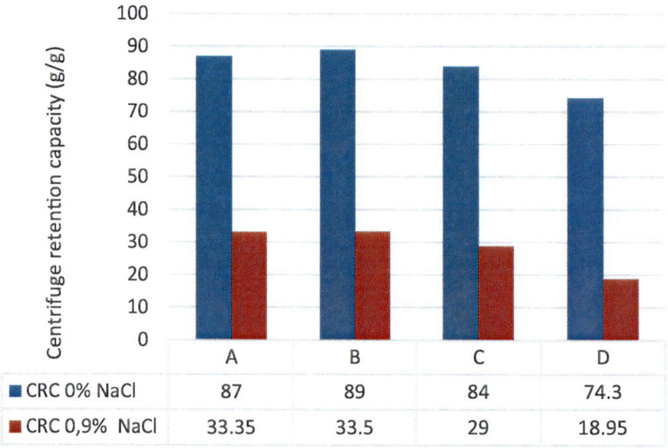


**Fig. 4** Histogram of CRC values

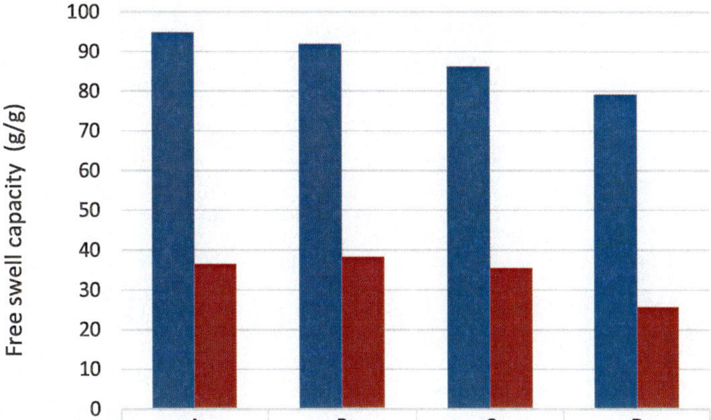


| | A | B | C | D |
|---|---|---|---|---|
| FSC 0% NaCl | 94.7 | 91.9 | 86.1 | 79 |
| FSC 0,9% NaCl | 36.45 | 38.15 | 35.4 | 25.65 |

**Fig. 5** Histogram of FSC values

The highest SAP CRC value was 33.5 g/g. It is that of diaper B and the lowest value was 18.95 g/g, which is that obtained for diaper D.

The difference in CRC values is due to the quality of the SAP used.

We can say as hypothesis that the smaller the CRC of the SAP, the higher the quantity of SAP in the diaper.

### 3.1.3 Free Swell Capacity (FSC)

Figure 5 shows the FSC values calculated from Eq. 2.

We note from the results shown in Fig. 5:

- The swelling of SAP varies from diaper to diaper.
- As the concentration of the solution increases, the swelling of SAP decreases.

We can say that the swelling behavior of SAP changes depending on the grade used.

### 3.1.4 Determination of the Quantity of SAP

From the CRC values found in the previous test, we calculated the amounts of SAP contained in a baby diaper using Eq. 3.

The results of this test are shown in the following Table 2.

The highest SAP value corresponds to diaper A (12.72 g) and the lowest to diaper B (12.01 g).

**Table 2** SAP quantity and centrifuge absorption capacity

| Baby diaper | Quantity SAP (g) | Centrifuge absorption capacity (%) |
|---|---|---|
| A | 12.72 ± 0.38 | 873 ± 21 |
| B | 12.01 ± 0.33 | 902 ± 27 |
| C | 12.66 ± 0.77 | 1121 ± 39 |
| D | 12.52 ± 0.64 | 1065 ± 40 |

The highest absorption capacity corresponds to diaper C (1121%) and the lowest to diaper A (873%).

### 3.1.5 Distribution of Fluff, SAP and NAM

Following the calculations effected in the SAP control test, we deduced the amount of fluff contained in each diaper, so that the distribution of pulp, SAP and NAM in the diaper structure can be analyzed.

Table 3 shows the quantities of SAP, pulp and NAM obtained.

As given in Table 3, the cellulose pulp values are different: diaper D contains a very small amount of cellulosic pulp (2.98 g) in contrast to diaper A and B, which contain a significant amount. As for the diaper C, the quantity of cellulose pulp it contains is average. The values of MNA are very close.

The absorbent core is the most functional part in the diaper. It is responsible for absorption and retention of liquids. The SAP absorbs and traps liquids and the cellulose pulp improves the absorption capacity and provides diffusion of liquids.

Table 4 gives the percentages of the SAP and pulp in relation to the absorbent core:

Table 4 gives us that the distribution of SAP and pulp in the AC is different:

**Table 3** Distribution of SAP, pulp and NAM in grams

| Baby diaper | SAP (g) | Pulp (g) | NAM(g) |
|---|---|---|---|
| A | 12.72 ± 0.38 | 23.65 ± 0.6 | 11.63 ± 0.4 |
| B | 12.01 ± 0.33 | 20.33 ± 0.3 | 11.27 ± 0.5 |
| C | 12.66 ± 0.77 | 13.17 ± 0.5 | 12.49 ± 0.4 |
| D | 12.52 ± 0.67 | 2.98 ± 0,33 | 10.26 ± 0.3 |

**Table 4** Percentage of SAP and pulp in relation to the absorbent core

| Baby diaper | SAP (%) | Pulp (%) |
|---|---|---|
| A | 35 | 65 |
| B | 37 | 63 |
| C | 49 | 51 |
| D | 81 | 19 |

**Table 5** Mass-thickness-density of baby diapers of SAP

| Baby diaper | Total mass (g) | Mass of AC (g) | thickness(cm) | Density (g/cm$^3$) |
|---|---|---|---|---|
| A | 48 ± 2.3 | 36.37 ± 2 | 0.88 ± 0.57 | 0.1 |
| B | 43.61 ± 2.2 | 32.34 ± 2 | 0.9 ± 0.38 | 0.07 |
| C | 38.32 ± 2.8 | 25.83 ± 0.38 | 0.85 ± 0.4 | 0.06 |
| D | 25.76 ± 2.4 | 15.5 ± 2 | 0.35 ± 0.56 | 0.11 |

Diapers A and B: 60% pulp, 40% SAP.
Diaper C: 50% pulp, 50% SAP.
Diaper D: 20% pulp, 80% SAP.

The results found in the CRC determination explain the percentages of SAP in the absorbent core, the lower the CRC of SAP, the higher the amount of SAP in the diaper.

The higher percentage of SAP does not mean an added value to this diaper. This is due to the poor distribution of the components that appear in the small amount of cellulose pulp put in the absorbent core, so it is necessary to have a balanced distribution and distribution of SAP and pulp.

## 3.2 *Analysis of the Fabrication Characteristics: Mass, Thickness and Density of the Baby Diapers*

Table 5 presents the results of masses, thickness and density of the baby diapers.

Table 5 gives that the highest mass (48 g) is for diaper A. On the contrary, diaper D is the lightest. Its mass is almost equal to half that of diaper A. This characteristic is very important because the lighter the diaper the more comfortable it is for the baby.

The diaper B is the thickest diaper and the diaper C is the thinnest (less thick) compared to the other diapers.

The density varies from 0.07 g/cm$^3$ to 0.11 g/cm$^3$.

The diaper B has the highest density, while the diaper C has the lowest density.

## 3.3 *Analysis of Comfort Characteristics*

### 3.3.1 Total Absorption Capacity

The results of the total absorption capacity of diapers in saline solution are given in Table 6:

Table 6 Total absorption capacity values

| Baby diaper | Total absorption (%) |
|---|---|
| A | 1605.73 ± 12.32 |
| B | 1741.44 ± 18.40 |
| C | 1774.34 ± 40.89 |
| D | 1725.73 ± 46.32 |

According to Table 6, the lowest total absorption capacity 1605.73% is in diaper A that is relatively low compared to the other values, this may be due to the low percentage of SAP contained in the diaper.

The total absorption capacity of the other diapers varies between 1774.34% and 1725.73%.

The total absorption capacity is related to the percentage of SAP contained in the diaper and the quality of SAP used from a CRC and FSC perspective.

### 3.3.2 PLUTO Absorption Capacity

The results of the diffusion length as well as the PLUTO absorption capacity are shown in Table 7.

From Table 7, we can see that the highest diffusion length is attributed to the diaper A and diaper B and then to the diaper C, this length is lower in the diaper D.

We have previously found that the percentages of pulp in diaper A and B are important, this explains the results of diffusion lengths found. We can therefore confirm that the cellulosic pulp is responsible for the diffusion of the liquid in the baby diaper.

Table 8 gives the values of the absorption capacity PLUTO.

Table 7 Values of diffusion length of the liquid

| Baby diaper | Diffusion length (%) |
|---|---|
| A | 18 |
| B | 18 |
| C | 17.5 |
| D | 8 |

Table 8 PLUTO absorption capacity values

| Baby diaper | Ca PLUTO |
|---|---|
| A | 166.36 |
| B | 200.56 |
| C | 142.02 |
| D | 79 |

According to the results found, PLUTO absorption is highest in the B diaper and lowest in the D diaper.

The PLUTO test is a test that describes the capillary rise of liquid in the diaper.

It is done to measure the mass of liquid absorbed by the diaper, which is maintained at a 30° angle of inclination in contact with the saline solution. We can say that the more the quantity of water is important, the more the material is hydrophilic.

The higher PLUTO absorption of diaper B can be explained by the fact that the pulp of this diaper has more hydroxyl groups on its macromolecular chain.

### 3.3.3 Fluid Runoff Quantification

To determine the surface effect of the baby diaper, which illustrates the absorption rate of the liquid, the quantification of liquid runoff was studied from a specific angle.

Figure 6 shows the results of quantitative evaluation of primary, secondary and tertiary fluid runoff.

The primary runoff values vary from 0 to 0.4 g.

The highest primary runoff value is attributed to the diaper D and the lowest to the diaper A. This means that, compared to the different diapers, the diaper A absorb liquids rapidly.

The secondary runoff values are between 0 g and 0.62 g. The tertiary runoff values are between 0 g and 0.74 g.

The best topsheet is the one with the lowest runoff values, i.e., topsheet A.

Therefore, the topsheet of the diaper A is the best, and it is designed to transfer liquids fast while remaining soft and dry.

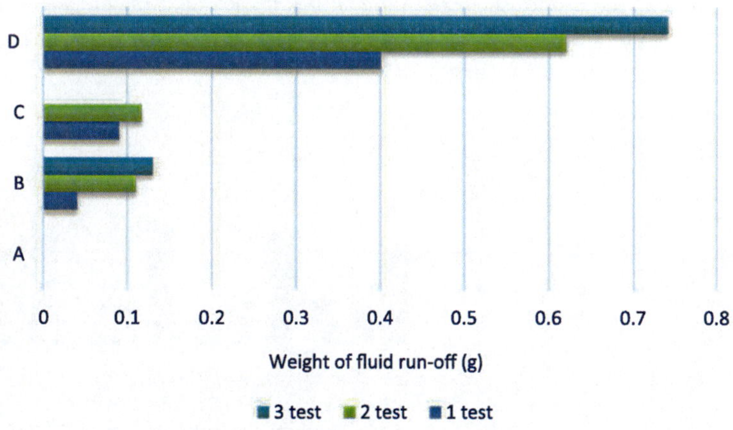


**Fig. 6** Fluid runoff quantification for baby diaper in saline solution

**Table 9** Rewet values of diapers

| Baby diaper | R1 | R2 | R3 |
|---|---|---|---|
| A | 0.02 | 0.04 | 0.31 |
| B | 0.03 | 0.08 | 0.31 |
| C | 0.04 | 0.09 | 0.14 |
| D | 0.03 | 0.03 | 0.99 |

### 3.3.4 Measurement of Diaper Rewets

The rewet test results are given in Table 9.

The 1st rewet values vary from 0.02 g to 0.04 g.

The 2nd rewet values vary from 0.04 g to 0.09 g.

The 3rd rewet values vary from 0.14 g to 0.99 g.

The purpose of this test is to evaluate the rewet of the saline solution. The lowest rewet values provide a very good ability to keep the baby dry.

According to Table 8, the diaper C has the best rewet values, and this is only a direct consequences of the results found previously. Indeed, this diaper has a better total absorption capacity and the liquid diffuses and propagates easily and rapidly in the structure of this diaper. So the liquid will systematically move away from the baby's skin to the inside of the structure where it is trapped and thus keep the baby dry.

Unlike the diaper D that has bad values of rewet, this is explained by the low values of runoff found as well as the low length of diffusion.

So the diaper does not absorb liquid easily and the liquid does not spread fast.

Figure 7 shows the values of the absorption rates calculated from Eq. 7.

According to the results obtained in Fig. 7, the diaper C has the highest value of absorption rate, that is to say that it absorbs the liquid slow unlike the other diapers.

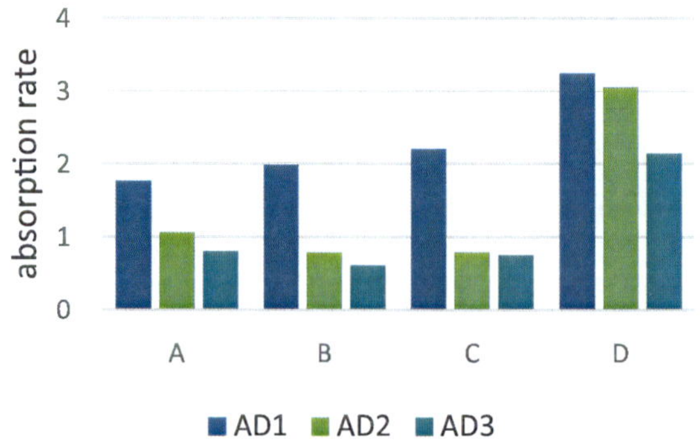


**Fig. 7** Values of absorption rates

## 4 Conclusion

In this paper, comparative study is carried out whose purpose is to know the best diaper.

First, the complete structure of the DBD was tested by measuring its dimensions as well as quantifying the weight of the SAP by using the centrifuge retention capacity (CRC) method.

The results indicated that even though the diapers were of the same size (size 5), they had different dimensions and content of SAP, cellulosic pulp and NAM, which in turn affects the performance of the diapers. This influences the performance of the diaper, especially in terms of total absorption capacity, PLUTO absorption capacity, and fluid runoff quantification as well as rewets values.

According to the results obtained in this present study, the diaper C is the best diaper, this diaper has a total absorption of 1774.34%, the distribution of the components is balanced (33.04% SAP/34.34% pulp/32.59 NAM), the structure of the absorbent core (homogeneous) contains 50% pulp and 50% SAP, the total mass is 38.32 g, the values of rewet, pluto, runoff are average.

We can deduce that the structure and the composition make it possible to provide better results from the point of view of absorption and comfort of the baby diapers.

## References

1. BabyDiapers-Worldwide, https://www.statista.com/outlook/cmo/tissue-hygiene-paper/babydiapers/worldwiderevenue, Last accessed 05 May 2023.
2. Davis, D.: Seven decades of disposable diapers. The Winthrop Group, Inc .on behalf of EDANA (European Disposable and nonwoven Association), (2005).
3. Couches-bébés-fonctionnalité-composants, https://www.grouphygiene.org/fr/couchesbebes-fonctionnalite-composants/#_ftnref1, Last accessed 24 Oct 2022.
4. EDANA: European Disposable and nonwoven Association, sustainability-report- 2015–4th EDITION, (2015).
5. Baby Diapers Global Market. The Business Research Company (2023).
6. Y. Qin, Superabsorbent polymers and their medical applications, Med. Text. Mater. (2016) 71–88. https://doi.org/10.1016/b978-0-08-100618-4.00006-6.
7. EDANA, Centrifuge Retention Capacity, No NWSP 241.0. R2 (19), Harmonized Nonwovens Standard Procedures: Polyacrylate Super absorbent Powders- Determination of the Fluid Retention Capacity in Saline Solution by Gravimetric Measurement Following Centrifugation, 2019.
8. EDANA, "Free Swell Capacity, No NWSP 240.0.R2 (19), Harmonized Nonwovens Standard Procedures: Polyacrylate Superabsorbent Powders - Determination of the Free Swell Capacity in Saline by Gravimetric Measurement, 2019.
9. ISO 11948-1. (1996). Urine-absorbing aids – Part 1: Whole-product testing/trans. Incontinence ISO/TC 173/SC 3. Aids for Ostomy.
10. EDANA, INDA, Fluid Run Off, No NWSP 080.9.R1 (19), Harmonized Nonwovens Standard Procedures: Non woven Run Off, equivalent to ERT 152.2 (2019).

# DMAIC Approach to Optimize the Production Capacities of a Tunisian Clothing Manufacture: A Case Study

**Nesrine Bhouri, Naoufel Bhouri, and Dorra Oualha**

**Abstract** Manufacturing issues have always been a hurdle for managers, engineers, and professionals. They can lead to low productivity, poor quality, high costs, and ultimately loss of customers. Problems should be avoided by fair means and following well-established methodologies of continuous process improvement. The main objective of this paper is to standardize a production line of T-shirts by optimizing material, human, and logistic resources using lean management tools. The DMAIC method as a full approach was used to improve the workstation design and to decrease human fatigue. The SWOT analysis followed by the QQOQCCP method was used to define the project; the Pareto analysis coupled with the IO method and the Ishikawa diagram have been explored the main causes of hazards. The improvements focused on the design of some workstations to change their operating modes lead to the productivity optimization. New machine arrangement method was made to develop more comfortable space and less flows disruption. The results lead to increase the productivity by 40% and decrease the number of needed staff from 22 to 19 workers. The number of daily outstanding was decreased from 254 pieces/day to 64 pieces/day.

**Keywords** DMAIC · Performance keys · Hazards · Human fatigue · Productivity · Daily outstanding rate

N. Bhouri (✉) · N. Bhouri
Textile Materials and Processes Research Unit, UR17ES33, University of Monastir, Monastir, Tunisia
e-mail: bhouri.nesrine@yahoo.fr

D. Oualha
Department of Forensic Medicine, Research Laboratory LR12SP14, Teaching Hospital Fattouma Bourguiba, 5000 Monastir, Tunisia



S. B. Abdessalem et al. (eds.), *Proceedings of the Second International Conference of Innovative Textiles and Developed Materials-ITDM'2; 05-06 May 2023; Tunisia*,
https://doi.org/10.1007/978-981-99-7950-9_21

## 1 Introduction

Nowadays, there are huge pressures on organizations to improve customer satisfaction and quality, at the same time to decrease ineffectiveness and reduce the number of errors [3] #70. Problems should be avoided by fair means and by following well-established methodologies of continuous process improvement. There are many different conceptions, methods, and tools that may be used to maintain a good quality level and help in the continuous development in the company [1] #72. DMAIC is among many different tools of quality management which may be considered methods of quality and process improvement. DMAIC consists of five stages that are connected to each other [3], #73. It is an acronym from the words Define–Measure–Analyze–Improve–Control. This method is based on process improvement according to Deming Cycle [4] #70. It aims to improve, optimize, or stabilize an existing process by detecting and removing the defects or inefficiencies in the process, specifically the output defects [2] #71.

The main objective of this paper was to standardize a production line of T-shirts by optimizing material, human, and logistic resources using lean management tools. The DMAIC method as a full approach has been used to improve the workstation design and to decrease human fatigue.

## 2 Research Methodology

This study presents a step-by-step DMAIC implementation road map to standardize a production line of T-shirts. The define phase starts with identifying a problem that requires a solution and ends with a clear standing of the problem. We have used the WWWWHHW method to identify the problem. The main purpose of this stage is to verify whether the actions are connected with the priorities in the work frame and that there is support from management and availability of required resources.

The measure phase consists of measuring the extent of the problem in the manufacturing process. In order to evaluate the productivity of the process, the daily yield of the investigated line was recorded for two weeks (six working days). To have an idea of the organizational system applied during the distribution of the work among the employees of the studied chain, we studied the saturation of the various posts defined according to the time of the line.

The objective of the analysis phase is to investigate the data to find the causes of the problem and propose solutions and improvements. In this phase, we have to use instant observation (IO) to list the fundamental causes that ensure fluctuations from the optimal rate. Indeed, a total of 900 observations were made on 22 stations during six successive days, with seven observations per post per day made during various intervals during the day. The activity of the company was estimated at 64% with a precision of 5%. Pareto analysis by the ABC method is used to identify priority hazards to be managed, and the Ishikawa diagram was used to analyze the causes of each major hazard.

The goal of improving phase is to take necessary information to create and develop an action plan in order to improve the functioning of the organization issues. In this phase, we are going to present the changes applied during standardization of the production of some posts as well as the new distribution of the works between the employees and the new arrangement of the machines to ensure a less encumbering flow in the line.

The main purpose of control phase is not only improving the process performance but also having the improved results sustained in the long run. For this phase, we have controlled the line productivity, the daily stock, and the workers saturation.

# 3 Case Study

## *3.1 Define*

The first step of the DMAIC methodology starts by defining the problem and identifying the objectives to be achieved. The results of WWWWHHW method are used to define the main objectives of this project.

- What?: The objective of this project is to standardize the production line of T-shirt.
- Who?: Services having a direct effect on production hazards in the quilting room.
- Where?: Within the method departments and the production line of T-shirts.
- When?: Synthesis of works performed by engineering and technical students dealing with the same problem between February 2020 and July 2021.
- How?:
  - Waste hunting
  - Pareto
  - Ishikawa
  - Flow analysis
- How much?: Make a better estimate of the workforce, have a better estimate of deadlines and costs, act on manufacturing costs and the necessary resources.
- Why?: Improve performance, reduce daily outstanding rate, reduce uncertainties, and meet production deadlines.

In this project, we aim to standardize the production line of the T-shirt and the production lines in order to avoid the difficulties of the manufacturing process in just time.

## 3.2 Measure

The evolution of the line's yields during two weeks of investigation shows that the average yield is about 60.7% ± 18%, and this does not reach the initial target of 80%. The saturation of various posts on this line has been shown that 41% of the workers were outside the standard range of production stability (six workers were found with saturations above 110% and three workers with saturations below 90%). This justifies the fluctuations in the evolution of the average daily output of the line. The work process observations at the GEMBA workstation showed a huge loss of time when repairing the defective parts detected during the quality control.

The ratio of defective legs is 5% ± 3.3%, and it can reach the value of 10.5% in some cases. These results show on one side that the rate of defect recorded at this post exceeds the level of quality accepted in this company considered at 2%, and there is a significant fluctuation in the daily values reflecting bad stability in the applied operating modes.

## 3.3 Analyze

In order to study the activity and the hazards in this company, the obtained results were divided in activity rate of 59.85% and hazard rate of 40.15%.This value, quite high, reflects the lack of organization and preparation of work in the stitching room. Pareto analysis by the ABC method is presented in Figs. 1 and 2.

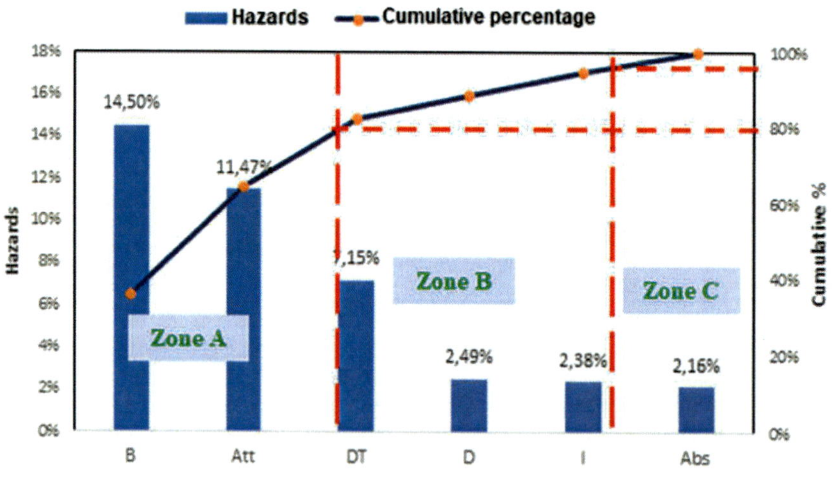


**Fig. 1** Pareto graph

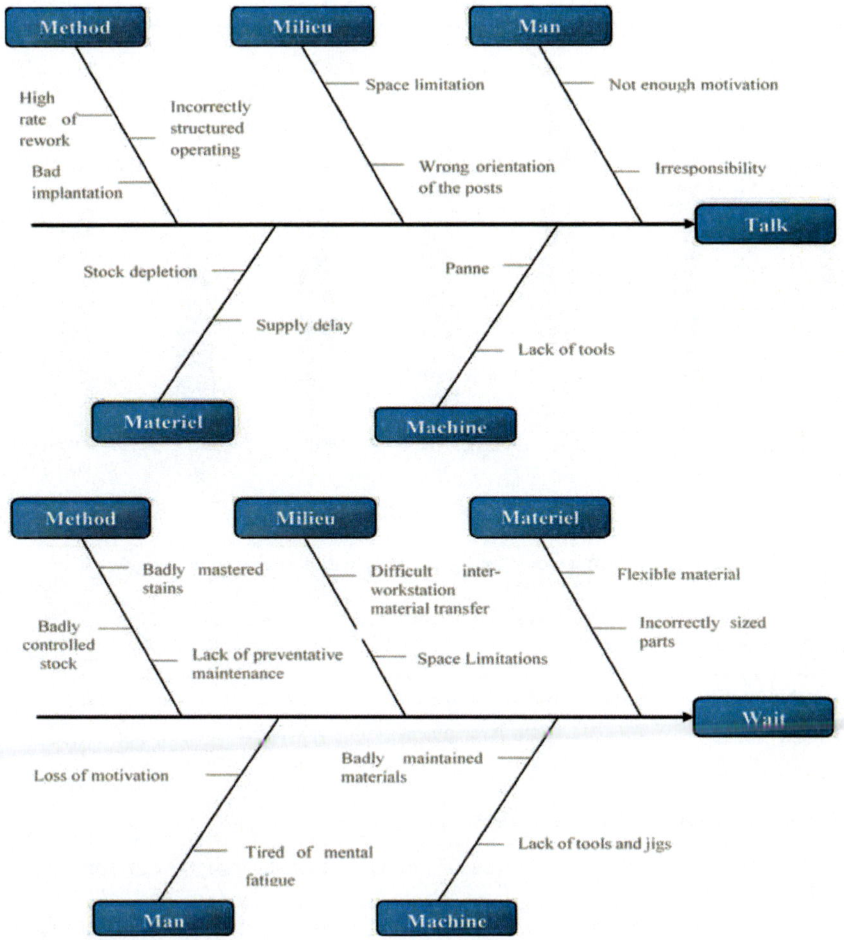


**Fig. 2** Ishikawa diagram

Figure 1 shows that the main hazards, identified by zone A, were the talk (B) with 14.5% of hazards and the wait (Att) with 11.47%. Figure 3 illustrates the Ishikawa diagram of talk and wait.

## *3.4 Improve*

The improvements focused on the design of some workstations to change their operating modes leading to productivity optimization. A new machine arrangement method was made to develop a more comfortable space and fewer flows disruption.

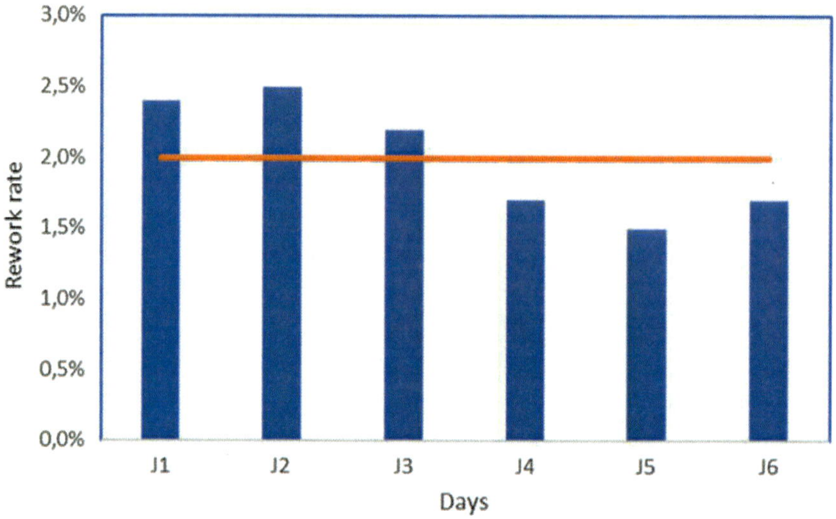


**Fig. 3** Evaluation of the rework rate at the leg assembly station after reorganization

**Table 1** Comparative study of effective time at the "leg assembly" workstation before and after reorganization

| | Average (%) | Gain |
|---|---|---|
| Before development | $375 \pm 7$ | 13% |
| After development | $325 \pm 4$ | |

The new implementation has allowed to:

- Eliminate the flow return because of the respect of the progress of conception T-shirt (Table 1).
- Reduce daily outstanding rate and the rework rate by a better distribution of tasks (Fig. 3).
- Reduce chatter by changing the stations orientations.
- Flow optimizing by removing the works need to move to the next lines.
- The three removed employees from the production line are transferred to other production lines and are considered as replacement workers in case of absence or major hazards.

## *3.5 Control*

The results show that the labor requirement of the line has decreased from 22 to 19 workers to produce 540 pieces/day with a gain of 40%. This was due to better control of the distribution of tasks, equal saturation between workers, and the lower daily outstanding rate. The average daily yield increased to 79.2%, thus meeting the company's overall target.

## 4 Conclusion

The development of a quality approach through the implementation of the five steps of the DMAIC method was carried out in order to standardize a T-shirt manufacturing chain. This approach allowed us to optimize the production costs thanks to the reduction in the staff needs, the optimization of flows and daily outstanding rates, the improvement of workstations productivities, and the reduction of the defects rate.

## References

1. Abbes, N., Sejri, N., Chaabouni ,Y., et al. : Application of Six Sigma in Clothing SMEs: A case study. IOP Conference Series: Materials Science and Engineering 2018; 460: 012009. DOI: https://doi.org/10.1088/1757-899x/460/1/012009.
2. Bhargava, M., and Gaur, S.: Process Improvement Using Six-Sigma (DMAIC Process) in Bearing Manufacturing Industry: A Case Study. IOP Conference Series: Materials Science and Engineering 2021; 1017: 012034. DOI: https://doi.org/10.1088/1757-899x/1017/1/012034.
3. Shankar, R.: Process improvement using Six Sigma : a DMAIC guide. American Society for Quality, Quality Press, 2009, p.127.
4. Smętkowska, M., and Mrugalska, B.: Using Six Sigma- DMAIC to Improve the Quality of the Production Process: A Case Study. Procedia - Social and Behavioral Sciences 2018; 238: 590-596. DOI: https://doi.org/10.1016/j.sbspro.2018.04.039. +

# The Impact of Acid Blue Dye Concentration on Color Strength and the Chromaticity Coordinates of Raw and Tannery Wool: A Comparative Study

**Olfa Abdellaoui, Harizi Taoufik, Feriel Bouatay, and Msahli Slah**

**Abstract** Globally, the management of wool waste employs a range of strategies and practical techniques to maximize the value of wool fibers in both clothing and non-clothing applications. Recent studies have emphasized the importance of wool waste valorization in establishing a new supply chain for tannery companies. To utilize wool fibers in clothing, interior textiles, and technical textiles, it is necessary to first dye and finish them. The aim of this research was to investigate the effect of acid dyes on tannery wool. In order to properly assess the colorfastness of tannery wool, this study compared the chromaticity coordinates and color strength with those of raw wool. Samples of both raw and tannery wool were dyed with an acid dye and subjected to analysis of *K*/*S* values. The results demonstrated that although the dyeing properties of tannery wool were lower than those of raw wool, the latter possessed considerable potential that could be developed for use in textile products. Tannery wool fibers are now being promoted as a sustainable and eco-friendly alternative material.

**Keywords** Tannery wool · *K*/*S* · Acid dye

O. Abdellaoui (✉) · H. Taoufik · M. Slah
Textile Engineering Laboratory, University of Monastir, Monastir, Tunisia
e-mail: olfa.abdellaoui_ghozzi@yahoo.fr

F. Bouatay
Research Laboratory of Environmental Chemistry and Clean Processes, Faculty of Sciences of Monastir, University of Monastir, Monastir, Tunisia

O. Abdellaoui · H. Taoufik · F. Bouatay
Higher Institute of Fashion of Monastir, University of Monastir, Monastir, Tunisia

 
S. B. Abdessalem et al. (eds.), *Proceedings of the Second International Conference of Innovative Textiles and Developed Materials-ITDM'2; 05-06 May 2023; Tunisia*,
https://doi.org/10.1007/978-981-99-7950-9_22

# 1 Introduction

The present research was therefore aimed to conduct a comparative dyeing study between raw and tannery wool fiber. The chromaticity of raw and tannery wool fibers was measured and compared. The hide of sheep was divided into two identical parts. One part was sheared and identified as raw wool, while the second part was pulled and recuperated as a tannery wool using the chemical tanning process [1]. Chemical deliming is a process used to remove fibers from animal skins in order to produce a usable raw material for the manufacture of leather or wool products. The process involves the use of a chemical agent, known as a deliming agent, to dissolve the keratin protein that makes up the hair or fiber. This deliming agent is typically a solution of sodium sulfide $Na_2S$ and lime [2], which is applied to the skin of wool for a set period of time. Once the deliming agent is applied, the hair or fibers are mechanically removed from the skin.

According to the results obtained, we were able to determine that the raw wool fibers exhibited a higher color strength (*K*/*S*) and better color fastness compared to the tannery wool fibers. This information can be used to optimize the processing of tannery wool fibers and improve their dyeability, leading to a higher quality end product.

# 2 Experimental

## *2.1 A Wool Sample*

Sheep pelts were obtained from slaughter. The selected pelts refer to male Barbary red face sheep (BTR). The average mean fiber diameter of the wool samples was $26.94 \pm 4.21$ μm [1]. We were used ten pelts in our investigation. Each hide was divided into two identical parts. The samples of raw and tannery wool fibers were first scoured with a sodium carbonate solution $Na_2CO_3$ and soap at a moderate temperature ($52 \pm 3$ °C).Then, they are transformed into wicks to be dyed (Fig. 1).

The dyeing was carried out in the pots of Ahiba (Datacolor, AHIBA IR®). The dyeing was conducted in the presence of (C.I. Acid Blue 25), using the Agent of unison ALBGAL (CIBA-GEIGY), sodium sulfate, acetic acid. The chemical structure of the applied dye is shown in Fig. 2. A Calibration Lab Benchtop pH Meter was used to measure the pH values of the dyeing bath.

## *2.2 SEM Analysis*

Scanning electron microscopy (SEM) analysis was carried out with a JEOL JSM5400 microscope at 15 kV acceleration voltage after gold coating.

**Fig. 1** Wicks of raw and tannery wool

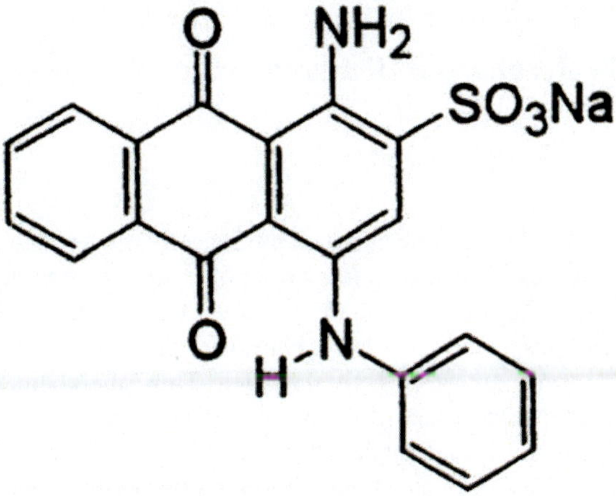


**Fig. 2** The chemical structure of C. I. Acid Blue 25, C.I 0.62055, molecular formula: $C_{20}H_{13}N_2NaO_5S$, molecular weight: 416.38 g/mol [3]

## 2.3 *Methods*

**Dyeing Method**

The wool samples were dyed with different concentrations of Acid Blue 25 in presence of acetic acid (pH = 4.5), with a liquor ratio L.R = 40:1.The raw and tannery wool yarns were introduced to the dyeing baths at 40 °C, and the temperature was raised to 100 °C at a constant rate and kept for 60 min. The dyed samples were thoroughly rinsed with cold water and dried at ambient temperature.

**UV–visible Spectrophotometer**

A UV–visible DR 6000 spectrophotometer was employed for absorbance measurements using a quartz cell of 1-cm path length. The maximum absorbance wavelength of the applied dye Acid Blue 25 was recorded at 605 nm using spectrophotometer.

**Colorimetric Measurements**

The color values of dyed samples were recorded on spectrocolorimeter. CIEL* *a* * *b* * color parameters and color strength (*K*/*S*) values were automatically calculated from reflectance data using Color iMatch software. The average color depth ($K/S_{AVG}$) over a wavelength range of 400–700 nm is reported in this research.

The *K*/*S* value was also calculated by Eq. (1):

$$K/S = (1 - R)^2/2R \tag{1}$$

where *R* is the measured reflectance, and *K* and *S* are the absorption and light scattering coefficients, respectively [4].

# 3 Results and Discussion

## 3.1 Morphological Study

Wool is a natural fiber, which intrinsically has a rough surface. The SEM images of examples of raw and tannery wool fiber are shown in Fig. 3. For raw wool, presented in Fig. 3a, it can be clearly observed that scales have sharp edges and the cuticle layer has a smooth surface. However, for the tannery wool, represented in Fig. 3b, the scales appear slightly damaged. Thus, the chemical treatment has an etching effect on the surface of the wool fiber. For wool fiber, it is evident that the damage is essentially caused by the attack of protein by the alkaline treatment [5].

At pH 4.5, the acid blue dye exists primarily in its anionic form, meaning that it has a negative charge. Wool, on the other hand, is amphoteric, meaning that it can have either a positive or a negative charge depending on the pH of the solution it is in.

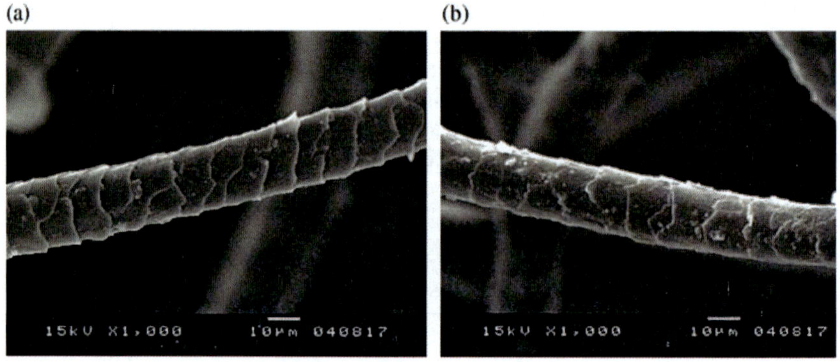


**Fig. 3** Scanning electron micrographs of **a** raw and **b** tannery wool fibers (×1000)

When wool is exposed to acid blue dye at pH 4.5, it can form several types of chemical bonds with the dye molecules. One of the primary types of bond is electrostatic attraction, which occurs between the negatively charged dye molecules and positively charged sites on the wool fibers. These positive sites on wool can come from several sources, including amino groups on the wool protein molecules and acidic functional groups present in the wool fibers.

Another type of bond that can form between wool and acid blue dye is hydrogen bonding. Hydrogen bonding occurs between the hydrogen atoms in the dye molecule and oxygen or nitrogen atoms in the wool protein molecules.

Finally, covalent bonding can also occur between the dye molecules and the wool fibers, although this is typically a less common type of bond. Covalent bonding involves the sharing of electrons between the dye molecule and wool fiber, resulting in a strong, permanent bond.

Overall, the specific types of chemical bonds that form between wool and acid blue dye at pH 4.5 will depend on several factors, including the pH of the solution, the concentration of the dye and wool, and the specific chemical structures of the wool protein and dye molecules.

## 3.2 *Color Strength ($K/S_{AVG}$)*

Figure 4 represents the effect of different concentrations of Acid Blue 25 on the color depth ($K/S_{AVG}$) values of raw and tannery wool yarns dyed. Significant variations were observed in the average $K/S$ values. The variations in the $K/S$ values depend upon the nature and the extent of the interaction developed between wool fiber and dye [6].

At lower dye concentrations, raw wool shows a higher adsorption capacity than tannery wool. This may be due to the more regular scales structures of raw wool and the presence of porous, which can facilitate dye adsorption. However, as the dye concentration increases, tannery wool begins to behave similarly to raw wool in terms of adsorption capacity.

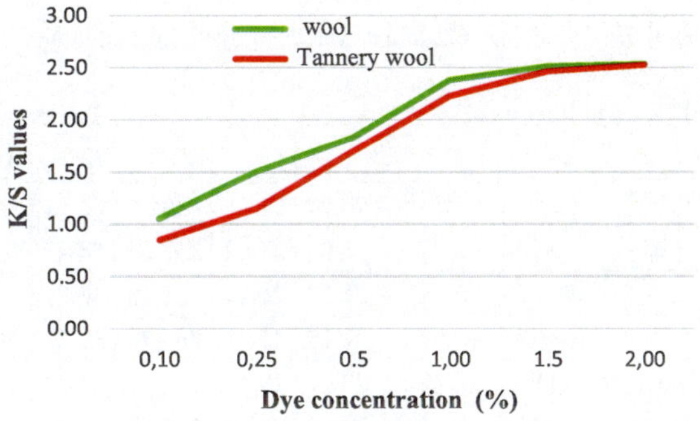


**Fig. 4** *K/S* values of the wool samples dyed with Acid Blue 25 in different concentrations

This change in behavior may be due to several factors. First, as the dye concentration increases, there are more dye molecules available to be adsorbed onto the wool fibers. This may mean that tannery wool begins to saturate with dye molecules at a higher concentration than raw wool, which increases its adsorption capacity.

In addition, the structure of tannery wool may be modified by the tanning process, which can affect its adsorption behavior. In fact, the tannery wool has a different wool fiber surface from that of raw wool, which could affect how dye molecules interact with it. In fact, damaged cuticles of tannery wool can make dyeing more difficult as they can act as a barrier to uniform dye absorption. This can also affect the regulation of dye absorption and diffusion into the fiber, which can have an impact on the speed and uniformity of dyeing.

## 3.3 Color Parameters ($L^*$ $a^*$ $b^*$ $C^*$ $h°$)

Table 1 provides measurement data for two types of wool, raw wool and tannery wool. It shows the values of the $L^*$, $a^*$, and $b^*$ color parameters for raw wool and tannery wool at different concentrations of dye.

The measurements include color values of $L^*$, $a^*$, $b^*$, $C^*$, and $h°$ for each sample. The values of $L^*$, $a^*$, and $b^*$ are color coordinates in three dimensions in the CIELAB color space, while the values of $C^*$ and $h°$ are measures of saturation and hue. $L^*$ represents the lightness values of the wool, with higher values indicating brighter or lighter colors, and lower values indicating darker colors. The $a^*$ value represents the red–green axis, with positive values indicating more red colors and negative values indicating more green colors. The $b^*$ value represents the yellow–blue axis, with positive values indicating more yellow colors and negative values indicating more blue colors.

In general, as the concentration of dye increases, the values of $L^*$ tend to decrease, indicating an increase in the darkness of the color. This is particularly evident for raw wool, where the values of $L^*$ decrease significantly as the dye concentration increases. However, for tannery wool, this trend is not as pronounced. From Table 1,

**Table 1** $L^*$, $a^*$, $b^*$, $C^*$, and $h°$ values of dyed raw and tannery wool at different concentrations of the Acid Blue 25

| D C (%) | Raw wool | | | | | Tannery wool | | | | |
|---|---|---|---|---|---|---|---|---|---|---|
| | $L^*$ | $a^*$ | $b^*$ | $C^*$ | $h°$ | $L^*$ | $a^*$ | $b^*$ | $C^*$ | $h°$ |
| 0.1 | 45.60 | – 7.53 | – 20.07 | 21.43 | 249.44 | 61.11 | – 11.21 | – 13.13 | 17.26 | 229.50 |
| 0.25 | 40.93 | – 5.97 | – 26.97 | 27.62 | 257.52 | 54.43 | – 9.94 | – 16.25 | 19.05 | 238.54 |
| 0.5 | 39.59 | – 6.69 | – 25.27 | 26.14 | 255.18 | 40.21 | – 7.07 | – 25.94 | 26.88 | 254.76 |
| 1 | 33.78 | – 3.92 | – 27.15 | 27.43 | 261.78 | 39.18 | – 8.23 | – 21.88 | 23.38 | 249.39 |
| 1.5 | 30.98 | – 0.37 | – 29.43 | 29.43 | 269.27 | 33.27 | – 3.35 | – 26.93 | 27.14 | 262.90 |
| 2 | 29.21 | 0.34 | – 29.18 | 29.18 | 270.66 | 31.05 | – 1.50 | – 27.98 | 28.02 | 266.92 |

it can be observed that at low concentrations of dye (0.1 and 0.25%), raw wool generally has lower $L^*$ values than tannery wool, indicating that tannery wool is brighter in color than raw wool. However, at higher concentrations of dye (0.5% and above), the difference in $L^*$ values between raw wool and tannery wool becomes smaller. This suggests that tannery wool may have a lower capacity for absorbing dye at lower concentrations, but at higher concentrations, it can behave similarly to raw wool.

Regarding the values of $a^*$ and $b^*$, there seem to be subtle but consistent variations in the hue of the color. For example, as the concentration of dye increases, raw wool tends to take on a more red hue. An increase in dye concentration results in an increase in color saturation ($C^*$) for both types of wool, while the hue angle ($h°$) varies depending on the type of wool.

Recovered wool from tanning waste can be different from raw wool in terms of texture, color, and chemical composition. This difference can impact how the wool is dyed. Firstly, the recovered wool from the tannery can be altered by the chemical treatment it underwent during tanning. For example, the recovered wool may contain residue of chemicals such as tanning agents, acids or solvents, which can affect how color is absorbed and fixed during the dyeing process. Therefore, it may be necessary to thoroughly clean the recovered wool before dyeing to eliminate all chemical residues.

Moreover, the tannery wool may have a different base color from that of raw wool due to the chemical treatment it underwent. The base color of the tannery wool may be more uniform than that of raw, as the chemical treatment may have eliminated natural color variations in the raw wool. However, this can also mean that the tannery wool is more difficult to dye, as it may have fewer dye-fixation sites.

Finally, the texture of the tannery wool may differ from that of raw wool. The tannery wool may be more compact and dense than raw wool due to the tanning process. This can make the recovered wool more difficult to dye as it may require more time to absorb and fix dyes.

## 4 Conclusion

The results of the present research study prove that raw wool has a higher adsorption capacity than tannery wool at lower dye concentrations, but both types of wool behave similarly at higher concentrations. This may be due to differences in the structure and properties of raw and tannery wool fibers, as well as the available dye concentration for adsorption.

In summary, wool from the tannery can differ from raw wool in terms of texture, color, and chemical composition, which can affect how it is dyed. It is important to take these differences into account during the dyeing process to achieve the desired results.

## References


1. Olfa, A., Taoufik, H., Riadh, Z., &Slah, M. The Valorization Potential of Tannery Wool Waste in the Textile Industry. Journal of Natural Fibers,1(20). (2023) DOI: https://doi.org/10.1080/15440478.2022.2146251.
2. Moghassem, A. R., & A. A. Gharehaghaji. Evaluating pile yarn characteristics in hand woven carpet using stress-strain behavior in compression. International Journal of Engineering, Transactions B: Applications 21 (3):303–12. (2008).
3. Abubakar, M., Yusuf, M., & Ismail, M. Equilibrium, kinetics and thermodynamic adsorption studies of acid dyes on adsorbent developed from kenaf core fiber. Adsorption Science & Technology, 1–19. (2017) DOI: https://doi.org/10.1177/0263617417715532.
4. Jahan, M.S., Chowdhury, D.A.N., Hasan, M.R., & Hossain, M.M. Utilization of jackfruit (Artocarpusheterophyllus) waste as a potential source of food fiber: a review. Journal of Food Processing and Preservation, 39(1), 11-20. (2015).DOI: https://doi.org/10.1111/jfpp.12244.
5. Helal, A., & M. M. Mourad. Pulled wool as a recycled material. World Applied Sciences Journal 7 (6), 693–98. (2009)
6. Safapour, S., Rather, L.J., &Mazhar, M. Coloration and functional finishing of wool via Prangos-ferulacea plant colorants and bioactive agents: Colorimetric, fastness, antibacterial, and antioxidant studies. Fibers and Polymers, 24, 1379-1388. (2023). DOI: https://doi.org/10.1007/s12221-023-0828-6.

# Smart Textile Industry: Threats and Opportunities

**Aref Meddeb**

**Abstract** Smart textiles, also known as e-textiles or electronic textiles, are fabrics that integrate sensors, actuators, and communication devices, to create new functionalities. These are used in a variety of applications, from sports, entertainment, and fitness, to healthcare, safety, and military mission critical applications. The integration of Industry 4.0 and smart textiles can lead to significant advances in manufacturing and product development. Smart textiles can enable new product functionalities, such as monitoring of vital signs or biometric data, and provide valuable insights into the collected information. While the integration of Industry 4.0 and smart textiles represents a significant opportunity for innovation and growth in various domains, it comes with some tricky security challenges. Thus, there is an urgent need for the development of new skill sets among stakeholders. This paper provides a comprehensive overview of the challenges encountered in the context of smart textiles. We advocate the need to adopt straightforward, streamlined, simple, and standard solutions to provide smart, seamless, scalable, safe, and secure user experience.

**Keywords** Smart · Cyber · Security · Textiles · 4.0 Industry

## 1 Introduction

Smart textiles can provide enhanced functionality, such as monitoring vital signs or providing location-based services, which can improve quality of life [1]. Smart textiles can also be used in safety applications, such as monitoring worker health in hazardous environments or providing alerts in emergency situations. Increased connectivity is a key feature of smart textiles, which can connect to devices such as smart phones or other wearables, enabling seamless integration and communication. Smart textiles can further provide a wealth of data that can be analyzed to gain insights into user behavior, preferences, and health conditions.

A. Meddeb (✉)
National Engineering School of Sousse, University of Sousse, Sousse, Tunisia
e-mail: aref.meddeb@eniso.u-sousse.tn.com



S. B. Abdessalem et al. (eds.), *Proceedings of the Second International Conference of Innovative Textiles and Developed Materials-ITDM'2; 05-06 May 2023; Tunisia*,
https://doi.org/10.1007/978-981-99-7950-9_23

The integration of Industry 4.0 and smart textiles can lead to significant advances in manufacturing and product development. For example, the use of sensors and data analysis can help to optimize production processes, reduce waste, and improve product quality. The combination of Industry 4.0 and smart textiles can enable the creation of more personalized products, which can be customized based on individual preferences and needs. This can lead to greater customer satisfaction and improved loyalty.

While the integration of Industry 4.0 and smart textiles represents a significant opportunity for innovation and growth in various industries, it also presents challenges, such as the need for increased security and the development of new skill sets among employees. Smart textiles can collect and transmit sensitive data, such as personal health or location information. The risk of unauthorized access, data breaches, and misuse of this information is a significant concern. Proper encryption, access controls, and data anonymization techniques should be implemented to protect user privacy.

Smart textile products often include embedded sensors, microcontrollers, and communication modules, which can be vulnerable to malware attacks and hacking attempts. Manufacturers should ensure that these devices have robust security measures, such as secure boot mechanisms and regular firmware updates, to mitigate the risk of unauthorized access or control during the entire product lifecycle.

Further, smart textiles rely on wireless communication protocols, such as Bluetooth or Wi-Fi to transmit data to other devices. These wireless transmissions can be intercepted if proper encryption and authentication mechanisms are not in place. Adequate encryption protocols and secure communication channels should be implemented to prevent data interception and tampering.

Furthermore, the smart textile industry involves complex supply chains with various components and suppliers. The risk of compromised components, counterfeit products, or insecure manufacturing processes can lead to vulnerabilities in the final products. Manufacturers need to ensure the security and integrity of the supply chain to minimize the risk of introducing security weaknesses. In fact, smart textiles are expected bring more comfort, safety, and security to our lives than threats and discomfort.

This paper provides a comprehensive overview of the threats encountered in the context of Smart Textile. In Sect. 2, we describe the enabling technologies used in smart textiles. In Sect. 3, we describe some applications of smart textile. Section 4 describes cybersecurity issues raised by smart textiles. Section 5 presents some standards and regulatory measures for smart textiles. We conclude the paper by summarizing the most important findings and by providing some future research directions.

## 2 Smart Textile Enabling Technologies

Smart textiles rely on various technologies to integrate electronics and functionality into fabric or textile materials [2]. These technologies work together to transform traditional textiles into intelligent and functional products. Some key enabling technologies used in smart textiles include:

- Conductive materials that enable the transmission of electrical signals and are used in smart textiles to create circuits and sensors. Conductive materials can be incorporated into textiles using conductive fibers or by coating the textile with conductive polymers or metals.
- Sensors used to detect and measure various physical parameters such as temperature, pressure, humidity, and motion. These sensors can be incorporated into textiles to create smart garments that can monitor the wearer's vital signs, activity levels, and other biometric data.
- Actuators that convert electrical signals into physical actions. These are used in smart textiles to create garments that can respond to environmental changes or user inputs. Examples of actuators include shape memory alloys, electroactive polymers, and piezoelectric materials.
- Power sources that operate the electronic components. Power sources can be integrated into the garment, such as batteries or solar panels, or they can be provided through external devices such as wireless charging pads.
- Wireless communication: Smart textiles can communicate with other devices wirelessly through technologies such as Bluetooth, Wi-Fi, or Radio Frequency ID (RFID). This allows for real-time data collection and analysis, as well as remote control of the garment's functions.
- Data processing and storage: Smart textiles generate a large amount of data that needs to be processed and stored. This can be done through microcontrollers or other embedded computing devices that are integrated into the garment.

The components of a smart textile fabric are illustrated in Fig. 1. By integrating electronics, sensors, communication capabilities, and data processing into textile fabrics, smart textiles can offer a wide range of applications, including healthcare monitoring, sports performance tracking, fashion-tech, and interactive textiles for various industries.

### 2.1 Sensors Used in Smart Textiles

Smart textiles rely on various types of sensors to collect data and enable functionalities [3]. The specific sensor types used in smart textiles depend on the desired applications and the data needed for the intended functionality. Common sensors used in smart textile include:

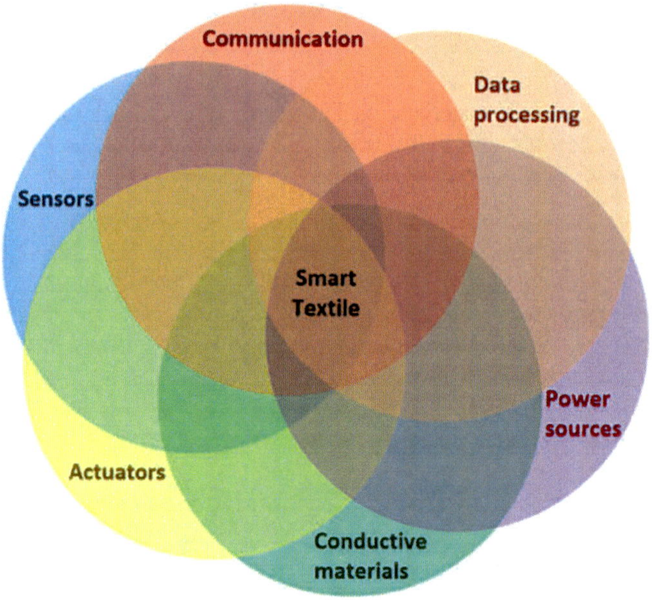


**Fig. 1** Smart textile fabric components

- Temperature sensors that can detect changes in temperature and can be used to monitor body and ambient temperature, or changes in temperature caused by physical activity.
- Pressure sensors that can detect changes in pressure and can be used to monitor posture, movement, and pressure points.
- Stretch sensors that can detect changes in fabric stretch and can be used to monitor movement and position.
- Accelerometers that can detect changes in acceleration and can be used to monitor motion and position.
- Electrocardiogram (ECG) sensors that can be used to monitor heart rate and rhythm.
- Electroencephalogram (EEG) sensors that can be used to monitor brain activity and detect changes in cognitive state.
- Light sensors that detect changes in light levels and can be used to monitor exposure to light or changes in ambient lighting.
- Moisture sensors that can detect changes in moisture levels and can be used to monitor things like sweat levels and hydration.
- Gas sensors that can detect changes in gas levels and can be used to monitor air quality or detect the presence of specific gases.
- Proximity sensors can detect the presence of nearby objects and can be used to monitor things like distance and movement.

These sensors, when integrated into smart textiles, enable various functionalities such as health monitoring, activity tracking, environmental sensing, immersive, and interactive experiences.

## 2.2 Conductive Materials Used in Smart Textile

Conductive materials play a key role in enabling the integration of electronics and electrical pathways within smart textiles [4]. These materials allow for the electricity flow and facilitate the connection of electronic components. Some commonly used conductive materials in smart textiles include:

- Conductive yarns, which are made with conductive fibers such as metallic fibers, carbon fibers, or conductive polymers. They can be woven, knitted, or embroidered into the fabric to create conductive pathways.
- Conductive inks, which contain conductive particles such as silver, copper, or carbon. They can be printed onto the fabric to create conductive traces, circuits, or sensors.
- Conductive coatings, which contain conductive materials such as metal oxides, carbon nanotubes, or graphene. They can be applied to the fabric to create conductive surfaces or electrodes.
- Conductive adhesives, which contain conductive particles such as silver or copper. They can be used to attach electronic components to the fabric or to create conductive joints between different parts of the fabric.
- Conductive foams, which contain conductive particles such as carbon black or metallic particles. They can be used to create soft and flexible electrodes or sensors.
- Conductive films, which are thin layers or sheets made from conductive materials, such as metals or conductive polymers. These films can be laminated onto textiles or incorporated as adhesive layers. Conductive films provide a flexible and conductive surface for attaching electronic components or creating conductive pathways.
- Conductive elastomers, which are stretchable materials infused with conductive particles, allowing them to maintain conductivity even under stretching or deformation. These materials are used in applications that require both flexibility and conductivity, such as stretch sensors or wearable electrodes.
- Conductive nanomaterials, such as carbon nanotubes or graphene, which offer exceptional conductivity and mechanical properties at the nanoscale. They can be incorporated into textiles through various methods, such as coating, printing, or composite fabrication. Conductive nanomaterials provide high-performance conductivity and enable advanced functionalities.
- Smart fibers and coated fibers are textile fibers that are infused or coated with conductive materials. These fibers can be integrated into textiles during the spinning or weaving process, providing conductive pathways, or sensing capabilities within the fabric structure.

These conductive materials enable the integration of electronic components, sensors, and conductive pathways into smart textiles, allowing for functionalities such as sensing, actuation, data transmission, and energy management. The choice of conductive materials depends on factors like conductivity requirements, flexibility, durability, as well as the manufacturing processes.

## 2.3 Actuators for Smart Textiles

Smart textiles typically incorporate a variety of actuators, which are devices that convert electrical, thermal, or mechanical energy into physical actions such as movement, sound, or light [5]. These actuators are responsible for enabling the smart textile to respond to its environment in various ways. These are components that execute physical actions or produce a response to external stimuli. Common actuators include:

- Shape memory alloys (SMAs)—These are materials that can change shape in response to heat or electrical currents. They are often used in smart textiles to create fabrics that can change shape or stiffness in response to a stimulus.
- Electroactive polymers (EAPs)—These are materials that can change their shape or size in response to an electrical current. They are often used in smart textiles to create fabrics that can contract or expand in response to an electrical signal.
- Piezoelectric materials—These are materials that generate an electrical charge when they are subjected to mechanical stress. They are often used in smart textiles to create fabric that can generate an electrical signal when it is stretched or compressed.
- Shape memory polymers (SMPs)—These are materials that can change shape in response to a stimulus, such as heat or light. They are often used in smart textiles to create fabrics that can change shape or stiffness in response to a specific environmental condition.
- Micro-electromechanical systems (MEMS)—These are small devices that can be integrated into smart textiles to provide various functions. They combine mechanical and electrical components on a very small scale. MEMS devices typically consist of microstructures, sensors, actuators, and electronics integrated onto a single chip.
- In some cases, smart textiles utilize pneumatic or hydraulic systems as actuators. Soft inflatable chambers or channels can be embedded within the fabric to enable controlled inflation or deflation, resulting in shape changes or mechanical movements. Pneumatic or hydraulic actuators can be controlled by valves and pumps to provide dynamic functionality.

The choice of actuators used in a particular smart textile depends on the desired functionality and the specific application of the textile.

### 2.4 Power Sources Used in Smart Textile

Power sources include batteries or solar panels, or they can be provided through external devices such as wireless charging pads [6]. Power can also be generated by movement or pressure. The power sources used in smart textiles depend on the specific application and the type of electronic components involved. Common power sources include:

- Batteries: Rechargeable or non-rechargeable batteries can be integrated into smart textiles to provide power for electronic components such as sensors, microcontrollers, and LEDs.
- Solar cells: Photovoltaic cells can be incorporated into the fabric to convert sunlight into electrical energy, which can power the electronic components.
- Piezoelectric materials: These materials can generate electrical energy when they are mechanically deformed. This energy can be used to power small electronic devices.
- Thermal energy: Thermoelectric materials can convert temperature differences into electrical energy, which can be used to power small electronic devices.
- Wireless power transfer: Smart textiles can be designed to receive power wirelessly from a nearby power source, such as a charging pad or a wireless power transmitter.

Often, a combination of power sources and energy harvesting techniques is used to provide continuous or sustainable power to smart textiles. The choice of power source depends on the specific requirements of the smart textile application, such as the amount of power needed, the duration of operation, and the environmental conditions.

## 3 Smart Textile Applications

Smart textiles have a wide range of applications across various industries [7]. They find extensive use in the healthcare sector [8]. They can be used for monitoring vital signs, such as heart rate, respiration rate, and body temperature. These textiles can incorporate sensors and conductive fibers to collect data and transmit it wirelessly to medical professionals or monitoring systems. Smart textiles can also be used for pressure sensing in bedsore prevention, postural monitoring, and drug delivery systems.

Smart textiles have also made significant advancements in sports and fitness applications [9]. They can monitor biometric data, such as heart rate, breathing rate, and movement, to provide insights into an individual's performance and optimize training regimes. Smart textiles can be integrated into sports apparel, such as shirts, socks, and compression garments, to enhance comfort, monitor hydration levels, or provide muscle support.

In [10], textile engineers have developed a fabric woven from ultra-fine nano-yarns composed in part of phase-change materials and other advanced substances that combine to produce a fabric that can respond to temperature changes to heat and cool its wearer as needed. By combining the yarns with electro-thermal and photo-thermal coatings that enhance the effect, they essentially developed a fabric that can both quickly cool the wearer and warm them up when conditions change. The T-shirt can also be controlled by the smart phone. A piece of fabric that can keep track of how much the wearer is sweating, and then "open" to cool them down is described in [11].

Moreover, smart textiles have a growing presence in the fashion industry, enabling innovative and interactive designs [12]. They can incorporate LED lights, flexible displays, or color-changing properties to create visually appealing and customizable garments. Smart textiles in fashion can also include temperature-regulating fabrics, moisture-wicking properties, or embedded sensors for enhanced comfort and functionality.

Safety and protection are other example of smart textile applications to enhance personal safety and monitor environmental conditions. For instance, they can integrate sensors to detect harmful gases, temperature extremes, or impacts, providing early warnings or activating safety mechanisms. Smart textiles are also used in firefighter suits, military uniforms, and protective clothing for industrial workers, offering enhanced functionality, comfort, and safety features.

Smart textiles are also employed in the automotive and aerospace industries to improve passenger comfort, safety, and functionality. They can be integrated into car seats to monitor the driver's vital signs, detect drowsiness, or adjust seating positions. In aerospace, smart textiles can be used for pressure sensing in seats, temperature regulation, or embedding antennas for wireless communication within the aircraft. For example, Dutch custom textile manufacturer By Borre (https://byborre.com/) has developed the fabrics for BMW's future interactive car interior. The cabin's interior upholstery was turned into a tactile three-dimensional knitted surface interface. The embedded smart materials offer the passenger an integrated, naturally intuitive, and interactive experience of control and demand over the car's interior [13].

Further, smart textiles can be utilized in home textiles and interior design applications to enhance comfort and functionality. They can include self-cleaning fabrics, energy-efficient curtains that adjust transparency based on light levels, or textiles with integrated sensors for home automation. Smart textiles can also be incorporated into bedding for sleep monitoring or temperature regulation.

Furthermore, smart textiles play a vital role in military and defense applications, providing enhanced capabilities and protection to soldiers. They can include fabrics with camouflage properties, ballistic protection, and integrated sensors for monitoring vital signs or detecting chemical agents. Smart textiles are used in military uniforms, body armor, and wearable communication systems.

Moreover, smart textiles enable new modes of human–computer interaction, such as gesture recognition or touch-sensitive interfaces. Fabrics with integrated sensors or conductive materials can turn clothing into interactive surfaces, allowing users to control devices or access information through simple gestures or touch interactions.

Smart textiles can also be designed to generate electricity through various mechanisms. These electricity-generating smart textiles are often referred to as energy harvesting textiles [14]. Further, a full-duplex enabled wireless power transfer system via textile for miniaturized IMD is presented in [15]. Finally, searchers from China's Fudan University built a fabric that can turn into a working keyboard, light up like a display, or even help send texts all while being bent, folded, and even washed like any other piece of clothing [16].

## 4 Cybersecurity in the Smart Textile Context

The smart textile industry has experienced rapid growth in recent years with the development of innovative fabrics that incorporate electronics, sensors, and other technologies. While these advancements offer exciting opportunities for enhancing functionality and connectivity, they also come with significant cybersecurity threats [17]. It is essential to prioritize cybersecurity in the design, development, and deployment of smart textiles to minimize the risks and maximize the benefits.

### *4.1 Cyber Threats*

Since unsuspected communication and storage devices can be disseminated in clothes, hair, garments, or even inside the body, this raises some new threats such as broadcasting malicious software or capturing and storing sensitive information on hidden writable devices. With the large number of connected devices that one may wear, it is very hard to keep track of their individual status and it is easy to forget some of them, leaving the door open to unsuspected attacks. One can be connected without even knowing it.

Attacks of the future will behave like "natural disasters" that might be predictable but unavoidable, and users will be asked to "get off the net" when these attacks are close to them. There will be seasons of wide scale attacks, like flu seasons, or isolated attacks, like headaches.

When someone travels with his wearable devices, he may catch a virus and bring it home, contaminating all the devices including his car, fridge, and smart TV. We may need agents which will act like antibiotics and security guards behaving like vaccines.

A smart shirt with embedded sensors could be hacked to steal the wearer's biometric data, such as their heartbeat and breathing rate. One could intercept the data transmission between the shirt and its companion app, and then use the data to create a fake biometric profile of the wearer. A smart glove with embedded sensors could be hacked to control a drone. One could manipulate the glove's sensors to mimic the movements of a person's hand, and then use those movements to control the drone's flight path.

Smart textiles collect and transmit a vast amount of sensitive data, such as personal health information or location data, making them attractive targets for cybercriminals. Malware can be embedded in smart textiles through compromised software or hardware, enabling cybercriminals to take control of the device, steal data, or damage the fabric's functionality. Smart textiles may be targeted by physical attacks, such as tampering or theft, which can compromise the integrity and security of the fabric. The smart textile industry is relatively new, and there is a lack of regulation around cybersecurity, making it challenging to ensure the security and privacy of end users.

Smart textile components may constitute several sources of vulnerabilities. Sensors, actuators, and conductive devices may pose serious threats. If a hacker takes control of a sensor, he/she might order an actuator to excessively increase pressure on the body or increase/decrease the temperature above/below normal. Conductive yarns, ink or foam may be physically altered, causing unexpected behavior of the wearable fabric. While it is very easy to get rid of a smart phone or tablet when it is hacked, it may be less straightforward to remove a "smart" T-Shirt in public places.

## *4.2 Privacy Versus Security*

Connected wearable devices raise an overwhelming number of security and privacy challenges [17]. Tiny and numerous connected devices raise obvious security and privacy concerns. While security issues are quite straightforward, mainly from background knowledge, privacy issues are more complex and may constitute challenging obstacles to large-scale deployment of connected devices. In fact, unprotected personal information may expose sensitive and embarrassing data to the public.

*Privacy may be an anomaly*, Vinton Cerf said [18]. Privacy constitutes a rather tricky and intriguing concept even for ITC experts. Technology have evolved far beyond any expectations, and we are not prepared to deal with it. McNealy further pointed out: "You have zero privacy anyway. Get over it!" [19].

Consumers may want to have the ability to control the security level they use. From an industry perspective, privacy is a matter of user conduct and responsibility. Consumers need to be trained to understand how by storing their personal data on various devices, they expose themselves to various types of attacks. Consumers may not be able to figure out if they are being "watched." Often, there is no mean for users to know whether their personal data are being tracked or "stolen" by third parties.

The contextual nature of consumers' comfort level with data sharing reveals that acceptance depends on device and on the perceived value in return for sharing the data. For instance, sharing location information with Google Maps may be accepted by travelers but not by residents. Also, in disaster situations, users are much more likely to divulgate their locations and identities.

## 4.3 Impact on Health

It is important to be aware of the potential health risks associated with the use of smart textiles and to take steps to minimize these risks. This might include choosing products made with safe and non-irritating materials, properly maintaining and calibrating devices, and limiting exposure to electromagnetic radiation. It is also important to use only standard and homologated materials and equipment. Common health threats include:

- Radio waves exposures, which may harm human tissues and cells if the specific absorption rate (SAR) is exceeded or if they are exposed for long periods of time [20]. This can be especially concerning if the smart textiles are worn close to the body, such as fitness clothing or wearable medical garments. SAR provides a straightforward means for measuring the RF exposure characteristics of on-body connected devices to ensure that they are within the safety guidelines.
- Skin irritation: Some smart textiles are made with materials that can cause skin irritation or allergic reactions, especially if they are worn for long periods of time. This is particularly true if the smart textiles contain chemical components or are made with synthetic materials.
- Disruption of sleep patterns: Smart textiles that emit light can disrupt the body's natural sleep patterns and lead to sleep disturbances.
- Psychological effects: The use of smart textiles can also have psychological effects, such as anxiety or stress related to constant monitoring or tracking.
- Inaccurate data: If smart textiles are not properly calibrated or maintained, they can provide inaccurate data that could lead to incorrect diagnoses or treatment recommendations.
- Malfunction or failure: Like any electronic device, smart textiles can malfunction or fail, potentially leading to inaccurate data or even injury. For example, a smart garment that malfunctions and overheats could cause burns.
- Addiction or dependency: The constant monitoring and tracking provided by smart textiles could lead to addiction or dependency on the technology, with negative consequences for mental and physical health.

## 4.4 Designing Adequate Security Solutions

Designing adequate security solutions constitutes a rather challenging task, even for the most skilled developer [21]. Since many consumers often use easy to remember and similar (and often identical) passwords in various systems, protecting data becomes somewhat a utopia. Vinton Cerf: "figuring out how to make a security system work well that doesn't require the consumer to be an expert is a pretty big challenge."

Nevertheless, designing adequate security solutions for smart textiles is essential to ensure the privacy and protection of the data and functionalities associated with these textiles. Some considerations and strategies for designing robust security solutions for smart textiles include:

- Authentication and access control to ensure that only authorized users can access and control the smart textile's features. This can involve user authentication through passwords, biometric data (such as fingerprints or facial recognition), or wearable authentication devices like smart watches or RFID tags.
- Data encryption to protect the data transmitted between the smart textile and external devices or networks. Encrypting the data ensures that it cannot be easily intercepted or accessed by unauthorized parties. Strong encryption algorithms and secure key management practices should be employed.
- Use secure communication protocols, such as HTTPS or MQTT with Transport Layer Security (TLS), to establish secure connections between the smart textile and external devices or networks. These protocols encrypt the data during transmission and provide authentication and integrity checks to prevent data tampering.
- Ensure that personal data collected by smart textiles is handled with utmost privacy. Implement privacy policies and practices that adhere to applicable data protection regulations. Minimize the collection and storage of sensitive data and provide transparency and control to users over their personal information.
- Regularly update the firmware or software of the smart textile to address security vulnerabilities and apply patches for known issues. Establish a secure update mechanism to prevent unauthorized or malicious updates that could compromise the system.
- Consider physical security measures to protect the smart textile from unauthorized physical access or tampering. This can include secure enclosures or tamper-evident features to deter unauthorized opening or manipulation of the textile's components.
- Implement intrusion detection and prevention systems to monitor and detect any unauthorized access attempts or malicious activities targeting the smart textile. This can involve technologies like anomaly detection, behavior analysis, or network monitoring to identify and respond to security incidents promptly.
- Educate users about the security features and best practices for using smart textiles. Provide guidelines on secure usage, such as avoiding connecting to untrusted networks, using strong passwords, and being cautious of phishing or social engineering attempts.
- Adhere to established security standards and guidelines specific to smart textiles and IoT devices, such as the IoT Security Foundation's IoT Security Compliance Framework or industry-specific standards like ISO/IEC 27001 for information security management.

By incorporating these security measures and best practices, the risks associated with smart textiles can be mitigated, ensuring the privacy, integrity, and security of both the user and the data transmitted or stored by the textile [21].

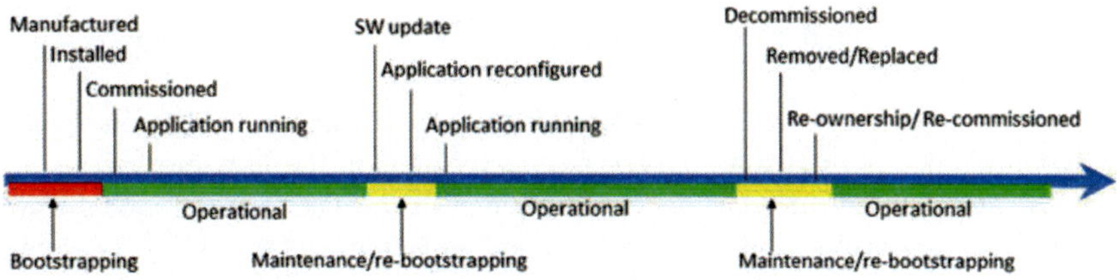


**Fig. 2** Lifecycle of a thing

### *4.5 Life Cycle of a Thing*

Some key new security features that go beyond current paradigms are to consider the lifecycle of a thing as shown in Fig. 2. Things such as fabrics and wearables go through various stages during their lifecycle: *Manufactured*, *Installed*, *Commissioned*, *Running*, *Updated*, and *Reconfigured*. In the manufacturing and installation phases, the thing is *Bootstrapped,* while during the commissioning and running phases, the thing is *Operational*. In each stage, security credentials and ownership information need to be updated.

Before we design, develop, or use smart textiles, we should understand their lifecycle. This suggests knowing what data was/is/will be collected, stored, processed, transmitted, shared, and deleted. Users and developers should be aware of the potential threats and vulnerabilities that can compromise the data at any stage, such as hacking, tampering, interception, unauthorized access, or loss. By understanding the lifecycle, we can identify the risks and implement the appropriate measures to mitigate them.

## 5 Smart Textiles Standards

Standards play a crucial role in the development, manufacturing, and integration of smart textiles. They provide guidelines, specifications, and protocols that ensure interoperability, safety, and quality of smart textile products. Standardization of smart textile is still in its infancy. Nevertheless, some very interesting actions are being taken to put in place a general framework for terminology, systems, and requirements for smart textile applications.

### *5.1 Mainstream Standards for Smart Textiles*

- **ASTMD8248-20**: *Standard Terminology for Smart Textiles.* Provides terminology and definitions related to smart textiles such as electrical textiles and wearable electronics are included as well as the fiber, yarn, and fabric that compose them, and end products.

- **BS EN IEC 63203-101-1:2021**: *Wearable electronic devices and technologies Terminology.* Provides terminology frequently used in the literature related to wearable electronic devices and technologies in the IEC 63,203 series. This list includes wearable electronic devices and technologies, near-body wearable electronics, on-body wearable electronics, in-body wearable electronics, and electronic textiles.
- **BS EN IEC 63203-201-1:2022**: *Wearable electronic devices and technologies Electronic textile. Measurement methods for basic properties of conductive yarns.* Specifies provisions and test methods for measurement of properties of conductive yarns. Conductive yarns covered in this document have conductivity of a level that can be used for transmission of electric signals, supply of electric power and electromagnetic shield. They do not include high-resistance conductive yarn used for antistatic and heating use. Conductive yarns are the basic material in electronic textiles and are mainly used as conductive traces in clothes-type wearable devices, as well as with secondary processing (woven, knitted, embroidered, non-woven, etc.) to provide conductive fabrics. This document does not define the required characteristics of the conductive yarn; rather, it specifies the handling and measurement methods for general and electrical properties of conductive yarn.
- **IEC 63203-201-2:2022**: *Wearable electronic devices and technologies Electronic textile. Measurement methods for basic properties of conductive fabrics and insulation materials.* Specifies the provisions for conductive fabrics and insulation materials used for electronic textiles and measurement methods for their properties. Conductive fabrics covered by this document are basic materials in electronic textiles and are mainly used as conductive traces, electrodes, and the like in clothes-type wearable devices. This document does not cover high-resistance conductive fabrics used for antistatic purposes and heater applications. Insulating materials handled in this document are materials used for electrical insulation of conductive parts in electronic textiles. They include materials for covering the conductive parts, and general fabrics constituting the basic structure of clothes-type wearable devices. This document does not define the required characteristics of the conductive fabric and insulation materials; rather, it specifies measurement methods for general and electrical properties of the conductive fabric and insulation materials.
- **BS EN IEC 63203-201-3:2021**: *Wearable electronic devices and technologies Electronic textile. Determination of electrical resistance of conductive textiles under simulated microclimate.* Specifies a test method for the determination of the electrical resistance of conductive fabrics under simulated microclimate within clothing. The microclimate is the climate of the small air layer between the skin and clothing having a specific temperature and humidity. This test method can be applied to conductive fabrics including multilayer assemblies for use in clothing.
- **BS EN IEC 63203-204-1:2023—TC: Tracked Changes**: *Wearable electronic devices and technologies Electronic textile. Test method for assessing washing durability of e-textile products.* Specifies a household washing durability test

method for e-textile products. This document includes testing procedures for e-textile products with electrically conductive components and sensors to collect the data of the user.

- **PD CEN ISO/TR 23383:2020**: *Textiles and textile products — Smart (Intelligent) textiles — Definitions, categorization, applications, and standardization needs (ISO/TR 23383:2020).* Provides definitions in the field of "smart" textiles and textile products as well as a categorization of different types of smart textiles. It briefly describes the current stage of development of these products and their application potential and gives indications on preferential standardization needs.
- **PD CEN/TR 17945:2023**: *Textiles and textile products. Textiles with integrated electronics and ICT. Definitions, categorization, applications, and standardization needs.* Provides definitions in the field of e-textiles and electronic textile systems, as well as the categorization of different types of electronic textiles and electronic textile systems. It briefly describes the current stage of development of these products and their application potential and gives indications on preferential standardization needs. This document also provides guidelines to determine general verification of claimed performance, innocuousness, durability of properties, product information, and environmental aspects of textile electronics. This document is not intended for products which are placed inside or are (permanently) attached to the human body. It also does not specifically address the electronics information communication link between the textile with integrated electronics and external data processing. This document therefore also does not focus on the design of software to be implemented in electronic textiles of textile systems.
- **PD IEC TR 63203-250-1:2021**: *Wearable electronic devices and technologies Electronic textile. Snap fastener connectors between e-textiles and detachable electronic devices.* This document reviews the use cases of conductive snap fasteners applied as electrical connectors for e-textile products available on the market and provides guidance on future standardization work.
- **PD IEC/TR 62899-250:2016**: *Printed electronics Material technologies required in printed electronics for wearable smart devices.* This part of IEC 62899, which is a Technical Report (TR), explores a new technological field to establish standardization activities in TC 119 (Printed electronics) in particular, and to contribute to the development and market expansion of wearable smart device (WSD) technology.
- **IEEE 360-2022:** *IEEE Standard for Wearable Consumer Electronic Devices — Overview and Architecture.* Provides an overview, terminology, and categorization of wearable consumer electronic devices (wearables). An architecture for a series of standard specifications that define technical requirements and testing methods for different aspects of wearables — from basic security and suitableness of wear to various functional areas, such as health, fitness, and infotainment — is further outlined.
- **Bluetooth SIG**: The Bluetooth Special Interest Group (SIG) sets standards for Bluetooth wireless communication technology. Bluetooth standards are crucial for the interoperability of smart textiles with other devices, such as smart phones, tablets, or wearable devices, enabling seamless connectivity and data exchange.

- **NFC Forum**: The NFC (Near Field Communication) Forum develops standards for NFC technology, which enables short-range wireless communication between devices. NFC standards are relevant to smart textiles with NFC capabilities, allowing them to interact with NFC-enabled devices for applications like authentication, payment systems, or data transfer.

## *5.2 Regulation of Smart Textiles*

- Specific regulations and standards may vary across countries and regions. Organizations such as international standardization bodies, industry associations, and government agencies often collaborate to develop harmonized standards and guidelines for smart textiles to facilitate global trade and ensure consistent product quality and safety. Here are some considerations regarding the regulatory aspects of smart textiles:
- **Privacy and Data Protection**: Smart textiles that collect and transmit personal data are subject to privacy and data protection laws. In the EU, the General Data Protection Regulation (GDPR) regulates the collection, processing, and storage of personal data. In the USA, laws such as the California Consumer Privacy Act (CCPA) and the Children's Online Privacy Protection Act (COPPA) provide similar protections.
- **Product Safety**: Smart textiles must also comply with product safety regulations. In the EU, the General Product Safety Directive (GPSD) sets safety requirements for all consumer products, including smart textiles. In the USA, the Consumer Product Safety Commission (CPSC) regulates consumer products to ensure their safety.
- **Electromagnetic Compatibility** (EMC): Smart textiles that emit or receive electromagnetic signals must comply with EMC regulations to ensure that they do not interfere with other electronic devices. In the EU, the EMC Directive regulates electromagnetic compatibility, while in the USA, the Federal Communications Commission (FCC) regulates the use of the electromagnetic spectrum.
- **Medical Devices**: If a smart textile is intended to be used as a medical device, it must comply with regulations set by the relevant regulatory agencies. In the United States, the Food and Drug Administration (FDA) regulates medical devices, including smart textiles that are intended for medical use. The EU has similar regulations in place, such as the Medical Device Regulation (MDR) and the In-vitro Diagnostic Medical Device Regulation (IVDR).
- **Energy Efficiency**: Smart textiles often incorporate batteries, energy harvesting systems, or power management technologies. Energy efficiency standards and regulations may be in place to ensure that smart textiles optimize energy usage and minimize environmental impact.
- **Labeling**: Regulations may require clear labeling and product information for smart textiles. This includes information about the functionality, limitations, maintenance requirements, and any potential risks associated with the product. Proper

labeling helps consumers make informed decisions and understand how to safely use and maintain smart textiles.

- **Intellectual Property**: Intellectual property rights, including patents and trademarks, play a role in the regulation of smart textiles. Companies and inventors may seek patents to protect their technological advancements, while trademarks can be used to differentiate their smart textile products in the market.

## *5.3 Risk Assessment*

The risk assessment should consider the entire lifecycle of the product, from design and development to manufacturing, use, and disposal. The development of smart textiles involves incorporating electronic components and advanced technologies into fabric, which can introduce new risks and challenges compared to traditional textiles. It is important to conduct a risk assessment when working with smart textiles to identify potential hazards and develop strategies to mitigate them.

Once the risks have been identified, strategies can be developed to mitigate or eliminate them.

This may involve design modifications, improved manufacturing processes, the use of protective equipment, or the development of user instructions and warnings. Regular monitoring and evaluation should also be conducted to ensure that risks are effectively managed throughout the lifecycle of the product. Combining electronics and textiles, which both are relatively short-lived mass consumer goods, would intensify product obsolescence and lead to even shorter life cycles and abandonment of products. Some key smart textile risk assessment includes:

- **Health and safety risks**: The use of certain materials or chemicals in smart textiles may pose health risks to workers and end users, such as skin irritation or allergic reactions.
- **Electrical and fire hazards**: Smart textiles may involve electrical components and batteries, which can pose risks of electrical shock, fire, and explosions if they are not designed and manufactured correctly.
- **Cybersecurity risks**: Smart textiles that are connected to the internet or other networks may be vulnerable to cyberattacks, which could compromise user privacy and data security.
- **Environmental risks**: The use of certain materials or chemicals in smart textiles may pose risks to the environment during manufacturing and disposal.
- **Regulatory and standard compliance**: Smart textiles may be subject to various regulations and standards, such as those related to electrical safety, electromagnetic compatibility, and environmental protection.
- **Functional reliability**: Smart textiles rely on the proper functioning of integrated electronic components and systems. Failure or malfunctioning of these components can impact the intended functionality, leading to potential safety risks or inaccurate data.

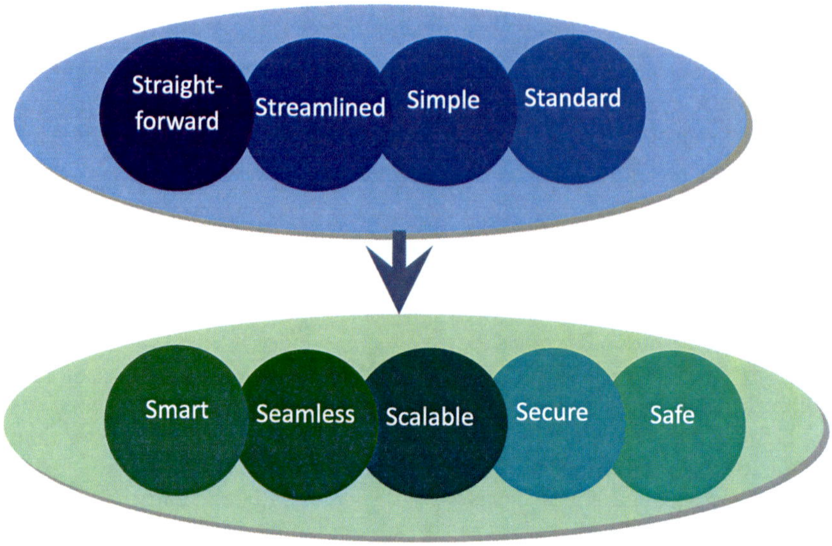


**Fig. 3** The 9S design paradigm

### 5.4 *The 9S Design Paradigm*

End users seek *smart*, *seamless*, *scalable*, *secure*, and *safe* solutions to widely adopt smart textile solutions. To this end, we need to consider only straightforward, streamlined, simple, and standard solutions as illustrated in Fig. 3. Such solutions are more viable options to provide a great user experience than proprietary ones, leading to acceptance and confidence among stakeholders.

## 6 Conclusion

Smart textiles have already started to revolutionize the textile industry, and they are expected to have a bright future. One of the major areas where smart textiles are expected to make a significant impact is in the healthcare industry. Smart textiles can be used to develop wearable medical devices that can monitor vital signs, detect health issues, and even deliver drugs or other treatments directly to the patient's body.

The future of smart textiles is promising, and we can expect to see a plethora of new and innovative applications emerging in the coming years as the technology continues to evolve and mature. Smart textiles have the potential to transform the fashion industry by offering new ways to personalize clothing and accessories.

With the help of embedded sensors and other technologies, clothing could be made to adapt to the wearer's body temperature, adjust to different lighting or weather conditions, or change colors in response to different stimuli.

Smart textiles are also expected to play a role in the automotive and aerospace industries. Textiles with integrated sensors could be used to monitor the structural integrity of vehicles, while fabrics with built-in heating elements could help to reduce energy consumption in cars and planes.

Nevertheless, smart textile raises some serious concerns with respect to safety, privacy, and security. All stakeholders including manufacturers, end users, providers, regulators, and standard bodies must cooperate to increase confidence in smart textile products and raise awareness of the potential risks when these are misused. Regular updates, usage of adequate security credentials, proper commissioning and decommissioning during the lifecycle of the product must be meticulously applied.

## References


1. Jansen, K.M.B.: Smart textiles: how electronics merge into our clothing. In Proceedings of the 20th International Conference on Thermal, Mechanical and Multi-Physics Simulation and Experiments in Microelectronics and Microsystems (EuroSimE), Hannover, Germany (2019).
2. Ruckdashel, R.R., Khadse, N., Park, J.H.: Smart E-Textiles: Overview of Components and Outlook. Sensors (Basel) 22(16), 6055 (2022).
3. Nesenbergs, K., Selavo, L.: Smart textiles for wearable sensor networks: Review and early lessons. In proceedings of IEEE International Symposium on Medical Measurements and Applications (MeMeA), pp. 402–406. Turin, Italy, (2015).
4. Gunnarsson, E.: Conductive Fabrics for Textile Electronic Interconnections and Capacitive Sensing - A Smart Textiles Perspective. 2017.
5. Persson, N.K., Martinez, J.G., Zhong, Y., Maziz, A., Jager, E. W. H.: Actuating Textiles: Next Generation of Smart Textiles. Adv. Mater. Technol. 3, 1700397, (.2018).
6. Beeby, S., Arumugam, S., Hillier, N., Li, Y., Shi J., Sun, Y., Wagih, M., Yong, S.: Power supply sources for smart textiles. In Smart Clothes and Wearable Technology (Second Edition), The Textile Institute Book Series, Woodhead Publishing (2023).
7. Wen, W.: A Study of the Intelligent Design of Smart Textile in Future Life. In proceedings of International Conference on Intelligent Design (ICID), pp. 172–175, Xi'an, China, (2020).
8. Kubiceket J., et al.: Recent Trends, Construction, and Applications of Smart Textiles and Clothing for Monitoring of Health Activity: A Comprehensive Multidisciplinary Review. IEEE Reviews in Biomedical Engineering 15, pp. 36-60 (2022).
9. Zewe, A.: Smart textiles sense how their users are movin, MIT News Office, July, 7, (2022).
10. Wu, J. Wang, M., Dong, L., Shi, J., Ohyama, M., Kohsaka, Y., Zhu, C., Morikawa. H.: A Trimode Thermoregulatory Flexible Fibrous Membrane Designed with Hierarchical Core–Sheath Fiber Structure for Wearable Personal Thermal Management. ACS Nano 16 (8), 12801 (2022).
11. Brian Rohrig.Don't Sweat It: How Moisture-Wicking Fabrics Keep You Cool and Dry, Chem Matters 2022 https://www.acs.org/education/resources/highschool/chemmatters/articles/dont-sweat-it-how-moisture-wicking-fabrics-keep-you-cool-and-dry.html.
12. Barrera, T.: Smart Textiles In Fashion: What They Are, Types & Exciting Examples. https://thetechfashionista.com/what-are-smart-textiles-and-examples/.
13. Textile Technology Source, "BMW unveils interactive textile interior," *ATA publications* (2020).
14. Chen et al.: Smart Textiles for Electricity Generation" Chem. Rev. 120, 8 ( 2020).
15. Lee, J., Bae, B., Kim, B. et al. : Full-duplex enabled wireless power transfer system via textile for miniaturized IMD. Biomed. Eng. Lett. 12, 295–302 (2022).

16. Cheng, K.: Chinese Scientists Invent Illuminating Fabric That Can Display Text. www.pandaily.com.
17. Özlem, K., Kuyucu, M.K., Bahtiyar, S., İnce, G.: Security and Privacy Issues for E-textile Applications. In proceedings of 4th International Conference on Computer Science and Engineering (UBMK), pp. 102–107, Samsun, Turkey (2019).
18. Ferenstein, G.: Google's Cerf Says, "Privacy May Be An Anomaly. Historically, He's Right", (2013). https://techcrunch.com/2013/11/20/googles-cerf-says-privacy-may-be-an-anomaly-historically-hes-right/
19. Sprenger, P.: Sun on Privacy: 'Get Over It' (1999). https://www.wired.com/1999/01/sun-on-privacy-get-over-it/.
20. Vallozzi, L., Hertleer, C., Rogier, H.: Latest developments in the field of textile antennas. In Smart Textiles and their Applications, Woodhead Publishing Series in Textiles, Woodhead Publishing (2016).
21. Textile Industry, "How do you protect the data and privacy of smart textile and wearable technology users? https://www.linkedin.com/advice/0/how-do-you-protect-dataprivacy-smart-textile?trk=cq.

# 4.0 Industry & Textile Processes

# Fabric Texture Reconstruction Via Multilayer Dictionary Learning

**Bo Xing, Qingqing Shao, Xianyi Zeng, and Jun Wang**

**Abstract** Fabric texture reconstruction is a critical task in the textile industry, with applications in computer-aided design, quality control, and production optimization. However, traditional methods suffer from limitations such as poor accuracy, slow convergence, and low robustness. To address these issues, this study proposes a novel approach that combines multilayer dictionary learning-based encoding and variational autoencoder (VAE) for fabric texture reconstruction. The proposed method utilizes multilayer dictionary learning-based encoding to extract deeper fabric texture features and generates new fabric images with similar texture features using VAE. Experiments on a publicly available dataset of fabric images demonstrate the superiority of the proposed method in terms of accuracy, robustness, and efficiency compared to state-of-the-art methods. The proposed method has significant implications for the textile industry.

**Keywords** Texture reconstruction · Multilayer dictionary learning · Variational autoencoder

## 1 Introduction

Fabric texture reconstruction [1] is a critical task in the textile industry with significant implications for computer-aided design, quality control, and production optimization. The complexity of fabric textures, which are characterized by intricate patterns and a high degree of variability, makes this task particularly challenging.

B. Xing (✉) · Q. Shao · J. Wang
College of Textiles, Donghua University, Shanghai 201620, China
e-mail: bo.xing@ensait.fr

J. Wang
e-mail: junwang@dhu.edu.cn

B. Xing · X. Zeng
GEMTEX-Génie Des Matériaux Textiles, Université de Lille, ENSAIT, 59000 Lille, France



S. B. Abdessalem et al. (eds.), *Proceedings of the Second International Conference of Innovative Textiles and Developed Materials-ITDM'2; 05-06 May 2023; Tunisia*,
https://doi.org/10.1007/978-981-99-7950-9_24

Traditional methods for fabric texture reconstruction have been largely based on image processing techniques. These methods often involve the use of filters or transform to extract texture features from fabric images, followed by a reconstruction process to generate a representation of the original fabric texture. However, these methods often suffer from limitations such as poor accuracy, slow convergence, and low robustness.

In recent years, machine learning techniques have been increasingly applied to the problem of fabric texture reconstruction. These techniques, which include deep learning and dictionary learning, have shown promise in improving the accuracy and efficiency of fabric texture reconstruction. Similarly, dictionary learning has been used for sparse representation of images and has shown potential in texture analysis. In light of the above, we propose a novel approach that combines multilayer dictionary learning-based encoding and VAE for fabric texture reconstruction. The proposed method overcomes the limitations of traditional methods and achieves high accuracy and reliability in fabric texture reconstruction.

## 2 Experiments Details

The fabric texture images were collected to form a dataset. We collected 300 images of pieces of fabric, subdivided into plain, satin, and twill weave fabrics. These images were resized to $56 \times 56$ pixel. Data augmentation techniques, including rotation, zooming, and horizontal flipping, were applied to generate a total of 18,000 samples. The dataset was split into 15,000 training images and 3000 testing images. Hyperparameters, including a learning rate of 0.001, a batch size of 32, and a dropout rate of 0.5, were selected based on empirical tuning and the literature recommendations (Fig. 1).

## 3 Methods

The proposed approach for fabric texture reconstruction consists of the following steps:

### *3.1 Multilayer Dictionary Learning-Based Encoding*

Multilayer dictionary learning is utilized to extract fabric texture features from input images. The K-SVD algorithm is used to learn the dictionary and represent the texture features as sparse codes. Multiple K-SVD layers [2] are stacked to extract deeper texture features, providing a more comprehensive representation of the underlying fabric texture features.

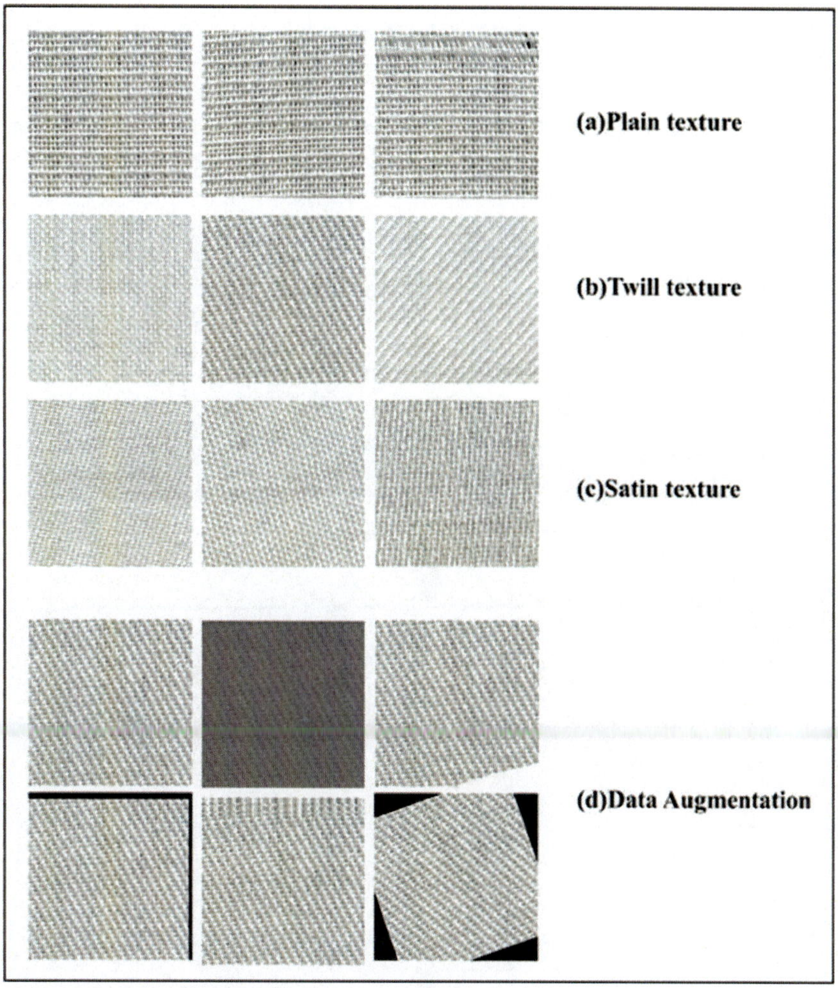


**Fig. 1** Few samples from datasets

A single level of dictionary learning plays an important role in latent representation of data and dictionary atoms which is provided by multilevel dictionaries in the deep dictionary learning structure. In classic dictionary learning, $X$, $D_1$, and $Z$ are data, dictionary, and sparse representation, respectively. In the synthesis, dictionary is firstly learned. Then, sparse representation is obtained according to the dictionary.

$$X = D_1 Z \tag{1}$$

Multilayer basic deep dictionary learning is shown in Fig. 2. Mathematical representation is given in Eq. 2:

$$X = D_1 D_2 Z \tag{2}$$

Following the suggested model and training algorithm proposed by Tariyal et al. [3] we took the gradient of the norm with respect to $\gamma$, require that it equals zero,

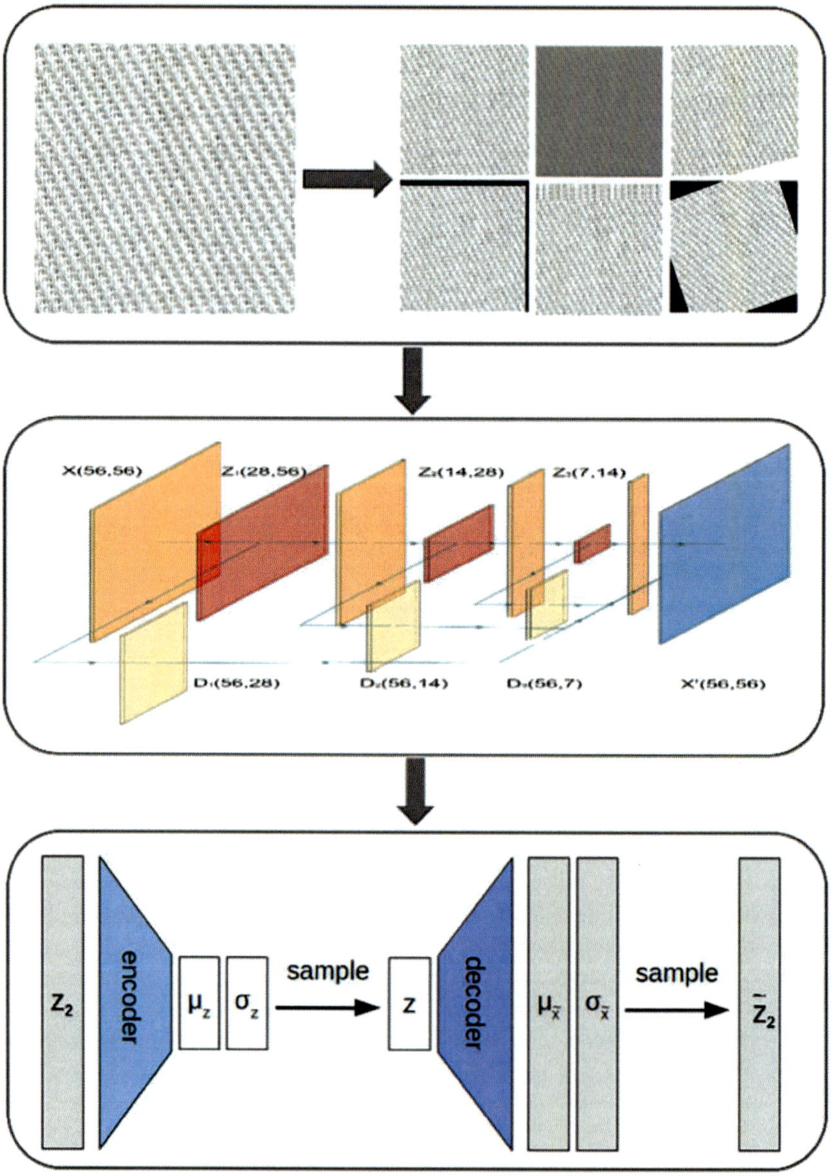


**Fig. 2** Proposed method for fabric texture reconstruction

and solve for the optimal $\gamma$. Therefore, solving the optimal $\gamma$ yields:

$$z_2 = \left(D_1 D_1^T\right)^{-1} D_1 z_1 \quad (3)$$

Similarly, we can solve for the optimal $D_1$ using least squares, and we obtain:

$$D_1 = \left(z_2 z_2^T\right)^{-1} z_2 z_1^T \quad (4)$$

In a similar manner, we can extend the above formulations to a batch of observed signals.

### 3.2 Variational Autoencoder-Based Texture Generation

After the encoding process, the texture features are fed into the decoder part, which consists of a VAE-based [4] texture generation network. The VAE-based texture generation process is composed of an encoder and a decoder network. The encoder network maps the input features into the latent space, where the mean and variance of the distribution of the latent features are computed. Then, the decoder network generates new fabric images by sampling the latent features from the distribution and decoding them into images.

### 3.3 Regularization Terms

To further improve the quality and robustness of the reconstructed fabric textures, regularization terms are introduced in both the dictionary learning and VAE training processes. The L1 norm of the sparse codes is used as a regularization term in the dictionary learning process, and the KL divergence [5] between the predicted and actual distributions of the latent features is used as a regularization term in the VAE training process.

The proposed approach is trained on a dataset of fabric images using a combination of the above steps. During the training process, the input fabric images are encoded using multilayer dictionary learning-based encoding, and the resulting texture features are used to generate new fabric images using the VAE-based texture generation network. The regularization terms are used to improve the performance and robustness of the reconstructed fabric textures.

## 4 Results

Experiments on a publicly available dataset of fabric images demonstrate that the proposed approach achieves superior performance in terms of accuracy, robustness, and efficiency compared to state-of-the-art methods. Multilayer dictionary learning-based encoding significantly improves the quality of the reconstructed fabric textures, as it captures deeper and more detailed fabric texture information. Additionally, the VAE-based texture generation process ensures the generated textures are realistic and accurate. The regularization terms introduced in the approach further improve the performance and robustness of the reconstructed fabric textures (Fig. 3).

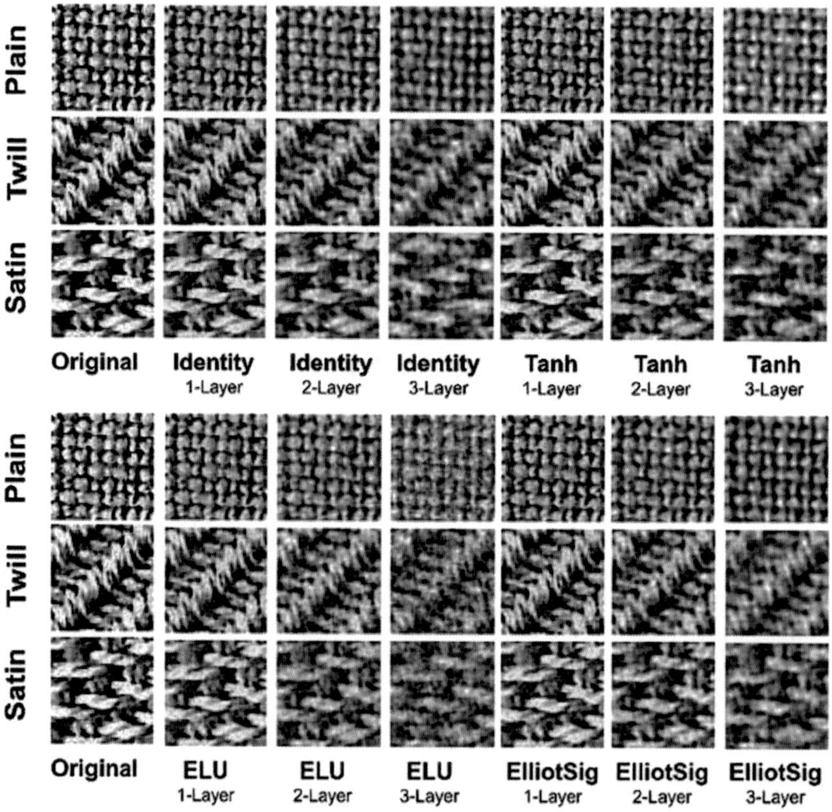


**Fig. 3** A visualization of the reconstructions for proposed models

**Table 1** Comparing reconstruction performance with varying activation functions and generative models

| Activation function-layers | Recons loss (RL) | AE-RL | GAN-RL | Proposed model-RL |
|---|---|---|---|---|
| Identify-1 | 1.122 | 1.153 | 1.345 | **0.861** |
| Identify-2 | 5.763 | 6.249 | 7.855 | 5.165 |
| Identify-3 | 30.786 | 14.256 | 10.569 | 8.149 |
| Tanh-1 | 2.367 | 3.101 | 3.879 | 1.876 |
| Tanh-2 | 5.964 | 6.501 | 7.865 | 7.254 |
| Tanh-3 | 10.452 | 15.219 | 12.328 | 9.357 |
| ELU-1 | 2.557 | 3.321 | 4.231 | 4.657 |
| ELU-2 | 8.621 | 9.315 | 10.418 | 10.234 |
| ELU-3 | 18.249 | 17.892 | 16.152 | 11.846 |
| ElliotSig-1 | 1.512 | 2.389 | 3.214 | 2.236 |
| ElliotSig-2 | 6.841 | 7.623 | 8.734 | 8.982 |
| ElliotSig-3 | 12.118 | 18.908 | 13.908 | 12.187 |

## 5 Discussion

The proposed approach effectively combines multilayer dictionary learning-based encoding and VAE to achieve high-quality and robust fabric texture reconstruction. The multilayer dictionary learning-based encoding extracts deeper fabric texture features and provides a more comprehensive representation of the underlying fabric texture features. By doing so, the proposed approach is able to capture more detailed fabric texture information and generate more accurate and realistic reconstructed fabric textures. Moreover, the regularization terms introduced in the approach improve the performance and robustness of the reconstructed fabric textures (Table 1).

The proposed approach has significant implications for the textile industry, as it can be used for computer-aided design, quality control, and production optimization. The accurate reconstruction of fabric textures can help designers create realistic fabric models, identify and correct fabric defects, and optimize fabric production processes.

## 6 Conclusion

This study presents a novel approach that combines multilayer dictionary learning-based encoding and VAE for fabric texture reconstruction. The proposed approach effectively overcomes the limitations of traditional methods and achieves high accuracy and reliability in fabric texture reconstruction. The results of experiments on a publicly available dataset of fabric images demonstrate the superiority of the proposed approach in terms of accuracy, robustness, and efficiency compared to state-of-the-art methods. The proposed approach has significant implications for the textile industry and can be applied in various applications such as computer-aided design, quality control, and production optimization.

**Acknowledgements** This study was funded by the China Scholarship Council Fund.

## References

1. Bernardini, F., I.M. Martin, and H. Rushmeier, *High-quality texture reconstruction from multiple scans*. IEEE Transactions on Visualization and Computer Graphics, 2001. **7**(4): p. 318–332.
2. Xing, B., et al. *Application of deep dictionary learning in automatic classification of woven fabric texture*. in *Machine Learning, Multi Agent and Cyber Physical Systems: Proceedings of the 15th International FLINS Conference (FLINS 2022)*. 2023. World Scientific.
3. Tariyal, S., et al., *Deep dictionary learning*. IEEE Access, 2016. **4**: p. 10096–10109.
4. Kingma, D.P. and M. Welling. *Stochastic gradient VB and the variational auto-encoder*. in *Second international conference on learning representations, ICLR*. 2014.
5. Bu, Y., et al., *Estimation of KL divergence: Optimal minimax rate*. IEEE Transactions on Information Theory, 2018. **64**(4): p. 2648–2674.

# Application of the Desirability Function for Optimizing the Use of Calcium Carbonate Used in the Production Process of PVC-Coated Textile

**Stambouli Mouna, Chaouech Walid, Gargoubi Sondes, Zouari Riadh, and Msahli Slah**

**Abstract** The application of calcium carbonate ($CaCO_3$) particles as functional fillers has been a judicious idea for many industrial plastic products in recent years. This paper deals with an evaluation of the quality of coated textiles used as a furnishing article, by changing the amount and the size of calcium carbonate particles using the desirability function. Referring to the literature studies, the desirability function remains an appropriate method that allows a level of flexibility over a graphical tool in defining and evaluating quality. It was concluded that the ideal mechanical properties of polymeric layers used for PVC furnishing articles (breaking load, tearing strength, elongation to break, and abrasion resistance) largely depend on calcium carbonate amount and particle size where the smallest particle size gave the highest values compared to the others. For the content effect analysis, the applied contents influenced mechanical properties significantly.

**Keywords** Coated textile · PVC · Synthetic leather · Layer · Mechanical properties

## 1 Introduction

The coating becomes increasingly important to add value to textiles. It aims to develop the functional properties of textiles and to improve certain characteristics [1].

The coating is used to guarantee that fabrics meet performance parameters that would not be achievable from uncoated fabrics [1].

Today, coated fabrics are essentially polymer-coated textiles, and the principles of polymers used for industrial products with textile substrates are polyvinyl chloride and polyurethane.

S. Mouna (✉) · C. Walid · G. Sondes · Z. Riadh · M. Slah
Textile Engineering Laboratory-LGTex, ISET Ksar Hellal, Monastir, Tunisia
e-mail: Mounastambouli@gmail.com



S. B. Abdessalem et al. (eds.), *Proceedings of the Second International Conference of Innovative Textiles and Developed Materials-ITDM'2; 05-06 May 2023; Tunisia*,
https://doi.org/10.1007/978-981-99-7950-9_25

Polyvinylchloride (PVC)-coated fabrics also called artificial leather fabrics are widely used in daily life [2], and they represent a group of lightweight materials currently used in a wide variety of industries, So far, the significance of their properties is much more accentuated due to their application in different domains such as footwear, automotive, upholstery, and clothing. These fabrics usually consist of a topcoat, middle coat, base coat, and backing cloth [3].

The advantage of PVC leather fabrics is easy processing, low cost, consistent appearance, and others [4]. Their main components are polymer (PVC), stabilizer, plasticizer, and filler. These components are uniformly mixed to form plastisol [5], and they are very important for the general behavior of PVC synthetic leather. The filler characteristics are a key issue in determining many technical properties of PVC. Fillers are also used to make final products cheaper [6].

Approximately 80% of the filler used in PVC is calcium carbonate ($CaCO_3$). Titanium dioxide is second at about 12%, followed by calcined clay at about 5%. The remains are other materials, including glass and talc [7].

Calcium carbonate is one of the most common fillers for the PVC leather industry [7]. The special properties of this product, such as its low cost as well as its availability, encourage industries to enhance its performance and optimize its use [8].

Traditionally, $CaCO_3$ filler was considered an additive, which, because of its reduced surface area and unfavorable geometrical features, was used to lower the cost, increase the melt viscosity, and moderately increase the modulus of the final product, whereas strength tensile and deformability remained unchanged or even decreased in some cases [9].

Recently, the particle shape and size distribution, the content of filler, and the degree of dispersion have been reported to exhibit markedly additional properties over PVC leather filled with calcium carbonate [10]. The outstanding properties of the final product are attributed to the small particle size and large interfacial area of the calcium carbonate particles, which lead to strong interfacial adhesion between the filler and the plastisol [5].

Extensive research work has been carried out to correlate the particle size and the content of calcium carbonate with the properties of PVC materials. Shuisheng et al. [8] showed that the tensile and impact strength of PVC greatly increases with decreasing $CaCO_3$ particle size. Nakamura et al. [9] observed that the yield stress of PVC composite decreases while increasing the filler content.

Conducted research has generally been carried out on PVC plastics and composites [1, 3, 5, 8], and very little work has been performed concerning the PVC-coated fabrics [11].

In order to find a suitable method for a practical application of this filler, this work aims to investigate the effects of calcium carbonate content and particle size on microstructural properties, breaking load, tearing strength, and elongation to break of PVC-coated fabrics. Desirability function is used to optimize this multiresponse system.

**Table 1** Different types of $CaCO_3$ used with different particle sizes

| Filler type | Particle size(μ) |
|---|---|
| *Fine natural calcium carbonate (Type 1)* | |
| 5S | 2.7 |
| 2S | 1.8 |
| *Ultra-fine coated calcium carbonate made from extremely white marble (Type 2)* | |
| 5MC | 2.5 |
| 2AMC | 1.4 |
| *Extremely fine natural calcium carbonate made from the marble of exceptional whiteness (Type 3)* | |
| 2M | 1.8 |
| 95M | 0.9 |

# 2 Materials and Methods

## 2.1 Raw Materials

PVC resin, plasticizer (DINP), stabilizer, $CaCO_3$ fillers with different types and particle sizes (Table 1), blowing agent (azodicarbonamide), kicker, transfer paper, and textile fabric were generously donated by PLASTISS Company (Monastir, Tunisia).

## 2.2 Synthesis of PVC Synthetic Leather

In this work, PVC plastisol has been used to create PVC foamed layers. 100 parts PVC resin, 80 parts DINP, four parts azodicarbonamide, two parts kicker, and 1.5 parts stabilizers were homogeneously combined in a mechanical stirrer to create PVC plastisol. The fillers (calcium carbonate) were then added at 25, 50, 75, 100, or 125% (by weight) and stirred until a uniform mixture was achieved.

To create PVC synthetic leather, we used the transfer coating technique. On the first coating head, a blade is used to spread the plastisol on the transfer paper while adjusting the thickness. The created layer called the superficial or skin layer is then dried at 140 °C for 20 s and chilled. On the second coating head, the plastisol of the second layer called foamed layer is applied to the superficial layer using a blade at a specific thickness. The knitted poly-cotton fabric is afterward attached to the resulting PVC layers, and the complex therefore created is then placed in the main furnace at 180 °C for 80 s where cellular expansion and gelling take place.

## 2.3 Desirability Approach

The desirability function is widely used to optimize an industrial process and especially to learn which combination of input variables provides the most satisfactory output.

This method is used to find the best formulation that can provide a product that can satisfy different properties.

In this case, the PVC synthetic leather layers used for furnishing products require the satisfaction of different characteristics at the same time.

These characteristics are essentially mechanical properties including tearing resistance, breaking strength, and elongation to break.

The desirability function approach transforms each response function $f_i$ to a desirability function di; it is a value between 0 and 1.

$$\begin{aligned} & f_i\colon [Y_{\min}, Y_{\max}] \rightarrow [0, 1] \\ & Y_i \rightarrow d_i(Y_i) \end{aligned} \tag{1}$$

Three types of desirability functions are proposed: desirability function to minimize, maximize, and achieve a target range of the property [10].

To maximize, we use the following:

$$d_i = Y_{\text{target}} \begin{cases} 0 \ \text{ if } Y_i < Y_{\min} \\ \left((Y_i - Y_{\min})/\left(Y_{\text{target}} - Y_{\min}\right)\right)^S \ \text{ if } Y_{\min} < Y_i \\ 1 \ \text{ if } Y_i > Y_{\text{target}} \end{cases} \tag{2}$$

To minimize this, we use the following:

$$di = Y_{\max} \begin{cases} 1 \ \text{ if } Y_i > Y_{\text{target}} \\ \left((Y_{\max} - Yi)/\left(Y_{\max} - Y_{\text{target}}\right)\right)^S \ \text{ if } Y_{\text{target}} < Y_i \\ 0 \ \text{ if } Y_i > Y_{\max} \end{cases} \tag{3}$$

To achieve a target range of the property:

$$di = \begin{cases} 0 \ \text{ if } Y_i < Y_{\min} \\ \left((Y_i - Y_{\min})/\left(Y_{\text{target}} - Y_{\min}\right)\right)^S \ \text{ if } Y_{\min} < Y_i < Y_{\text{target}} \\ \left((Y_{\max} - Y_i)/\left(Y_{\max} - Y_{\text{target}}\right)\right)^S \ \text{ if } Y_{\text{target}} < Y_i < Y_{\max} \\ 0 \ \text{ if } Y_i > Y_{\max} \end{cases} \tag{4}$$

Here, $Y_{\min}$ is the lower value, $Y_{\max}$ is the upper value, $Y_{\text{target}}$ is the target value, and $S$ is the exponent determining how important it is to attain the target value. $Y_i$ is the input vector.

The overall desirability function is made by multiplying all other functions by the power of the inverse of $R$, where $R$ is the number of functions.

$$\mathrm{DG} = \left(\prod_{r=1}^{R} \mathrm{d}r\right)^{1/R(1)} \tag{5}$$

## 3 Results and Discussions

### 3.1 Quality Assessment

To meet the customer demands of the furniture sector, the main properties that characterize PVC synthetic leather layers have been defined, under the direction of the PLASTISS company managers, which are mainly mechanical properties (strength at break, elongation at break, abrasion resistance, and tear resistance).

### 3.2 Choice of the Tolerance Interval

The standard PLASTISS furnishing article is considered to have "good" mechanical properties, which practically satisfies customer demands; nevertheless, improvement of its mechanical properties is still desired.

Twenty tests for each layer, still using the standard PLASTISS formulation, were carried out. The different mechanical tests were then carried out, the smallest value among the results found was considered a minimum value, and the average value found was considered a target value.

It has been found that for the superficial layer, the elongation at break must have a value greater than 201% but not more than 270%, as extra flexibility may affect the quality of the finished article.

Any value above 92 N is acceptable for breaking strength, any value above 5 N is approved for tear strength, and a loss of mass not exceeding 25% of the initial mass of the sample is allowed for abrasion resistance.

For the expanded layer, any value above 82% for the elongation at break is acceptable, but any value above 150% is rejected because an excessively soft expanded layer affects the quality of the final article, with 94% being the target value.

Any value above 0.52 N is acceptable for tearing strength, and any value above 17 N is allowed for breaking strength.

### 3.3 Optimization of the Production Process of Coated Textiles Used in Furniture Manufacture

Our study aims at experimenting with the process of valorization of the use of calcium carbonate in the manufacturing process of PVC synthetic leather used for furnishing articles.

For this purpose, three types of calcium carbonate with two different particle sizes each were used, and the amount of calcium carbonate was varied from 0 to 125 g in the superficial and expanded layer formulations. Thus, 31 samples for each layer loaded with the different types of calcium carbonate at different particle sizes and amounts were studied.

**Individual Desirability**. The desirability function is based on the idea that the quality of a product that requires several features is unacceptable if any of them is outside a "desirable" limit. Its objective is to find a good compromise between the different characteristics to provide the right solution [12].

Thus, to define the overall desirability of each layer, the desirability of each separate mechanical property, called individual desirability, must, first of all, be studied [13].

*Individual Desirability for the Superficial Layer*. In our case, for the elongation at the break, we have to reach a certain target, and the breaking strength and the tear resistance should be maximized. On the other hand, the abrasion resistance expressed in terms of mass loss must be minimized.

dAll, dFr, dRd, dRab, the series of graphs showing the individual desirability of each property characterizing the superficial layer are shown in Fig. 1.

*Individual Desirability for the Expanded Layer*. In our instance, for the elongation at the break we have to reach a certain target, the breaking strength and the tear resistance should be maximized.

dAll, dFr, dRd, the series of graphs showing the individual desirability of each property characterizing the expanded layer are shown in Fig. 2.

**Overall Desirability**. Once the $n$ variables (factors and response levels) are transformed into individual desirability functions, they are combined into a single function to find the best solution called overall desirability (D) [14].

$$D = \left(d_1^{r1} \times D_2^{r2} \times \cdots d_n^{rn}\right)^{1/\sum n} = \left(\prod_1^n d_i^{ri}\right)^{1/\sum n} \quad (6)$$

With $r_i$ is the importance of each variable to the others, according to Luciana Vera Candioti et al. [15]. This value is established by the analyst and can vary from 1 for the least important variable to 5 for the most important variable.

In our studies, the values of $r_i$ were assigned for the different mechanical properties to be checked (set according to the specifications of the furniture article) by

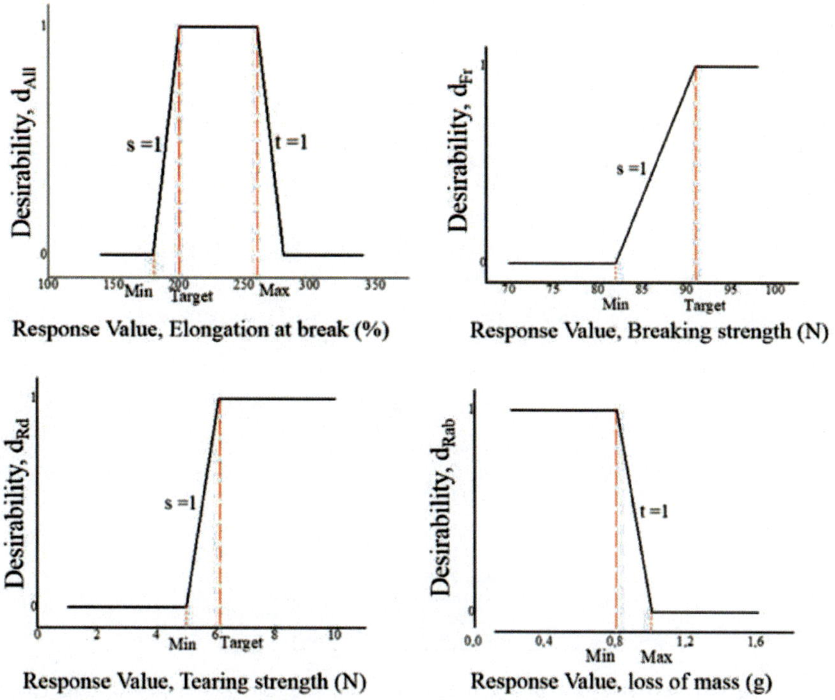


**Fig. 1** Graphical representation of the desirability functions for the different superficial layer properties

PLASTISS with the help of the company's experts and were established as follows regardless of the type of the layer (Table 2).

The value of *D* gives the overall assessment of the desirability of all the product characteristics; it lies in the interval [0–1] [16]. *D* increases if the balance of properties becomes more favorable, so it can be seen that the value of *D* increases when the desirability of the corresponding quality characteristic increases. In this way, a high *D* value means that the investigated sample fulfills almost all the required quality requirements [17]. Taking into consideration the opinions of PLASTISS experts, these conditions were set to find the best-performing product:

- If $D = 0$, the product quality is unacceptable.
- If $0.1 \le D \le 0.6$, this value is considered to be far from the target value 1, reflecting a lack in one or more properties that must necessarily be improved.
- If $D \ge 0.7$, the product quality is considered acceptable.

The value of *D* closest to 1 expresses the most desirable quality.

*Overall Desirability of the Superficial Layer*. The overall desirability *D* of the superficial layer varies between 0 and 0.95, and the diagram of the overall desirability of each sample is shown in Fig. 3 (Table 3).

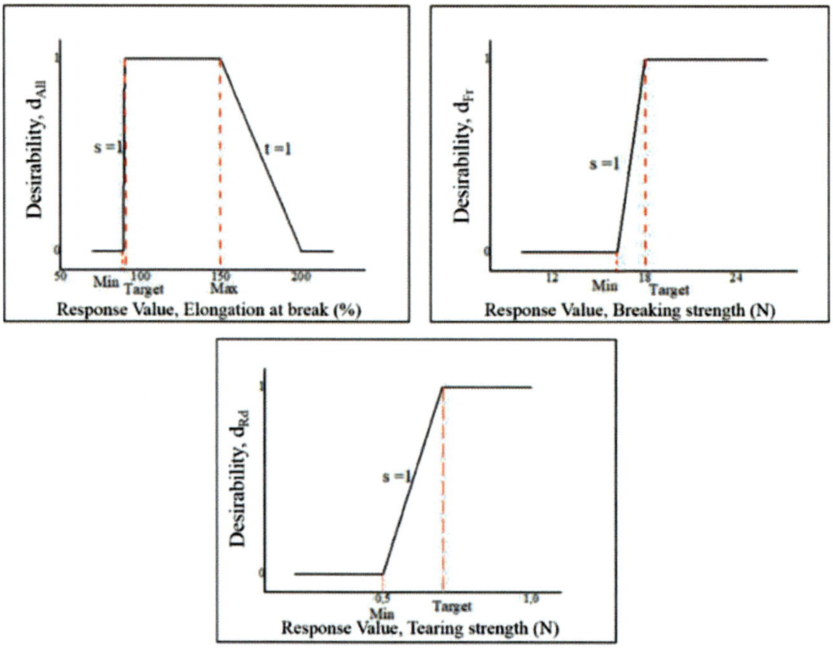


**Fig. 2** Graphical representation of the desirability functions for the different surface layer properties

**Table 2** Attributing weight to the different properties studied

| Tests | Weight ($r_i$) |
|---|---|
| Breaking load (N) | 5 |
| Elongation at break (%) | 3 |
| Abrasion resistance (Pete de masse (g)) | 5 |
| Tearing strength (N) | 5 |

Only the quality of the superficial layers Ps2, Ps26, and Ps32 is acceptable, so it can be concluded that for each type of calcium carbonate, only the amount of 25 g of small particle size filler in the superficial layer formulation will give the best final quality (Fig. 4 and Table 4).

*Overall Desirability of the Expanded Layer*

Only the quality of the expanded layers Pe2, Pe3, Pe4, Pe9, Pe15, Pe21, Pe26, and Pe33 is acceptable. Thus, it can be concluded that for each type of calcium carbonate, the amount of 50 g in the formulation of the foamed layer will give almost the best final quality.

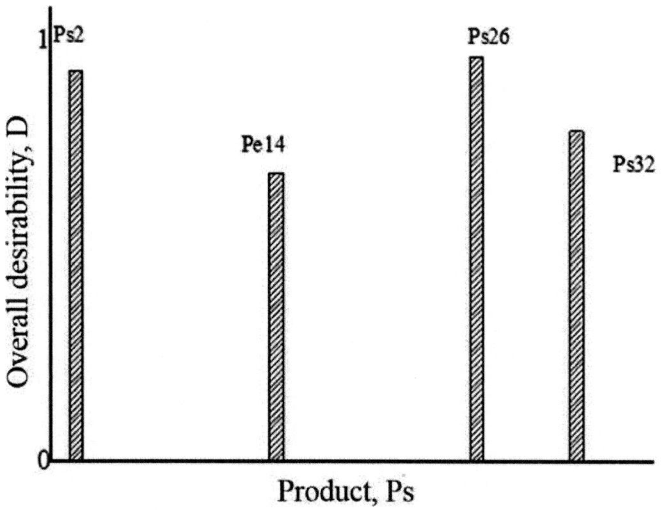


**Fig. 3** Diagram of the overall desirability of superficial layer samples

**Table 3** The overall desirability values for each sample of the superficial layer

| Particle size (μm) | | Product | dAbr | dAll | dRd | dFr | *D* |
|---|---|---|---|---|---|---|---|
| TYPE M | 0.9 | Ps1 | 0 | 0 | 0 | 0 | 0 |
| | | Ps2 | 1 | 0.9 | 1 | 1 | 0.93 |
| | | Ps3 | 0 | 0.4 | 0 | 0.4 | 0 |
| | | Ps4 | 0 | 0 | 0 | 0 | 0 |
| | | Ps5 | 0 | 0 | 0 | 0 | 0 |
| | | Ps6 | 0 | 0 | 0 | 0 | 0 |
| | 1.8 | Ps7 | 1 | 0 | 0 | 0 | 0 |
| | | Ps8 | 1 | 0 | 1 | 0 | 0 |
| | | Ps9 | 0 | 0 | 0 | 0 | 0 |
| | | Ps10 | 0 | 0 | 0 | 0 | 0 |
| | | Ps11 | 0 | 0 | 0 | 0 | 0 |
| | | Ps12 | 0 | 0 | 0 | 0 | 0 |
| TYPE MC | 1.4 | Ps13 | 1 | 0 | 0 | 0 | 0 |
| | | Ps14 | 1 | 0.83 | 1 | 0.83 | 0.69 |
| | | Ps15 | 0 | 0 | 0 | 0 | 0 |
| | | Ps16 | 0 | 0 | 0 | 0 | 0 |
| | | Ps17 | 0 | 0 | 0 | 0 | 0 |
| | | Ps18 | 0 | 0 | 0 | 0 | 0 |

(continued)

**Table 3** (continued)

| Particle size (μm) | | Product | dAbr | dAll | dRd | dFr | *D* |
|---|---|---|---|---|---|---|---|
| | 2.5 | Ps19 | 1 | 0 | 0 | 0 | 0 |
| | | Ps20 | 1 | 0.99 | 1 | 0 | 0 |
| | | Ps21 | 0 | 0 | 0 | 0 | 0 |
| | | Ps22 | 0 | 0 | 0 | 0 | 0 |
| | | Ps23 | 0 | 0 | 0 | 0 | 0 |
| | | Ps24 | 0 | 0 | 0 | 0 | 0 |
| TYPE S | 1.8 | Ps25 | 0 | 0 | 0 | 0 | 0 |
| | | Ps26 | 1 | 0.95 | 1 | 1 | 0.96 |
| | | Ps27 | 1 | 0 | 0 | 1 | 0 |
| | | Ps28 | 1 | 0 | 0 | 1 | 0 |
| | | Ps29 | 0 | 0 | 0 | 0 | 0 |
| | | Ps30 | 0 | 0 | 0 | 0 | 0 |
| | 2.7 | Ps31 | 0 | 0 | 0 | 0 | 0 |
| | | Ps32 | 1 | 0.73 | 1 | 1 | 0.79 |
| | | Ps33 | 1 | 0 | 0 | 1 | 0 |
| | | Ps34 | 0 | 0 | 0 | 0 | 0 |
| | | Ps35 | 0 | 0 | 0 | 0 | 0 |
| | | Ps36 | 0 | 0 | 0 | 0 | 0 |

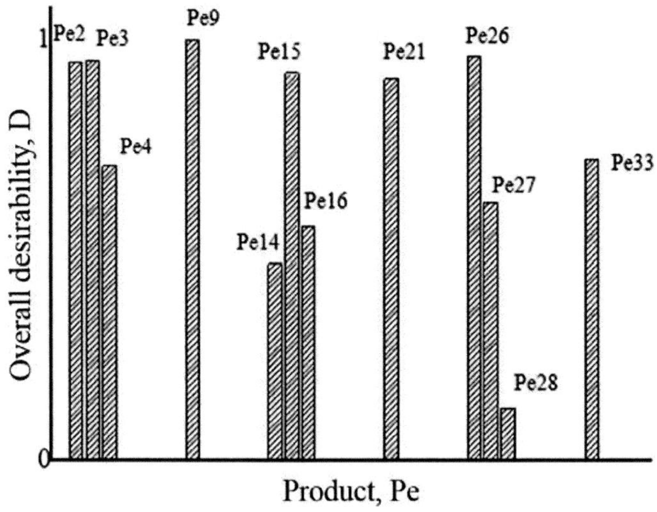


**Fig. 4** Diagram of the overall desirability of expanded layer samples

**Table 4** The overall desirability values for each sample of the expanded layer

| Particle size (μm) | | Product | dFr | dRd | dAll | *D* |
|---|---|---|---|---|---|---|
| TYPE M | 0.9 | Pe1 | 0 | 0.16 | 0 | 0 |
| | | Pe2 | 1 | 1 | 0.95 | 0.95 |
| | | Pe3 | 1 | 1 | 0.85 | 0.95 |
| | | Pe4 | 1 | 1 | 0.7 | 0.7 |
| | | Pe5 | 0 | 1 | 0.9 | 0 |
| | | Pe6 | 0 | 0 | 0.86 | 0 |
| | 1.8 | Pe7 | 0 | 0.16 | 0 | 0 |
| | | Pe8 | 0.15 | 0.5 | 0.4 | 0.01 |
| | | Pe9 | 1 | 1 | 1 | 1 |
| | | Pe10 | 0 | 1 | 0.79 | 0 |
| | | Pe11 | 0 | 0.69 | 0.66 | 0 |
| | | Pe12 | 0 | 0 | 0 | 0 |
| TYPE MC | 1.4 | Pe13 | 0 | 0.16 | 0 | 0 |
| | | Pe14 | 1 | 1 | 0.47 | 0.47 |
| | | Pe15 | 1 | 1 | 0.92 | 0.92 |
| | | Pe16 | 0.79 | 1 | 0.82 | 0.56 |
| | | Pe17 | 0 | 1 | 0.98 | 0 |
| | | Pe18 | 0 | 1 | 0.98 | 0 |
| | 2.5 | Pe19 | 0 | 0.16 | 0 | 0 |
| | | Pe20 | 0.15 | 1 | 0 | 0 |
| | | Pe21 | 1 | 1 | 0.91 | 0.91 |
| | | Pe22 | 0 | 1 | 0.95 | 0 |
| | | Pe23 | 0 | 1 | 0.82 | 0 |
| | | Pe24 | 0 | 1 | 0 | 0 |
| TYPE S | 1.8 | Pe25 | 0 | 0.16 | 0 | 0 |
| | | Pe26 | 1 | 1 | 0.96 | 0.96 |
| | | Pe27 | 1 | 1 | 0.61 | 0.62 |
| | | Pe28 | 0.77 | 0.65 | 0.39 | 0.13 |
| | | Pe29 | 0 | 0.05 | 0.9 | 0 |
| | | Pe30 | 0 | 0 | 0.94 | 0 |
| | 2.7 | Pe31 | 0 | 0.16 | 0 | 0 |
| | | Pe32 | 0 | 0.51 | 0 | 0 |
| | | Pe33 | 1 | 0.96 | 0.77 | 0.72 |
| | | Pe34 | 0 | 0.38 | 0.14 | 0 |
| | | Pe35 | 0 | 0.04 | 0 | 0 |
| | | Pe36 | 0 | 0 | 0 | 0 |

A desirability function method is a decision support approach. Thus, the present results leave it to the company's experts to choose from the best values found the most suitable formulation for the two polymeric layers to produce the final leatherette article.

## 4 Conclusion

The results of the present study showed that the PVC synthetic leather layer's formulations can be filled with different types of carbonate calcium at different levels, to provide an increase in the mechanical properties of the final product. Increase in $CaCO_3$ concentration in the mixture of both superficial and internal layers affected their quality than the final product significantly; decrease in the $CaCO_3$ particle size affected positively the quality of the layers more than the final product.

The desirability approach showed that the different carbonate calcium particle sizes and content can be used effectively for the PVC synthetic leather, but depending on the overall acceptability of the final product, only 95 M, 5 MC, 2 AMC, 5 S, and 2 S at a level of 25 phr can be used for the best result for the superficial layer.

For the internal layer, 95 M at 50 phr and 75 phr, 2 M at 50 phr, 2 AMC at 50 phr and 75 phr, 5 MC at 50 phr, and 2 S and 5 S at 50 phr for each one represent the best results.

The desirability function method is a decision support approach. Thus, the present results allow the company's experts to choose from the best values found the most suitable formula for the two polymeric layers to produce the final leatherette article.

**Acknowledgements** The authors would like to express their deepest gratitude to PLASTISS Company and especially to Mr. Aweb Baccar for their participation and technical support. This project is carried out under the MOBIDOC (mobilization of PhDs for the realization of collaborative research work in the socioeconomic environment); the scheme funded by the EU through the EMORI program and managed by the ANPR.

## References

1. Azimipour. B. and Marchand. F: Effect of calcium carbonate particle size on PVC foam. Journal of Vinyl and Additive Technology 12(2), 55–57 (2006).
2. Kang. G. et al.: Method for removing odor of artificial leather and artificial leather manufactured using the same. Google Patents (2017).
3. Matthews. G. and G.S. Plemper: Effects of calcium carbonate fillers on the behaviour of PVC in fires. British Polymer Journal 13(1), 17–21(1981).
4. Syabani. M.W. et al.:Silica from Geothermal Waste as Reinforcing Filler in Artificial Leather. Key Engineering Materials 849, 78–83 (2020).
5. Gargoubi. S. et al.: Getting rid of the unpleasant odor in new artificial leather using natural and synthetic fragrances. Chemical Industry and Chemical Engineering Quarterly 25(2),141–151 (2019).

6. Ramaraj. B.: Mechanical and thermal properties of ABS and leather waste composites. Journal of applied polymer science101(5), 3062–3066 (2006).
7. Rahman. G.S. et al.:Enhanced physico-mechanical properties of polyester resin film using $CaCO_3$ filler. Fibers and Polymers17(1), 59–65 (2016).
8. Liu. P, Zaho. M. and Guo. J.:Thermal stabilities of poly (vinyl chloride)/calcium carbonate (PVC/$CaCO_3$) composites. Journal of Macromolecular Science, Part B: Physics 45(6),1135–1140 (2006).
9. Chaturvedi. S. and Dave. P.N.: Role of Nanofillers in Blends, Interpenetrating Polymer Networks, and Gels of Unsaturated Polyester Resin Polymers, in Unsaturated Polyester Resins. Elsevier,173–180 (2019).
10. Stambouli. M, et al.: Effect of calcium carbonate particle size and content on the thermal properties of PVC foamed layer used for coated textiles. Turkish Journal of Chemistry 47(1), 40–46 (2023).
11. Mouna. S. et al.:The effect of calcium carbonate content and particle size on the mechanical and morphological properties of a PVC foamed layer used for coated textiles. Industria Textila 73(5), 580–586 (2022).
12. Costa. N.R., lourençon. J, and Pereira .Z.L.: Desirability function approach: a review and performance evaluation in adverse conditions. Chemometrics and Intelligent Laboratory Systems 107(2), 234–244 (2011).
13. Wu, F.-C.:Optimization of correlated multiple quality characteristics using desirability function. Quality engineering17(1),119–126 (2004).
14. Trautmann. H. and Weihs .C.: On the distribution of the desirability index using Harrington's desirability function. Metrika 63, 207–213 (2006).
15. Candioti. L.V., et al., Experimental design and multiple response optimization. Using the desirability function in analytical methods development. Talanta 124,123–138 (2014).
16. Guedri. W, Jaoudi.M., and Msahli.S., Evaluating farmer's satisfaction of different agrotextile bunch covers using desirability function. Textile Research Journal 92(17–18), 3337–3350 (2022).
17. Guedri. W., J. Mounir, and M. Slah, Dual desirability function and RSM approaches for modeling the ideal nonwoven cover bunch for protecting date fruit. The Journal of The Textile Institute,1–10 (2022).

# Ergonomic Study on Ironing Stations in Tunisian Clothing Companies

**Slim Boussaadoun, Nejib Sejri, and Mohamed Hamdaoui**

**Abstract** Human health at work mostly consists of physical health. The majority (60%) of musculoskeletal disorders (MSDs) experienced by employees in Europe. According to a study by the CNAM in Tunisia, the country's MSD rate was almost 80%. According to the American Bureau of Labor Statistics (BLS), MSDs were about 30% in 2020; however, they are around 85% in France. The OWAS approach and the stratified method will be used to achieve this study's goal in this situation. This sample technique evaluates the organizational risk factors and musculoskeletal risk factors of Tunisian staff in northern Tunisia within a group of four ISO 45001 and SA8000-certified businesses using the responsibility approach: corporate social responsibility (CSR) in the textile and clothing sector and implementation of standards in the ironing trades. The overall average total percentage of MSDs in these companies is 65%, and the areas of discomfort observed in the workers studied are as follows: legs (54%), right hand (24%), back (20%), hand left (19%), shoulders (19%), and neck (17%). There have been improvements made to this task, such as reducing one position's stop time by 19.75% and the average stop time for all other positions by 15%, increasing performance by 10%, and improving work quality by 10%. Appropriate solutions must be implemented in order to promote employee well-being, avoid occupational illnesses, improve recruitment procedures, eliminate sources of non-compliance, correct errors and defects, and try to lower the percentage of MSDs.

**Keywords** MSDs · Posture · Repetitive gestures · OWAS · CSR

S. Boussaadoun (✉) · M. Hamdaoui
UR MPTex, National Engineering School of Monastir, Monastir, Tunisia
e-mail: slimbinz@gmail.com

N. Sejri
Textile Engineering Laboratory, HITS of Ksar Hellal, Monastir University, Monastir, Tunisia



S. B. Abdessalem et al. (eds.), *Proceedings of the Second International Conference of Innovative Textiles and Developed Materials-ITDM'2; 05-06 May 2023; Tunisia*,
https://doi.org/10.1007/978-981-99-7950-9_26

# 1 Introduction

The prevalence of musculoskeletal disorders (MSDs), also known as "troubles musculosquelettiques" (TMS), is alarmingly high in Tunisia, highlighting the need for effective ergonomic diagnostic tools to address this issue. Iterative surveys on working conditions conducted in France and Europe since the 1990s indicate that preventive measures have helped reduce exposure to traditional hazards such as physical constraints, chemicals, and psychosocial and organizational stressors [1]. In addition, occupational diseases in Tunisia according to the latest statistics were represented, in 2020, 1860 cases including 1496 suffering from musculoskeletal disorders (MSDs). In addition, 762 cases were reported in companies in the textile–clothing sector, which represent 41% of the percentage of the population affected by MSDs. The last few years have seen a considerable increase of around 80% in the number of occupational diseases declared to the National Health Insurance Fund (CNAM) [2]. According to SUMER in 2017, 97% of workers experience at least two constraints: organizational and physical (45%). A study conducted in 2006 by occupational physicians in Tunisia on MSDs in the garment sector showed cases of illness mainly affecting the back (75.8%), neck (60.4%), and shoulders (65.4%) [3].

Despite the changes that Tunisia has undergone in recent years and the resulting impact on its economic growth, exports, and particularly the products manufactured in Tunisia for the European community, the sector has demonstrated remarkable adaptability [4].

As a result, today's society requires operators, who play a crucial role in the production process, to perform their tasks efficiently while adhering to the "quality, cost, time" triad to ensure compliant products or services. Fortunately, the search for improvement continues to have a positive impact on working conditions and the physical well-being of employees who are subject to muscular physical constraints, optimizing their health and safety [5, 6].

Musculoskeletal disorders (MSDs) resulting from poor working conditions are defined as periarticular conditions that affect the soft tissues such as muscles, tendons, nerves, vessels, and cartilage of the upper and lower limbs in workers. Regardless of their location, MSDs can result in permanent damage and even lead to serious disabilities [7].

According to the "Global Burden of Disease" study, musculoskeletal disorders are the second largest contributor to potential overall disability resulting from the mechanisms involved in these pathologies. Workplace risk factors are associated with the onset of health problems. It can be directly responsible for the onset of a health problem and can be considered a trigger for an MSD.

The factors contribute to the appearance of musculoskeletal disorders (MSDs) in the workplace. These factors can be biomechanical, personal, psychosocial, or additional such as environmental conditions [8, 9]. According to a study that was done in 2020 in Tunisia in the clothing sector shows that ironing stations in the clothing manufacturing process are one of the most stressful 15.4% [10]. Participatory ergonomics is considered a basis for studying professional activity and preventing

MSDs [11]. Participatory ergonomics involves the participation of workers in both the technological and organizational aspects of the process [12, 13]. This was the case in a recent study conducted on a sample of Tunisian employees working in ironing sections who have been applying ergonomic standards and corporate social responsibility principles for the past five years [14].

## 2 Methods of Work and Materials Used

In this study, we used a participatory approach as follows: mobilization, investigation, mastery, evaluation, and data collection regarding confidentiality:

(1) To mobilize and investigate indicators related to the company's operation, we considered factors such as absenteeism rates of less than 3%, an 8-h work schedule, defect percentages of less than 5%, a monitoring rate of around 10%, average performance levels of around 75%, versatility, average age, and average experience in the industry. We also evaluated indicators related to employee health and safety, such as the prevalence of musculoskeletal disorders (MSDs), which were reported by around 10% of workers in the ironing sections of the company, exposure time, and other related factors.
(2) To achieve mastery, we constructed a list of indicators that should be understood and followed by all employees in the company. We also collaborated with the company's staff to establish an ergonomic questionnaire to assess and improve working conditions.
(3) To achieve mastery, we constructed a list of indicators that should be understood and followed by all employees in the company. We also collaborated with the company's staff to establish an ergonomic questionnaire to assess and improve working conditions.

In this study, we utilized various methods, tools, and standards, including:

The first tool used was stratified sampling. This method involves dividing a heterogeneous population into subgroups or strata. The goal is to obtain the same proportions of each stratum in the sample as in the target population, based on the characteristics selected for the study. [15] For this study, a confidence level of 95% and an accuracy rate of 0.05 were used. The second tool used in this study was the OWAS method, which is a technique used for observing and evaluating the static working postures of all body parts [16]. The method is used for evaluating the constrained postures of the human body during work and can be applied to the neck, back, upper limbs (shoulders, elbows, and hands), chest, and legs. The method was first introduced in 1977 and has undergone several improvements, including the one made in 1992 [17]. The BORG scale is a tool for judging the level of subjective effort, graduated from 0 to 10: The method consists in explaining this scale to the operators who themselves evaluate the level of force for each operation carried out, relating to the body area concerned (Fig. 1).

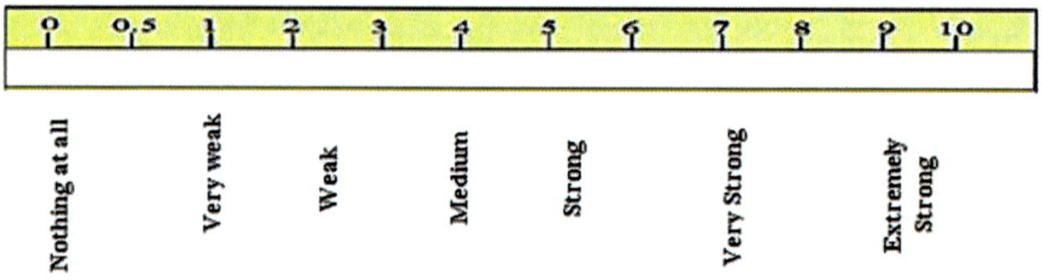


**Fig. 1** BORG scale [18]

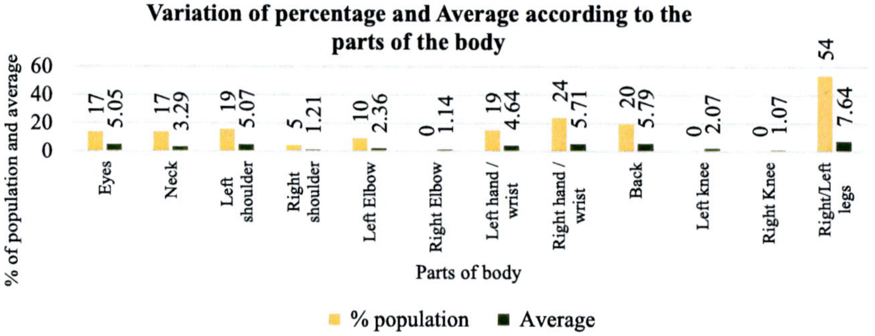


**Fig. 2** Distribution of discomfort in the ironing section

Moreover, we have devised a questionnaire to investigate the discomfort experienced by the workers while performing tasks such as ironing and sections. The main objective is to pinpoint the areas of discomfort and analyze them subsequently to devise preventive and remedial measures. The weights of these quizzes will be noted as shown in Fig. 2.

0–1–2: no pain/4–5: moderate pain/6–7: severe pain 7–8: very severe pain/9–10 extremely severe

To evaluate the scores, we focused on severe and extreme pain, which corresponded to scores of 7, 8, 9, and 10. By analyzing these scores, we were able to assess the extent of discomfort experienced by participants and gain insights into the severity of the pain. One commonly used method to define obesity is through the calculation of the body mass index (BMI), which measures the ratio of an individual's weight in kilograms to their height in meters squared [19]. In addition to IT tools like XLSTAT Premium and Excel version 2016, we used video and photo software like CyberLink PhotoDirector and Movie Maker. The CSR method was employed with the aim of not only attracting and retaining employees but also fostering a sense of pride among them and improving performance. This method involves developing a policy that encourages employee involvement and participation by selecting a line of work that aligns with their interests and goals. To determine the timing of different work phases involving bent postures, we measured the duration of each phase within well-defined time intervals such as [9:30 a.m. to 10:30 a.m.], [11:30:30 a.m. to 12:30 p.m.], and [2:30 p.m. to 3:30 p.m.]. We also accounted for the standard 30-min breaks that companies typically offer, which occur on alternating days at either 10 a.m., 10:30 a.m., or 11 a.m., as well as at 11:30 a.m.

To conduct our study, we relied on several recognized standards, including the international standards IN ISO 12228-3 and 12,226, which provided guidelines for studying postures and repetitive gestures. We also consulted the Deutsch Institute Standard, developed by the Deutsch Institute of Normalization, to obtain anthropometric measurements that informed the design of workstation heights. Furthermore, we applied the NF EN ISO 14738 standard for designing workstations, as well as the international standard NF IN 1005-5, to ensure compliance with established ergonomic principles.

# 3 Results

This study involved a participatory approach aimed at assessing ergonomic and organizational factors, which consisted of two parts. In the first part, we conducted a detailed analysis of the risk factors to identify problem areas. To achieve this, we used statistical tools and the OWAS method to evaluate the work environment in the ironing sections of four companies. The second part of the study was focused on proposing solutions to address the issues identified in the first part. Our analysis yielded the following results.

## *3.1 Discomfort Quiz and MSD Risk Factors*

The first set of results was obtained through a quiz, which was administered to each operator and is presented in Table 1. By analyzing the responses, we were able to gain insights into the level of discomfort experienced by workers in the ironing sections of the four companies included in the study and identify specific areas for improvement.

Following were the findings of the body discomfort quiz and survey, which measured the degree of discomfort felt by workers in the ironing sections while standing on an ergonomic mat: 54% of the participants said they had musculoskeletal symptoms in their legs, 24% in their right hand, 20% in their back, 19% in their left hand and shoulders, and 17% in their neck and eyes. The necessity for interventions to address the ergonomic issues experienced by workers in these fields is highlighted by these self-assessments. Based on the self-assessments acquired through the body discomfort quiz, Fig. 2 provides a clear visual picture of the distribution of discomfort experienced by workers in the ironing sections. The graph displays the frequency of musculoskeletal problems in various regions of the body, providing valuable insights for developing targeted interventions.

**Table 1** Body discomfort quiz

| | | | | | | | | | | | | | |
|---|---|---|---|---|---|---|---|---|---|---|---|---|---|
| **Example 1** | | | | | | | | | | | | | |
| **Date: 12/09/ 2022** | **weight** | | **74 kg** | **Stature: 1.78 m** | | | | | **Experience: 22 years** | | | | |
| **First name** | **Monia** | | | **Degree of versatility: 3 operations** | | | | | | | | | |
| **Station** | | **Ironing shirt** | | | | | | **BMI** | **42** | | | **Age** | **48 years** |
| **Job specification** | | **Table + fer + chaudière T 160 °C** | | | | | | | | **Station height** | **93.5 cm** | | |
| **Gender** | **Female** | | **weight** | **74 kg** | **Stature** | **1.78 m** | | | | **Experience** | **22 years** | | |
| **Parts of the body** | **0** | **1** | **2** | **3** | **4** | **5** | **6** | **7** | **8** | **9** | **10** | **Total** | **Spot of discomfort** |
| **Eyes** | | | | 3 | | | | | | | | | |
| **Neck** | | | | | | | | | 8 | | | **8** | |
| **Left shoulder** | | | 2 | | | | | | | | | **2** | |
| **Right shoulder** | | | | | | | | | | 9 | | **9** | |
| **Left Elbow** | | | 2 | | | | | | | | | **2** | |
| **Right Elbow** | | | 2 | | | | | | | | | **2** | |
| **Left hand/ wrist** | | | | | | 5 | | | | | | **5** | **Repetitive gestures** |
| **Right hand/ wrist** | | | | | | | | | | 9 | | **9** | |
| **Back** | | | | | | | | | 8 | | | **8** | |
| **Left knee** | | | | | 4 | | | | | | | **4** | |

(continued)

**Table 1** (continued)

| | | | | | | | | | | | | | |
|---|---|---|---|---|---|---|---|---|---|---|---|---|---|
| **Example 1** | | | | | | | | | | | | | |
| **Date: 12/09/ 2022** | **weight** | | **74 kg** | **Stature: 1.78 m** | | | | | **Experience: 22 years** | | | | |
| **First name** | **Monia** | | | **Degree of versatility: 3 operations** | | | | | | | | | |
| **Station** | | **Ironing shirt** | | | | | | **BMI** | **42** | | | **Age** | **48 years** |
| **Job specification** | | **Table + fer + chaudière T 160 °C** | | | | | | | | **Station height** | **93.5 cm** | | |
| **Gender** | **Female** | | **weight** | **74 kg** | **Stature** | **1.78 m** | | | | **Experience** | **22 years** | | |
| **Parts of the body** | **0** | **1** | **2** | **3** | **4** | **5** | **6** | **7** | **8** | **9** | **10** | **Total** | **Spot of discomfort** |
| **Right Knee** | | | | | 4 | | | | | | | **4** | |
| **Right/Left leg** | | | | | | | | | | 9 | | **9** | |

*BMI* Body mass index [20]

In this study, which focused specifically on ironing clothes, we used 30 participants with an average age of 33.8 years, an average difference in study size of 7.6, and averages for weight, stature, and mass index of 78 kg, 161 cm, and 30.09 kg, respectively. Body mass index, or BMI, is determined as follows: [20].

$$\text{BMI} = \text{weight (kg)/height} \times \text{height (cm)} \tag{1}$$

Table 2 presents risk factors such as BMI, age, and degree of versatility according to MSDs parameters.

This table's findings show that there is a proportional association between age and work adaptability, particularly among women. The greater their level of work adaptability, which lowers their risk of MSDs, the earlier they begin learning. The accompanying table, which summarizes the prevalence of risk factors for MSDs at work by job category, demonstrates that when BMI rises, so does the likelihood of developing MSDs. A rise in the average age of female workers may possibly be a factor in the emergence of TMS.

Additionally, it is apparent that a correlation exists between these factors, with a lesser degree of adaptability being associated with a higher percentage of female workers being targeted by MSDs. As a result, these criteria as well as additional organizational aspects are a priority for resolution, as they can influence both production and work quality.

Additionally, we have obtained the following results through Table 3 to assess the neck posture of three female workers.

**Table 2** Risk factors according to parameter

| N | Risk factors parameters | Parameters | Prevalence and percentages<br>Men and women (of the population affected by MSD) |
|---|---|---|---|
| 1 | MSDs | BMI | It is around 30% among women in the target population<br>It increases with time, around 10% in women |
| 2 | MSDs | Ages | |
| | | < 20 years | * MSDs is 1% in women |
| | | 20–30 years | *MSDs is around 4% in women |
| | | 30–40 years | * MSDs is around 15% of the affected female population |
| | | > 40 years | * MSDs is around 45% in women |
| 3 | MSDs | Degree of versatility | |
| | | *< 3 operations | *The prevalence of MSDs is around 55% among women |
| | | *3–15 operations | *The prevalence of MSDs is around 5% |
| | | *> 15 operations | *The prevalence of MSDs is about 5% |

**Table 3** Neck posture summary

| Posture / Name of the worker | Extension/flexion (frontal axis) % of time | Lateral bending in % | Rotation |
|---|---|---|---|
| « H» | **39** | **4** | **14** |
| « I» | **44.6** | 29 | 21 |
| « S» | **53** | **5** | **34** |

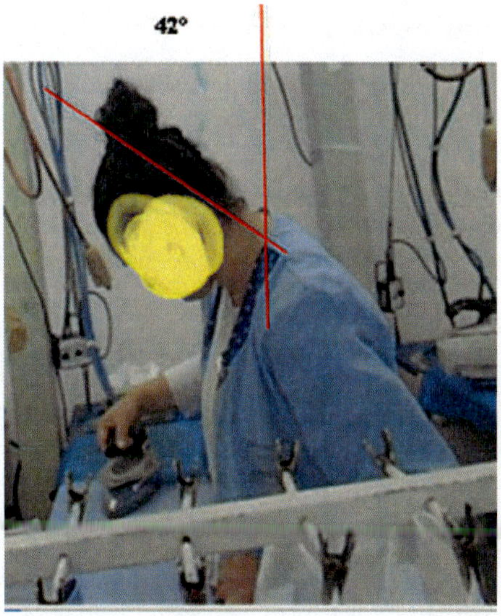


**Fig. 3** Bent posture of the worker's neck

**Interpretation**

We observe that all workers have a percentage higher than 30%, and this posture can lead to a risk of MSDs (ISO 11228, 2007) (Fig. 3).

## *3.2 Monitoring of Leaning Posture and Measurement of Arduousness*

The significance of constantly checking the workers' posture is to spot unsafe postures and gathering posture data to spot risky tasks. Surveys on the pain or discomfort experienced by employees can be used to assess whether preventative measures have been successful. The evaluation of work-related strain is valuable for evaluating the effectiveness of preventive measures. Finally, regular posture checks and strain measurements are necessary to effectively prevent MSDs. The following findings were made before correcting the operator's hunched posture and assessing the current situation:

**Before Improvement**

Ten measurements of the bent position (flexion/extension) were taken to get the numbers in Table 4 above. We discovered a maximum duration that was 22.22 h per week, or more than 20 h per week. (5.5 days in a week) Stress at work results from this. Additionally, Fig. 4 provides a clear explanation of the lean time threshold.

**Interpretation**

Figure 4 demonstrates that the percentages of controls (×1, ×4, ×5 ..., ×10) ranging between 48 and 57% surpass the ceiling for the longest permissible length, which is only 45%. Only the values of × 2 and × 3 are acceptable, and roughly 65% of the work is produced while 80% of it is of high quality. This demonstrates that the employee's posture needs to be corrected right away. Keep in mind that a 44-h workweek equates to 20 h every week, or 45.45%.

**Table 4** Measurement of leaning posture (ironing)

| Control numbers | X1 | X2 | X3 | X4 | X5 | X6 | X7 | X8 | X9 | X10 | Average % of time | % of an hour in min | % en heure/semaine |
|---|---|---|---|---|---|---|---|---|---|---|---|---|---|
| % time in leaning posture > 20 h/week | 50 | 45 | 40 | 49 | 51 | 55 | 48 | 54 | 57 | 56 | 50.5 | 32.9 | 22.22 h |

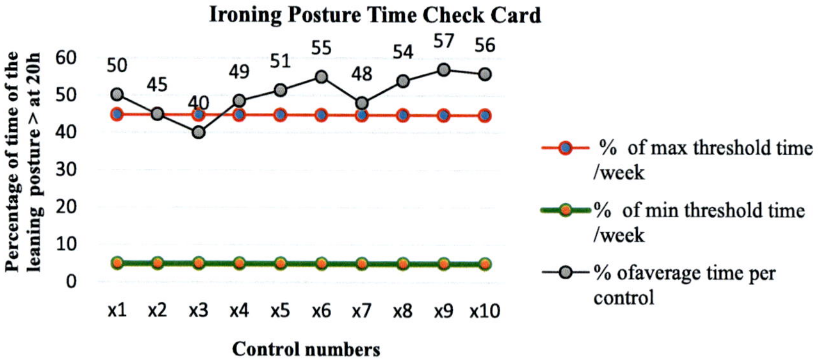


**Fig. 4** Neck bent posture time control chart

### After Improvement

The following outcomes were attained after raising awareness among the workers to break the habit of leaning forward and enhancing and changing the height of the ironing table to the worker's stature (Table 5).

This results in a gain of 19.75% ((% hours/week before improvement – % hours/ week after improvement)/%hours of work/week) * 100 = ((22.22 h – 13.53 h)/44 h/ week). Additionally, Fig. 5 depicts the 20° correction angle.

So, Fig. 6 explains well control improvement by reducing time of leaning posture ironing.

### Interpretation

The analysis of altering the operator's posture to attain more pertinent percentages of time (between 28 and 42%) of the duration of each control is summarized in the graph (above). It demonstrates that the majority of posture percentages fall below the 45% threshold. Performance has improved by 85%, while quality has improved

**Table 5** Control and measurement of leaning posture (ironing)

| Control numbers | X1 | X2 | X3 | X4 | X5 | X6 | X7 | X8 | X9 | X10 | Average % of time | % of an hour in min | % en heure/ semaine |
|---|---|---|---|---|---|---|---|---|---|---|---|---|---|
| % time in leaning posture > 20 h/ week | 38 | 28 | 39.5 | 40 | 38.5 | 30 | 25 | 32 | 28 | 42 | 30.75 | 20 | 13.53 h |

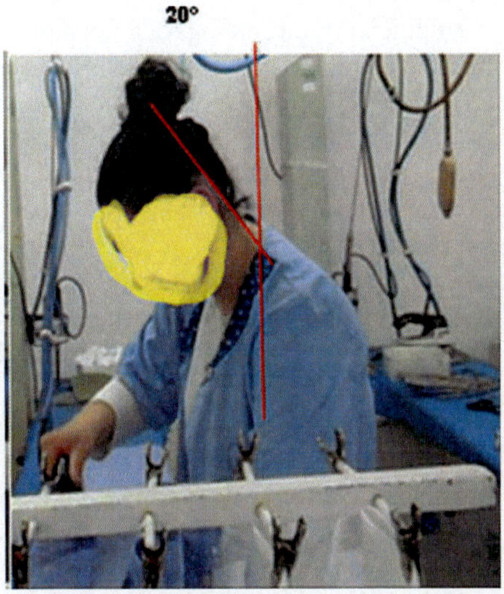


**Fig. 5** Correction of the bent posture of the worker's neck

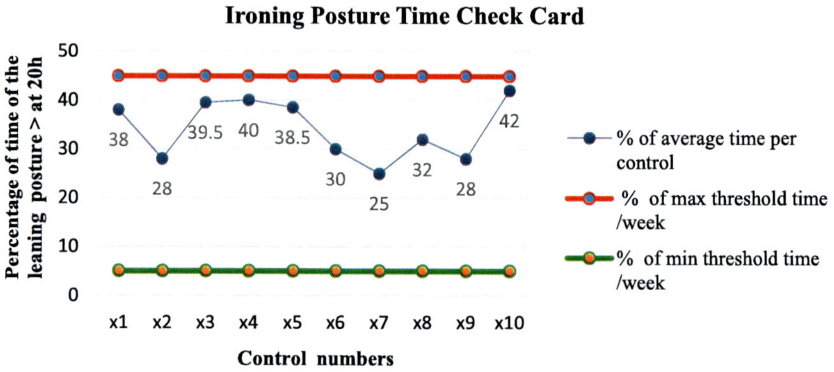


**Fig. 6** Neck bent posture time control chart after improvement

by 95%. This enables us to draw the conclusion that the condition of the operator is better and less exhausting.

## 4 Discussion on the Results

Comparison of what was found with other studies in Tunisia's apparel industry: The new figures we discovered in terms of MSDs sensation (54% for legs, 24% for right hand, 17% for neck ...) are quite relevant compared to the national scale which is around 80% and the European scale which is around 60%. Ergonomic studies in Tunisia are quite recent and require more effort, equipment, human, and financial resources.

Additionally, the findings of this study show that the average BMI is 30.09 9.4, and that 30% of the population was found to have BMIs that were higher than the standards advised, with "normal" body weight being between 18.5 and 24.9 kg/m$^2$, especially for women under the age of 35, who had a BMI of 25 kg/m$^2$ and above. Additionally, gaining weight and having an unhealthy BMI might make musculoskeletal pain from work and work-related psychosocial stress worse. One of our goals in this topic is to ascertain how BMI affects musculoskeletal problems associated with the workplace. Furthermore, this study revealed that the designed ergonomic ironing arrangement causes more discomfort and work-related stress for older individuals with a BMI > 30, who make up 30% of the population. Additionally, TMS has less of an impact on younger workers under the age of 30. The only persons who are affected are those with high BMI ratios compared to approved standards. People are more likely to develop TMS than those who participate in a wider range of activities if they engage in repetitive physical activity or spend a lot of time in uncomfortable positions. They move and position themselves in a wider range of ways during the day. By limiting the abuse of specific body components and allowing for appropriate recovery, greater variety can aid in the prevention of TMS. In this study, we have found that the workers who are most negatively impacted are those whose level of variety is less than three procedures.

## 5 Recommendation

It is impressive in this context to lessen and eliminate MSDs by taking preventative measures and incorporating them into the company's occupational risk management. The effectiveness of preventive strategies in lowering the risks of MSDs should be regularly assessed for their relevance. Workstation customization: Stretching should be promoted to employees on a regular basis, as needed, before, during, and after work. It is essential to educate and train staff members, managers, and supervisors about MSDs and how to prevent them. This will help them spot the early warning symptoms of these conditions and motivate them to adopt healthy postures and motions. Regular observation: to assess in real time the efficacy of adopted measures and adjust them appropriately, it is critical to routinely monitor MSD indicators, such as sick leave due to MSDs and complaints of pain and discomfort. Alternating between jobs will allow workers' muscles to rest; selecting the proper instruments (a light iron for ironing, for example); adjusting the ironing tables' height to fit the stature of each operator; setting up a different ironing station for a seated position between other workstations. This study revealed that some workers were not strictly employing ergonomic mats, which supports a rise in the proportion of leg discomfort along with other factors, including a body mass index that is higher than the threshold level. Out of a total of eight hours (one workstation/five operators), each operator spends two hours every day sitting down, alternating between sitting and standing postures during apprenticeship training. For instance, they would sew for the first part of the day and then swap ironing duties with other trainees for the second. In conclusion, it is critical to take preventative steps and promote consideration of postures and gestures.

## 6 Conclusion and Perspectives

To conclude, these factors have played an important role in calculating the risk factors. It is critical to familiarize yourself with studies on work parts, pinpoint the root problems, and weigh the importance of potential solutions. Companies in the worker/employee category must do more for their staff members. To further the goals of the CSR project, the organization must forge international partnerships. 65% of people have MSDs, and women over 40 are the most likely to have one because of things like age, BMI, and a lack of flexibility, among other things that need more research. Moreover, they (women between the ages of 20 and 40) represent 15% and 4% of the population, while young women under 20 years' old who start their careers early are less affected if the working conditions are favorable (better ergonomic conditions). The results obtained with these ergonomic methods and tools, such as reducing the posture of leaning and decreasing monotony in ironing positions by 15% through alternating positions, have resulted in positive improvements in performance by 5–10% and in work quality by 10%. These outcomes show that the staff is making an effort to meet the goals. Based on these results, this group of companies is still working on how to develop its managerial and organizational strategy for sustainable development projects with SMART objectives at various levels, including health, safety, ergonomics, quality, and productivity, in accordance with priorities and the urgency of problems as they arise. To reach a maximum of 5% of our goals, efforts should be undertaken to lessen the limitations that MSDs cause by 5% annually.

## References


1. Said Ben Sedrine and Mongi Amami, October 2021, Future of Textile, Clothing, Leather, and Footwear Industries for a Sustainable Industrial Policy Focused on Decent Work, ©Friedrich-Ebert-Stiftung Tunisia 2021.
2. J.-P. Bourgeois, 1997Project Management, Techniques de l'Ingénieur.
3. A. Mhamdi and R. Ghram, "Ergonomics in Tunisia: A Historical Overview," in 48th International Congress: French Language Ergonomics Society (SELF), 2013.
4. Vos, T., Abajobir, A. A., Abate, K. H., Abbafati, C., Abbas, K. M., Abd-Allah, F., ... & Aboyans, V. (2017). Global, regional, and national incidence, prevalence, and years lived with disability for 328 diseases and injuries for 195 countries, 1990–2016: a systematic analysis for the Global Burden of Disease Study 2016. The Lancet, 390(10100), 1211–1259.
5. Natacha Fouquet et al. Work-related musculoskeletal disorders: Number of avoidable cases by applying a theoretical prevention scenario // Troubles musculo-squelettiques liés au travail : nombre de cas évitables par l'application d'un scénario théorique de prévention. BEH Journal, February 16, 2021, France.
6. Rahul Pradeep Bhosale and M. Sunil Kumar, June 2022 "Analysis and Improvement of Working Postures in Cargo Securing Process During Outbound Shipment by Using Different Ergonomics Tools and Software,"Research Gate, https://doi.org/10.1007/978-3-030-94277-9_68, In book: Ergonomics for Design and Innovation (pp. 795–808).
7. Nahon, & Arnaud; 2011,Work situation analysis, Chain interactions. Health and Work, 35, 48.
8. Leplat, J. (2006). The notion of regulation in activity analysis. Interdisciplinary perspectives on work and health, 8(1).
9. Liliana & all (2005) "Simulation in ergonomics: Factor of innovation in product design" National Polytechnic Institute of Lorraine - France.
10. Sejri Nejib, Boussaadoun Slim, Faouzi Sakly, Ghazi Elbiche, 2020. Stress studies for clothing companies applying the social relationship of textile companies, vol. 71, no. 1 Tunisia. https://doi.org/10.35530/IT.071.01.1656
11. M.-È. Chiasson, D. Imbeau, K. Aubry and A. Delisle, "Comparing the results of eight methods used to evaluate risk factors associated with musculoskeletal disorders" International Journal of Industrial Ergonomics, pp. 478–488, 2012.
12. S. Dockrell, E. O'Grady, K. Bennett, C. Mullarkey, R. Mc Connell, S. Twomey and C. Flannery, 2012. "An investigation of the reliability of Rapid Upper Limb Assessment (RULA) as a method of assessment of children's computing posture," Applied Ergonomics, pp. 632–636.
13. K. Dohyung and K. Waldemar, "A Comparison of Three Observational Techniques for Assessing Postural Loads in Industry," International Journal of Occupational Safety and Ergonomics, pp. 3–14, 2007.

14. Vincent Maymo, Geoffroy Murat (2017). Sustainable Development Management Environmental Protection Corporate Social Responsibility Edited by Dunod. Paris - 2017 ISBN 978-2-10-058497-0
15. Lacroix, S., Four Sampling Methods, Available on http://www.sylvainlacroix.ca/ESW/Files/306_SamplingMethod.pdf (Accessed on 2018)
16. The OCRA index is described in the European standard EN 1005-5 (2007) Safety of machinery - Human physical performance - Part 5: assessment of risk related to repetitive handling at high frequency.
17. M.-È. Chiasson, D. Imbeau, K. Aubry and A. Delisle, 2012, "Comparing the results of eight methods used to evaluate risk factors associated with musculoskeletal disorders," International Journal of Industrial Ergonomics, pp. 478–488.
18. P. Meyer, Department of Human Work, INRS, Subjective evaluation of workload, use of Borg scales, reference in occupational health 2014, No. 139.
19. F. Laclergeriea, B. Jacquemet, G. Guicharda, S. Bernardini, E. Chabannes, L. Martina, J. Pastori, V. Baillya, H. Bittarda, F. Kleinclau, Flexible ureterorenoscopy with Holmium-YAG laser in the management of urinary lithiasis in obese patients: results of a monocentric cohort Published by Elsevier Masson SAS. https://doi.org/10.1016/j.purol.2014.03.005 1166-7087/© 201
20. Filipa Rafael , Maria Dias Rodrigues, Jose Bellver , Mariana Canelas-Pais March 2023 journal recherche gate The combined effect of BMI and age on ART outcomes. https://doi.org/10.1093/humrep/dead042

# An Algorithm at Incoming Quality Control of Raw Woven Fabrics—Cottonade Jacquard

**Tashka Ivanova Koleva and Ivelin Rahnev**

**Abstract** The quality control procedure of the fabrics between the weaving production and the subsequent dyeing and finishing processes is mostly intermediate in nature. From the removal of the raw fabric from the loom, to the final grading and storage, fabrics have the properties of semi-finished products. At the output of each process, the fabric exits with output lab results, which are sometimes checked at the input of the next process. The intermediate nature of the previous quality control consists in the speed of obtaining a quality assessment and adopting a technological decision for the upcoming processing. The subject of this article is a complex of laboratory tests and qualitative analysis arranged algorithmically for the selection of raw fabrics. The purpose of the development is to compile an operative procedure for the incoming control of the QMS for the correct direction of the raw fabrics during dyeing, printing and finishing.

**Keywords** Textile · Metrology · Fabrics

## 1 Introduction

Like any organization, a textile enterprise with a vertical structure conducts intermediate quality control in the transition between the weaving mill and the finishing. Two factors determine the proper routing of the raw fabric to subsequent dyeing, printing, or finishing operations. The first one is the complex of physico-chemical properties of each fabric roll. The properties and overall quality of the fabric are a consequence of the combination of the input parameters of the fibrous raw materials,

Supported by organization E. Miroglio EAD.

T. I. Koleva
Dobri Zhelyazkov Vocational High School of Textiles and Clothing, 8800 Sliven, Bulgaria

I. Rahnev (✉)
E. Miroglio EAD, Industrial District, 8800 Sliven, Bulgaria
e-mail: Ivelin.Rahnev@emiroglio.com

 
S. B. Abdessalem et al. (eds.), *Proceedings of the Second International Conference of Innovative Textiles and Developed Materials-ITDM'2; 05-06 May 2023; Tunisia*,
https://doi.org/10.1007/978-981-99-7950-9_27

the construction of the threads and the fabric, as well as the general machine mode. And the slightest change in any of the input parameters leads to a deviation of the properties of the fabric, such as areal mass and its distribution, background colour of the surface, and isotropy of mechanical resistance. Knowledge of the properties predetermines the general mode of subsequent treatments. For example, for a fabric roll with the same composition and construction, the shades of the raw surface determine whether the fabric will be dyed in dark, saturated colours or – clear, light tones. The second factor is the effectiveness of the chain of custody implemented, which tracks each fabric roll from the incoming weaving threads to the off-loom raw fabric. Even in the case of a complete match of both factors, an additional evaluation of the fabric rolls is required. The goal is to avoid accidental deviations that would lead to a deterioration in the quality of the finished fabric. The non-admission of inappropriate processing during the refinement of the raw fabric requires a mandatory intermediate control between the weaving loom and the finishing.

The subject of this article is a complex of laboratory tests and qualitative analysis arranged algorithmically for the selection of raw fabrics. The purpose of the development is to compile an operative procedure for the incoming control of the QMS for the correct direction of the raw fabrics during dyeing, printing, and finishing. Tasks: literature research, technological observations, sampling, laboratory tests, and qualitative analysis.

## 2 Theoretical Prerequisites

The declared construction parameters of the raw fabric can serve as a starting point above all for the partial rejection of some laboratory tests, such as checking the fibrous composition and, the type of weaving threads and the weave.

As appropriate, the intermediate quality control covers the following indicators: width and areal mass in the middle and both ends of the fabric, surface fastness, and mechanical resistance on warp and weft. The final quality assessment is based on the laboratory determined colour shade of each roll of raw fabric against a suitable benchmark. The width of the fabric is measured under ISO 3801 [1] conditions, where one to three consecutive measurements are recorded and then averaged. Clause No. 6.7, method 5 for determining the mass per unit area by means of small samples according to ISO 3801 is applied for the determination of the areal mass.

The average surface mass of the fabric in one roll is important when grouping several rolls in one group to achieve uniformity in dyeing and finishing. Of greater importance is the uniform distribution of the surface mass across the width of the fabric. The evaluation of the non-uniformity of the weight distribution is universally represented by the coefficient of variation by mass relative to the middle and both ends of the fabric. The limit value of the coefficient of variation ($Cv(weight) < 5.0\%$) above which the fabric is declassified was established by a long experimental work by Andonova and Rimini in 2020 [2].

Another main indicator for the selection of raw fabric with relatively uniform properties is the colour shade of the face and the reverse surface. Colour differences are usually applied in case of inconsistencies in the fastness of dyes and textile materials. It is a complex of prescriptions and recommendations collected in the series of standards ISO 105. Part A02 (ISO 105-A02) [3] describes the "5-level grey scale" as a means of assessing colour deviations between a standard and a test specimen. The primary purpose of grey scale is to allow evaluations and judgement of match/mismatch between reference and sample. The means and method are suitable for operational quality control of a discrete nature (0/1). The duration of visual observation, comparison and subjective inference is small, but does not have a detailed numerical value of the deviation. Therefore, grey scale is not suitable for mass selection and classification in groups of large volume of samples or rolls of fabric.

The colorimetric system CIE 1964 10° [4] of CIELAB colour space represents colour by three values: L—for the light/dark characteristic of colour and (a* and b*) for the four single colours of human vision: red, green, blue, and yellow. Despite the receptor differences, the system is extremely useful in industry for detecting small colour differences. This system uses the standard St-D65 daylight [8]. The total colour difference is calculated according to the well-known formula from 1976 with known parameters of saturation and colour deviation, respectively— 〚ΔL〛 *, 〚Δa〛 * and 〚Δb〛 *:

$$\Delta E_{ab}^{*} = \sqrt{\Delta L^{*2} + \Delta a^{*2} + \Delta b^{*2}} \tag{1}$$

The essential contribution of the application of the $\Delta E$ CieLab colorimetric system and the datacolour spectrophotometer is obtaining a numerical estimate of the colour difference between the selected standard and each of the samples row.

The presence of a numerical row of single values for each sample allows gradation from minimum to maximum difference, as well as grouping of samples according to acceptable limit values. The values of each group are limited to the interval of $1^{(\pm 0.20)}$, which corresponds to an evaluation level of 4–5 of the 5-point grey scale. The evaluation and application of numerical data from spectrophotometers acquire the meaning of a practical guide from the experimental studies of Staneva and his team [5]. In some cases, the weaving threads or the raw fabric go through a treatment that is the deposition of a thin layer of optical bleachers/brighteners. The fluorescent properties given to the fabrics sharply distinguish them from the others. Mixing optically bleached with untreated fabrics during dyeing or printing risks failure of an entire production batch. Practically useful for the identification of fluorescent substances are the experimental studies of Staneva and colleagues [6].

Various complex fabric grading systems are known. Such a system is the Tiger of Sweden Standard Trading Procedure (STP) point system, [7]. In this system, as in others, the focus is on the correspondence between the declared and established quality of the fabrics. Derivative results from these systems relate to the final ranking of fabrics and the awarding of corresponding trade values. These systems do not consider the properties of fabrics as a prerequisite for future technological adjustment. The advantage of the presented algorithm consists in the targeted application of the established quality indicators to the corresponding machine mode of processing and equalization of the quality of the general population.

## 3 Experimental Work

The experimental work was carried out in the production conditions of E. Miroglio EAD and reflects the working procedures for the application of the QMS in the internal quality control.

### *3.1 Experimental Set*

The subject of the experimental work is raw jacquard woven fabric—cotton type. The fabric consists of 79% cotton, 18% polyamide, and 3% elastane in the weft. The width of the raw fabric is 145 cm with a surface mass of 190 g/m$^2$.

The warp has a density of 900 threads/dm and consists of polyamide monofilament with a linear density of Tt 3.6 tex. The wefts have a density of 300 threads/dm and consist of cotton core spun yarn with a linear density of Tt 37.2 tex. Figure 1 shows microscopic photographs of the fabric face.

A standard circular scissors and an electronic scale provide the main laboratory equipment for the weight selection of the fabric. The optical properties of the fabric face were tested with a Leica MS5AmScope microscope and a MU100 3B camera, a Datacolour 600 spectrophotometer, and a very vide SAC 120 UV light camera.

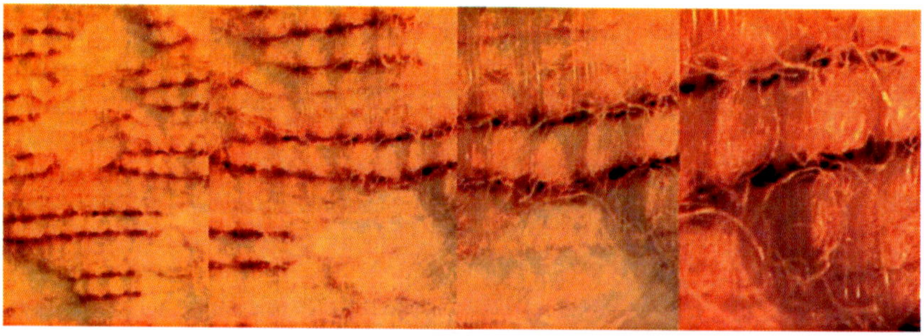

**Fig. 1** Jacquard cottonade face view

## 3.2 Description of Experimental Tests

From the available general population of the reportable fabric, individual samples from each roll are separated, labelled, and sent for testing.

The measurements are arranged in the general procedure as follows:

- measurement of the area mass in the middle and near the two selvedges of the fabric;
- determination of fluorescence response for use of optical bleachers on fabric;
- measurement of the surface colour of the face of the fabric compared to a reference white colour (Datacolour—white plate) on the spectrophotometer.

## 3.3 Experimental Results

The results of the fluorescence reaction are discrete in nature (0/1). The performed inspection facilitates the separation of the optically bleached rolls of fabric for the appropriate subsequent treatment. Figure 2 shows the light response of an optically bleached sample and an untreated sample under the action of a UV lamp.

The selection and grouping of the population is based on areal mass measurements and photometric measurements. According to the order of the general procedure, the measurements take place in two stages:

**Fig. 2** Optical bleachment UV identification

**Table 1** DeltaE CieLab levels

| Groups levels of DeltaCieLab | | | | | | | |
|---|---|---|---|---|---|---|---|
| Light | | Medium Light | | Medium Dark | | Dark | |
| Min | Max | Min | Max | Min | Max | Min | Max |
| 2.88 | 3.88 | 3.89 | 4.86 | 4.94 | 5.88 | 5.89 | 6.89 |

**Table 2** Surface mass levels

| Groups levels of surface mass, g/m$^2$ | | | |
|---|---|---|---|
| Light | | Heavy | |
| Min | Max | Min | Max |
| 177.0 | 188.0 | 189.0 | 199.7 |

First, the photometric properties of the fabric face are measured. Numerical values from colour measurements—DeltaCieLab—are presented in Table 1.

At the second stage, the weights of the small samples are measured across the width of the fabric. The measurement data are given in Table 2.

As a result of the measurement of one property—surface mass after processing, two main indicators are obtained. The first indicator is the average surface weight of the fabric roll, and the second indicator is the unevenness of the surface weight. Both indicators are equally important when selecting the rolls and grouping them into relatively uniform lots.

## 4 Analysis and Discussion

From the obtained results, three numerical series are obtained—of the colour differences, the average mass, and the unevenness in the distribution of the surface mass. These three numerical series: colour, weight, and unevenness become clear after their arrangement in ascending order and graphical visualization. In Fig. 3, the distribution functions of the three indicators are given in three graphs.

First of all, it should be pointed out that this arrangement only shows the increasing nature of the distribution functions. That is, these functions are not equally related to the fabric roll, but according to their value. Without direct relevance to the purpose of this paper, the distribution functions are linear in nature. This means that among the rolls of fabric there are no extraneous rolls that accidentally fell into the general assembly. Furthermore, the linear nature of the property variation suggests a harmonious grouping of the rolls. In this way, the smooth transition of properties between individual groups will be preserved.

The groups will be a prerequisite for adjusting the subsequent treatments so that finally a general population with relatively uniform quality is obtained.

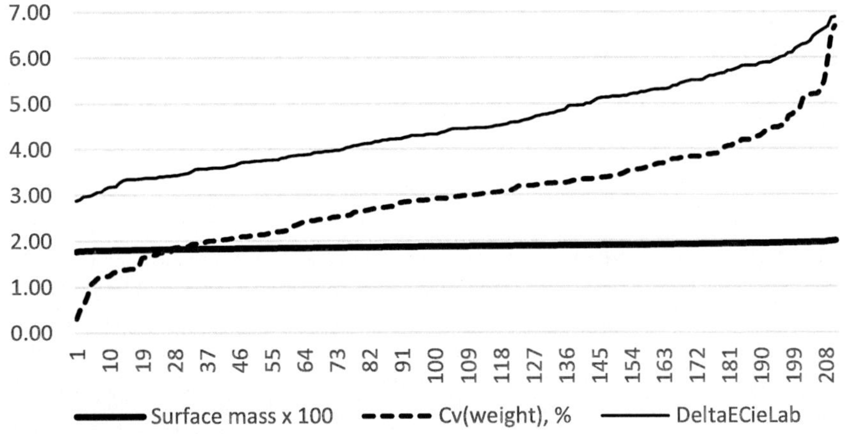


**Fig. 3** Numeric series development

The first grouping selection is by colour. Colour differences range from 2.88 to 6.89 DeltaCieLab. Considering that the permissible deviations are within the limits of $\Delta$CieLab $= \pm 1^{(\pm 0.20)}$, then all the rolls are grouped into four groups: light, medium light, medium dark, and dark. This is how all the rolls from the general population are initially grouped.

The second selection is the rejection of rolls with weight unevenness Cv(weight) $> 5\%$ from their colour groups and setting them aside as potentially non-conforming product.

The third selection involves the rolls, grouped into four colour groups. The general assembly consists of a common batch of one type of fabric with the same raw materials, construction, and properties. From the weight distribution functions, it is found that the margin varies from $-10$ to $+ 10\%$ relative to the surface mass average value. The requirement not to allow deviations greater than 5% necessitates the regrouping of the rolls into two subgroups: light and heavy. Thus, each of the colour groups is divided into two subgroups: light and heavy.

Thus, eight groups are formed with internally uniform properties in terms of colour and weight. In order to practically ease the technological flow in the production program, the fourth colour group "dark" is combined and includes both the light and heavy rolls, as well as the uneven ones from the second stage of selection. The seventh group formed in this way contains rolls of fabric with properties and quality that are borderline to the non-conforming product.

Algorithmic sequence of the work procedure.

- Sampling;
- Labelling;
- UV optical bleaching identification;
- DeltaE CieLab colour differences measurement;
- Small sections weight measurement;

- Fabric face colour selection and groupment;
- Unevenness mass selection and groupment;
- Surface mass selection and groupment.

## 5 Conclusion

The present paper is predominantly experimental in nature. The development reflects a practical solution to optimize incoming quality control of raw or finished fabrics. The proposed algorithmic sequence of operations can constitute a work procedure from the QMS of a textile or sewing enterprise. As a direction for future research, the need to determine the correlation dependences between the weight characteristics and the optical properties of the face of the fabrics is outlined.

## References

1. ISO 3801:1977, Textiles—Woven fabrics—Determination of mass per unit length and mass per unit area (1977).
2. Andonova, S.A., Rimini, G., Analysis and practical criteria for assessing the uniformity of fabric mass per unit area, Vlakna a Textil, ISSN: 1335-0617, Volume 27, Issue 2, Pages 13–17, (2020).
3. BDS ISO 105-A02:2005, Textiles—Tests for colour fastness—Part A02: Grey Scale for assessing change in colour, (2005).
4. Staneva, D., Vasileva-Tonkova, E., Yordanova, S., Kukeva, R., Stoyanova, R., Grabchev, I., Spectral characterization, antimicrobial and antibiofilm activity of poly(propylene imine) metallodendrimers in solution and applied onto cotton fabric, (2020) International Journal of Polymer Analysis and Characterization, (2020), 25 (5), 374–384, (2020).
5. H. Manov, D. Staneva, E. Vasileva-Tonkova, P. Grozdanov, I. Nikolova, S. Stoyanov, I. Grabchev Photosensitive dendrimers as a good alternative to antimicrobial photodynamic therapy of Gram-negative bacteria, Journal of Photochemistry & Photobiology, A: Chemistry (2021), 419, 113480, (2019).
6. Staneva D., Vasileva-Tonkova E., Grabchev I., A new bioactive complex between Zn(II) and a fluorescent symmetrical benzanthrone tripod for an antibacterial textile, Materials, 2019, (21) 3473, (2019).
7. Standard Trading Procedure, 1st Edition 2021, Legal & Compliance Part. https://www.tigerofsweden.com/se/ (2021)
8. Datacolor 650/600/400 User's Guide, Part No. 4230-0395M Rev 1, January, 2007, (2007).

# Application of Multi-objective Ant Colony Algorithm to Resolve Textile Color Matching Problem

**Sabrine Chaouch, Ali Moussa, and Neji Ladhari**

**Abstract** Color formulation is a crucial aspect of colorimetry, especially in the dyeing and finishing industries where it poses a significant challenge. To address this issue, this study proposes a multi-objective ant colony algorithm designed to resolve textile color matching problem by optimizing the reproduction step of requested target color. The algorithm achieves this by minimizing both color difference and metamerism between the proposed and the target color. To test the developed algorithm, reactive and direct dyestuffs were used for dyeing cotton samples, resulting in impressive results. All the obtained color differences were either below or within the acceptable threshold of 1, and approximately 75% of the samples displayed minimal metamerism, indicating a good match.

**Keywords** Color recipe prediction · Ant colony algorithm · Color matching · CMC color differences · Multi-objective function · Metamerism index

## 1 Introduction

Color has become a critical factor in the selection and quality of many products today, especially in textiles, where colors are obtained through dyeing or printing. The challenge in textile industries is to formulate the right color, which involves predicting the optimal recipe of compatible dyestuffs and their respective quantities

S. Chaouch (✉) · A. Moussa · N. Ladhari
Textile Engineering Laboratory, University of Monastir, Ksar-Hellal, Tunisia
e-mail: chaouchsabrine@hotmail.fr

S. Chaouch
Higher Institute of Technological Studiesof Ksar-Hellal, Ksar-Hellal, Tunisia

A. Moussa
National Engineering School of Monastir, University of Monastir, Monastir, Tunisia

N. Ladhari
Higher Institute of Fashion of Monastir, University of Monastir, Monastir, Tunisia



S. B. Abdessalem et al. (eds.), *Proceedings of the Second International Conference of Innovative Textiles and Developed Materials-ITDM'2; 05-06 May 2023; Tunisia*,
https://doi.org/10.1007/978-981-99-7950-9_28

required to attain an identical color match. Over the years, various methods have been developed to predict color recipes [1, 2], with the majority derived from the Kubelka–Munk approach [3]:

$$\frac{K(\lambda)}{S(\lambda)} = \frac{[1 - R(\lambda)]^2}{2 \times R(\lambda)} \tag{1}$$

where $R$, $K$, and $S$ are the reflectance, the absorption, and the scattering coefficient at a given wavelength $\lambda$, respectively.

However, assumptions are needed to formulate differential equations when employing the Kubelka–Munk theory [2]. These assumptions dictate that the absorption and the scattering coefficients of single dyes are proportional to their quantities and are additive in a blend. Moreover, each dyestuff in a blend behaves independently, with the substrate contributing significantly to the scattering. There is also a linear relation between the concentration $C_i$ of every dyestuff $D_i$ and its $\left(\frac{K}{S}\right)_i$ valor, which makes it possible to determine the concentration of each dyestuff in any specific mixture [2] as follows:

$$\left(\frac{K}{S}\right)_{\text{mix}} = \left(\frac{K}{S}\right)_{\text{sub}} + \sum_{i=1}^{n} C_i \times \left(\frac{K}{S}\right)_i \tag{2}$$

where mix, sub, and $i$ stand for mixture, substrate, and dye $D_i$, respectively, and $n$ is the total number of mixed dyes.

Various approaches and techniques have been elaborated to tackle the problem of textile color matching, including colorimetric and spectrophotometric methods [4–7]. For example, Agahian and Amirshahi have proposed a new color matching algorithm based on the equalization of the principal component coordinates of sample and target. Results indicate some types of improvements in comparison with similar previous algorithms [4]. Shams-Nateri has also proposed in this context an algorithm based on the derivative of the principal components as well as the derivative of the Kubelka–Munk function [5]. Moussa has proposed a new approach based on linear programming optimization to solve the textile color formulation problem which showed very good results with very small values of errors and color differences [6]. Furferi and Carfagni have proposed a mathematical assessment of spectrophotometric color matching which showed satisfactory results [7]. Other researchers have proposed artificial intelligence techniques for color formulation problem [8–12]. In 1991, Bishop et al. have developed a neural network algorithm for recipe prediction. The application of this method represented a radical departure from conventional mathematical treatments, and this initial study suggested that the approach may be successful [8]. This technique was subsequently improved by several researchers, to improve the performance of the prediction system, and Almodarresi et al. have combined the neural network and the scanner technique [9]. Jawahar et al. have also developed an artificial neural network model to predict color recipe based on tristimulus values given the concentration of dyes. Results showed that this method has the potential to give better predictive performance than the conventional Kubelka–Munk model [10]. Genetic algorithm was also used in color prediction. It is a metaheuristic technique inspired by the process and the evolutionary ideas of natural selection and genetics (selection, crossover, and mutation) [12]. Another artificial intelligence technique was introduced by Chaouch et al. who proposed a new method of predicting recipes by applying the ant colony algorithm [11].

However, as far as we know, no research has been carried out on the use of multi-objective ant colony optimization algorithm to resolve the textile color matching problem. Additionally, no one of cited studies have combined both color difference and metamerism as optimizing criterion within the same multi-objective function for selecting the optimal dyeing recipe, during the recipe selection process. Instead, color differences were calculated under a second illuminant after the recipe had already been chosen, which often required corrections to the recipe.

So, to sum up in all the mentioned methods researchers have developed some color formulation algorithms whose objective was to optimize the color recipe prediction step in reproducing the desired shades by minimizing only the color differences between the target color and the color obtained by the proposed recipe. They did not introduced the phenomenon of metamerism as an objective criterion simultaneously with color differences from the start and before choosing the recipe. In fact, after the choice of recipe, some researchers calculated the color difference under the second illuminant that generated sometimes some corrections in the proposed recipe.

Given that metamerism phenomenon is a crucial problem in all textile color matching steps, we introduce an ant colony optimization algorithm with a multi-objective function allowing the prediction of the most suitable color dyeing recipe that reduces both color difference and metamerism from the beginning of color reproduction process. This way, the developed algorithm selects the recipe that minimizes both criteria simultaneously from all available dyes. The algorithm efficiency was tested and proven.

## 2 Experimental

### *2.1 Materials*

The textile samples utilized in this study were 100% bleached cotton fabrics. Two sets of dyes were employed to dye these samples, including three reactive dyestuffs (CI Reactive Blue 235, CI Reactive Red 238, and CI Reactive Yellow 145) and four direct dyestuffs (CI Direct Black 22, CI Direct Orange 34, CI Direct Blue 85, and CI Direct Red 227) obtained from Huntsman Company without any further purification.

To ensure compatibility and uniform dyeing properties among the dyestuffs blended in the same dyebath, their compatibility was evaluated and confirmed using the same methodology as described in the literature [13].

**Table 1** Optimal recipes of dyeing

| Type of dyes | Stage | Product | Value |
|---|---|---|---|
| Reactive dyestuffs | A | Reactive dyes | X%[a] |
| | | Wetting agent | 3 mL/L |
| | B | Electrolyte | Y[b] |
| | C | Caustic soda | Z[b] |
| Direct dyestuffs | A | Direct dyes | X%[a] |
| | | Wetting agent | 3 mL/L |
| | | Sodium carbonate | Y[b] |
| | B | Sodium sulfate | Z[b] |

[a]X% is the concentration of used dyes
[b]Y and Z are determined depending to the concentration of used dyes

## 2.2 *Dyeing*

The optimal dyeing recipe and procedure, for the two range of used dyes, are presented in Table 1 and Fig. 1, respectively.

Cotton fabrics were dyed with reactive and direct dyestuffs in a laboratory dyeing machine (Ahiba Nuance Top Speed) maintaining the material and liquor ratio of 1:10, at different dye concentrations from 0.05 to 4%. To test color reproducibility, each dye was performed three times.

## 2.3 *Color Evaluation*

The color of the dyed cotton fabrics was evaluated using a spectrophotometer (Spectraflash 600 Plus) under specific measuring conditions (measurement range: 400 to 700 nm, illuminants: D65 and A, standard observer: 10°). To reduce errors in reflectance measurements, three measurements were done at different locations for each sample [14].

To assess the color difference between the proposed and the target color, the CMC equation was used [15]:

$$\Delta E_{CMC(l:c)} = \left[ \left( \frac{\Delta L^*}{l S_L} \right)^2 + \left( \frac{\Delta C^*}{c S_C} \right)^2 + \left( \frac{\Delta H^*}{S_H} \right)^2 \right]^{1/2} \tag{3}$$

where *SL*, *SC*, and *SH* are weighting functions that adjust the CIE differences depending upon the location of the standard in CIE 1976 color space. ($\Delta L^*$) is the lightness difference, ($\Delta C^*$) is the chroma difference, and ($\Delta H^*$) is the hue difference. The *l* (lightness weight) and c (chroma weight) were included to allow different

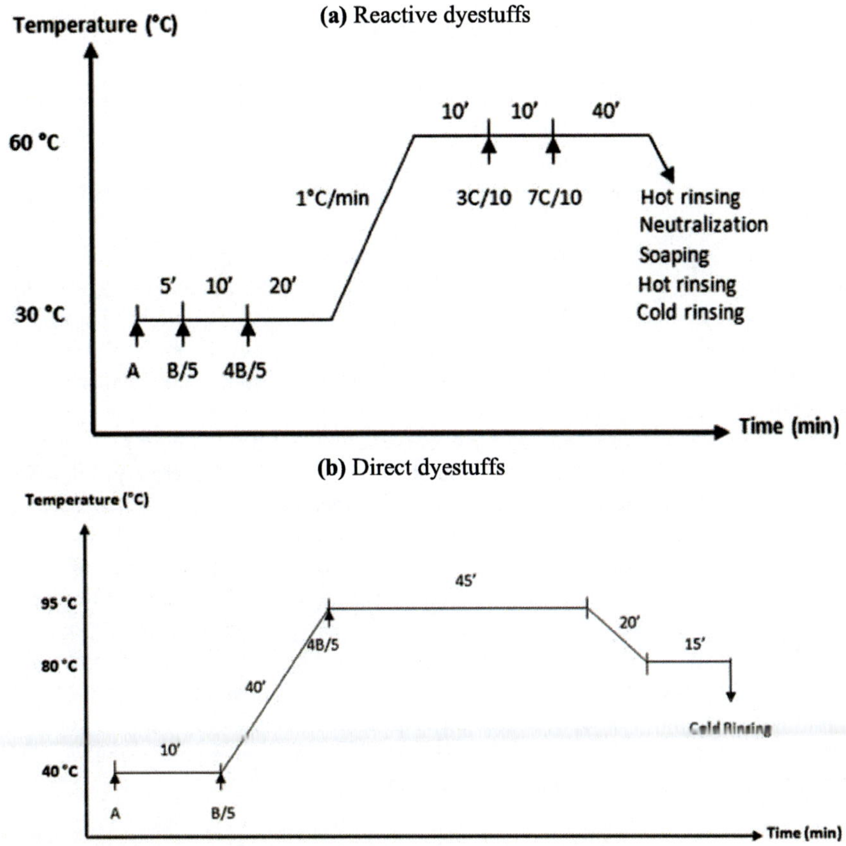


**Fig. 1** Optimal process of dyeing. **a** Reactive dyestuffs; **b** direct dyestuffs

weights to be used depending on the circumstances. The best $l$ and $c$ values, in textiles, have been found to be 2 and 1, respectively [16].

To calculate the index of metamerism $MI$ in this study, the following equation was used [17]:

$$MI_{D65/A} = \left[\left(DL^*_{D65} - DL^*_{A}\right)^2 + \left(DC^*_{D65} - DC^*_{A}\right)^2 + \left(DH^*_{D65} - DH^*_{A}\right)^2\right]^{1/2} \quad (4)$$

where $DL^* = \frac{\Delta L^*}{lS_L}$, $DC^* = \frac{\Delta C^*}{cS_C}$ and $DH^* = \frac{\Delta H^*}{S_H}$ calculated under illuminants D65 and A.

## 2.4 Multi-objective Ant Colony Algorithm

The ant colony optimization (ACO), an optimization algorithm that was introduced by Dorigo and Caro in the 1990s [18], is purely inspired from the foraging behavior of real ants. In their search for food, the process begins with ants exploring the surrounding area randomly while leaving a chemical pheromone trail as they move. The pheromone acts as a guide for other ants, as they are able to smell it and are more likely to choose paths with higher pheromone concentrations. If an ant locates a source of food, it will assess the quality and quantity of that food before bringing it back to the ants nest. On the way back, the ant may leave a stronger pheromone trail if the food is of higher quality or quantity. Through indirect communication via pheromone trails, ants are able to discover the shortest itinerary between that source of food and their nest [19].

In a single-objective ACO, we try to solve a problem by optimizing one single objective or criterion. While, in multi-objective ACO, we try to solve a problem by optimizing several objectives or criterions simultaneously.

To solve color matching problem in textile industries, our developed algorithm operates under the assumption that the reference color acts as the source of food, while the various dyestuff concentrations constitute the paths discovered by ants in looking for this source. The optimal dyeing recipe, which minimizes both the CMC color differences and the metamerism between the proposed and the reference colors, is likened to the shortest path between the ants nest and the source of food. The algorithm's multi-objective function *f* consists in our work to minimize both the CMC color difference and the index of metamerism at the same time, using weights ($\Omega$ and 1-$\Omega$) for each one of the two objectives:

$$f = \Delta E_{CMC(2:1)}.\Omega + MI_{D65/A}.(1 - \Omega) \tag{5}$$

where $f$ is the multi-objective function, $\Delta E_{CMC(2:1)}$ is the CMC color difference, $MI_{D65/A}$ is the index of metamerism, and $\Omega$ is the weight of factor to balance $\Delta E_{CMC(2:1)}$ and $MI_{D65/A}$ $(0 \leq \Omega \leq 1)$. $\Omega$ was fixed to 0.5 to have the same importance of the two criteria.

In effect, in the developed algorithm, each ant in the colony is required to randomly select a concentration $C_{ij}$ for each available dye in the first iteration and follow well-defined probabilities in subsequent iterations. As the ant moves through the path, which is a solution, it selects different concentrations of dyes to form a dye recipe. By implementing each predicted solution in dyeing, the resulting color is compared to the desired color, and some color differences and metamerism are obtained. The best solution is the one that minimizes the multi-objective function $f$, which aims to minimize both $\Delta E_{CMC(2:1)}$ and $MI_{D65/A}$ simultaneously.

At each iteration *t*, the choice of the concentrations is done randomly, based on the probability value $P_{ij}$:

$$P_{ij}(t) = \frac{\tau_{ij}^{\alpha}(t).{}_{ij}^{\beta}}{\sum_{l \in N_i^k} \tau_{il}^{\alpha}(t).{}_{il}^{\beta}} \tag{6}$$

where:

$P_{ij}(t)$ is the probability to choose the concentration $j$ of the next dye after choosing the concentration $i$ of the current dye;
$\tau_{ij}$ is the value of pheromones between the concentration $i$ of the current dye and the concentration $j$ of next dye;
$\eta_{ij} = \frac{1}{f}$ is the visibility between these two concentrations $i$ and $j$;
$N_i^k$ is the set of concentrations of dye that the ant has not yet chosen at the moment $t$;
α and β are the parameters controlling the relative importance between the rate of pheromones and visibility.

The updating of pheromone values is done as follows:

$$\tau_{ij}(t) = \rho.\tau_{ij}(t-1) + \Delta\tau_{ij} \tag{7}$$

where $\rho$ is the evaporation parameter of pheromones and $\Delta\tau_{ij} = \frac{1}{f}$.

The process of the proposed ant colony algorithm consists of several steps as shown in Fig. 2.

The performance of our proposed algorithm depends on the choice of its parameters; they have an impact on the quality of the obtained results. So, in order to optimize the ACO parameter values, a full factorial design was performed. The variation levels considered for each parameter are presented in Table 2. The values were selected according to previously researches. The result analysis was achieved using MINITAB statistical software with 0.05 significance level.

## 3 Results and Discussion

To test the efficiency of the proposed multi-objective ACO, 40 reference samples were prepared using various mixtures of direct and reactive dyestuffs. The objective is to predict the optimal recipe for each standard color using both reactive and direct dyestuffs simultaneously. The optimal recipe is the one that minimizes both the $\Delta E_{CMC(2:1)}$ and $MI_{D65/A}$ criteria between the proposed and the reference color, combined in $f$. To achieve this, the developed ACO is applied to select the most suitable dyes from all available options and estimate the ideal dyeing recipe.

The ACO algorithm parameters have been set depending to the results of the full factorial design. Indeed, the optimization plot presented in Fig. 3 proposes that those factors including $N_{iter}$, $N_{ants}$, α, β, and ρ should be at 2000, 81, 0.5, 0.5, and 0.1, respectively.

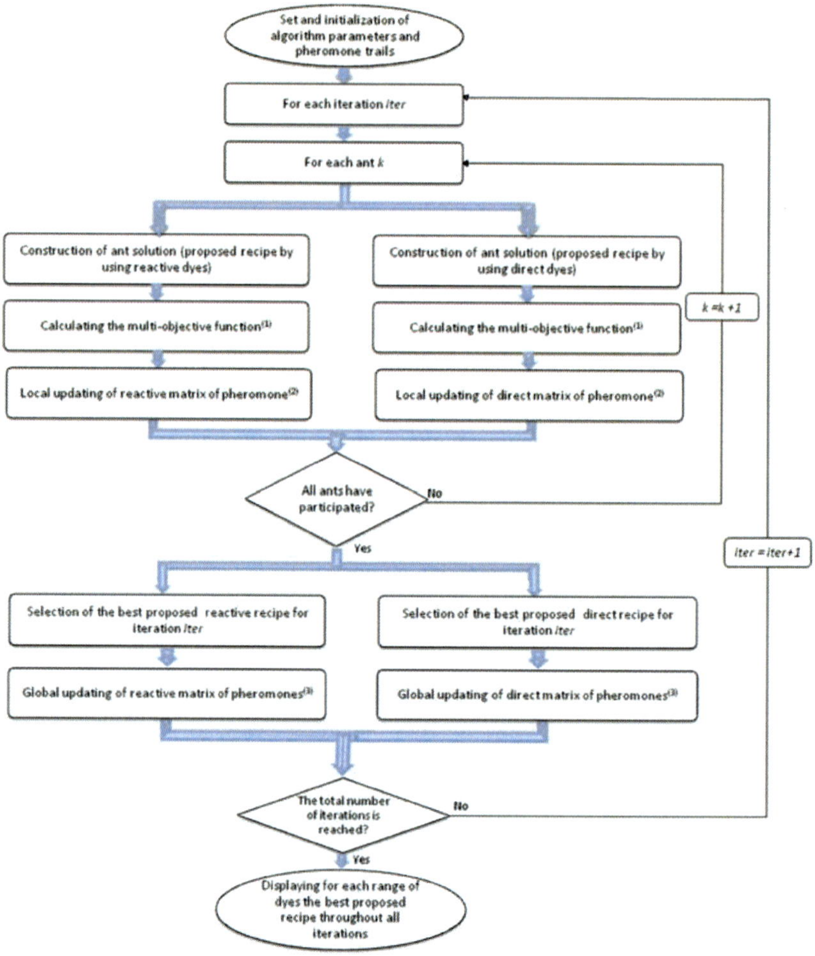


(1) Multi-objective function values using equation 5;

(2) Local updating of pheromone values using equation 7;

(3) Global updating of pheromone values using equation 7.

**Fig. 2** Process of the proposed multi-objective ACO

**Table 2** ACO parameters and variation levels

| Parameter | Variation levels |
|---|---|
| Number of iterations $N_{iter}$ | 500; 1000; 2000 |
| Number of ants $N_{ants}$ | 2; 20; 81 |
| Pheromone rate $\alpha$ | 0; 0.5; 1; 2; 5 |
| Visibility rate $\beta$ | 0; 0.5; 1; 2; 5 |
| Evaporation rate $\rho$ | 0; 0.1; 0.5; 0.9; 1 |

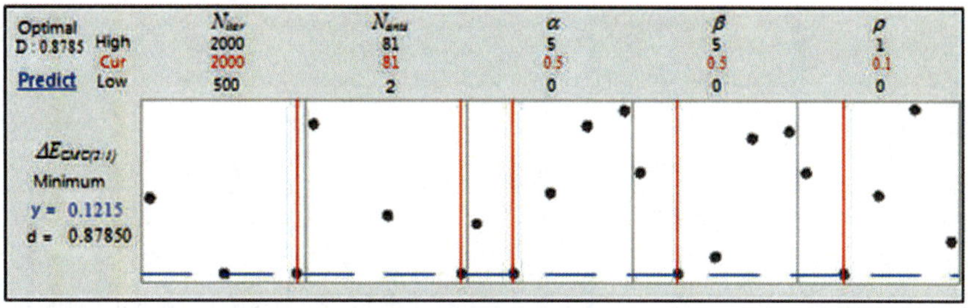


**Fig. 3** Optimization plot of algorithm parameters

Figures 4 and 5 show the results of applying the developed multi-objective ACO algorithm in proposing dyeing recipes for the 40 standard samples. For each standard sample, five simulations were carried out, so the obtained values of $\Delta E_{CMC(2:1)}$ and $MI_{D65/A}$are the results of five different simulations. We observed in Fig. 4a great color match between all the predicted color recipes and the reference colors. In fact, all $\Delta E_{CMC(2:1)}$ values are significantly less than the textile threshold of 1 [20], except samples 25 and 36 which its maximal values of $\Delta E_{CMC(2:1)}$ are close to the limit of textile threshold.

Concerning the metamerism index shown in Fig. 5, we notice that about 75% of predicted color recipes present a good color match, it has low$MI_{D65/A}$ values (less than 1). The highest $MI_{D65/A}$ values for other samples (greater than 1) can be related to the use of dyestuffs different from those used in reference sample preparation. It therefore needs further analysis.

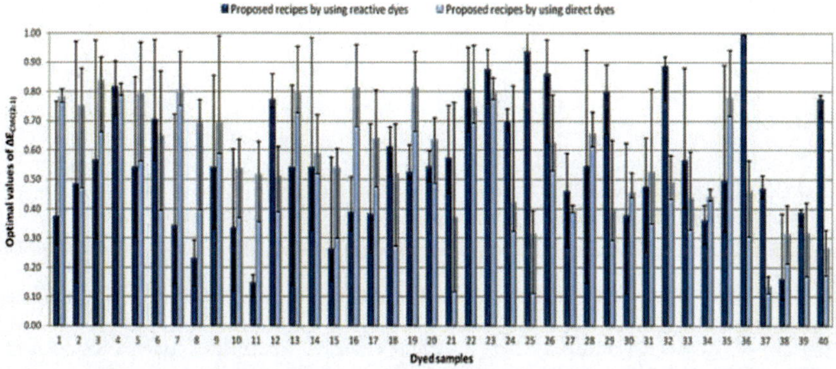


**Fig. 4** Optimal values of CMC color differences $\Delta E_{CMC(2:1)}$ between colors reproduced by the ACO and the reference colors

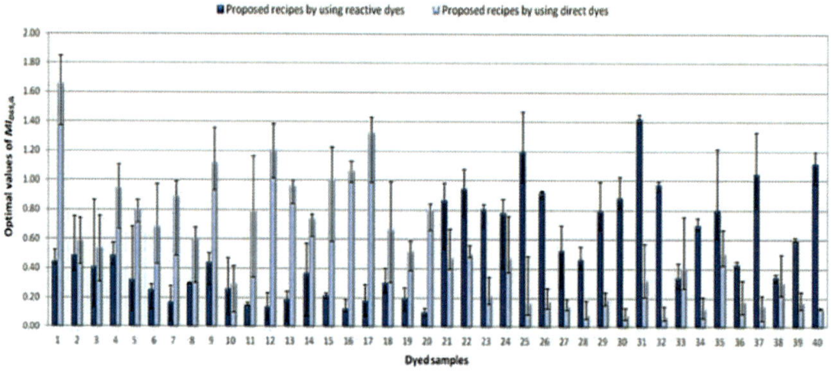


**Fig. 5** Optimal values of metamerism index $MI_{D65/A}$ between colors reproduced by the ACO and the reference colors

# 4 Conclusion

This study describes a new algorithm for solving textile color matching problem, which uses multi-objective function to reduce both $\Delta E_{CMC(2:1)}$ and $MI_{D65/A}$ between the obtained and the reference color. The technique of combining $\Delta E_{CMC(2:1)}$ and $MI_{D65/A}$ in the same algorithm function $f$ is helpful in achieving good results in color matching process. Application of this algorithm to reformulate standard colors results in satisfactory outcomes, along with most proposed recipes having small values of color differences and minimal metamerism. These initial findings demonstrate the potential of the developed multi-objective ACO and suggest further investigation is warranted.

# References

1. Agahian, F., Amirshahi, SH.: Design of Virtual Illuminants to Control the Colors Under Multiple Illuminants. Color Research & Application 34(3), 205–209 (2009).
2. Wyszecki G., Stiles WS.: Color Science: Concepts and Methods, Quantitative Data and Formulae. 2nd ed., New York: Wiley (2000).
3. Kubelka, P., Munk, F.: Ein beitragzuroptik der farbanstriche. Zeitschrift für Technische Physik.12, 593–601 (1931).
4. Agahian, F., Amirshahi, SH.: A new matching strategy: Trial of the principal component coordinates. Color Research & Application 33(1), 10–18 (2008).
5. Shams-Nateri, A.: Dye concentrations determination in ternary mixture solution by using colorimetric algorithm. Iranian Journal of Chemistry & Chemical Engineering 30(4), 51–61 (2011).
6. Moussa, A.: Textile color formulation using linear programming based on Kubelka-Munk and Duncan theories. Color Research and Application, 1–11 (2021).
7. Furferi, R., Carfagni, M.: An As-Possible Mathematical Assessment of Spectrophotometric Color Matching. Journal of Applied Sciences 10(18), 2108–2114 (2010).

8. Bishop, JM., Bushnell, MJ., Westland, S.: Application of neural network to computer recipe prediction. Color Research & Application 16(1), 3–9 (1991).
9. Almodarresi, ESY., Mokhtari, J., Almodarresi, SMT., Nouri, M., Shams-Nateri, A.: A scanner based neural network technique for color matching of dyed cotton with reactive dye. Fibers and Polymers 14(7), 1196–1202 (2013).
10. Jawahar, M., Babu, C., Kannan, N., Kondamudi-Manobhai, M.: Artificial neural networks for colour prediction in leather dyeing on the basis of a tristimulus system. Coloration Technology131(1), 48–57 (2015).
11. Chaouch, S., Moussa, A., Ben Marzoug, I., Ladhari, N.: Colour recipe prediction using ant colony algorithm: principle of resolution and analysis of performances. Coloration Technology 135(5), 349–360 (2019).
12. Chaouch, S., Moussa, A., Ben Marzoug, I., Ladhari, N.: Application of genetic algorithm to color recipe formulation using reactive and direct dyestuffs mixtures. Color Research & Application 45(5), 896–910 (2020).
13. Chaouch, S., Moussa, A., Ben Marzoug, I., Ladhari, N.: Study of C.I. Reactive Yellow 145, C.I. Reactive Red 238 and C.I. Reactive Blue 235 dyestuffs in order to use them in color formulation. Part 1: Characterization and compatibility. Journal of the Textile Institute110(7), 1042–1050 (2019).
14. Moussa, A., Dupont, D., Steen, D., Zeng, X.:Colour change as a result of textile transformations.Coloration Technology124(4), 234–242 (2008).
15. Clarke, FJJ., McDonald, R., Rigg, B.: Modification to the JPC79 colour-difference formula. JSDC100, 128–132 (1984).
16. Kuehni, RG.: The CMC(l:c) colordifference formula and the values for the weights l and c. AATCC Review 5, 25–26 (2005).
17. Steen, D., Dupont, D.: La métamérie. Euroforumconferences, Paris (2001).
18. Dorigo, M., Caro, GD.:New Ideas in Optimization. Chapter: The Ant Colony Optimization Meta-Heuristic, pp. 11–32. London: McGraw-Hill Ltd. (1999).
19. Blum, C.: Ant colony optimization. Introduction and recent trends. Physics of Life Reviews 2(4), 353–373 (2005).
20. Berger-Schunn, A.: Practical Color Measurement.. New York: John Wiley & Sons (1994).

# Deep Learning for DENIM Defects Detection

**Khaled Hammami, Imed Feki, and Saber Ben Abdessalem**

**Abstract** Fabric defects can be attributed to various factors such as raw material, spinning, weaving, knitting, and dyeing processes. These factors are the main causes of fabric defects, which are a major quality issue in the garment industry. The demand for high-quality fabrics has increased, and customers are now more aware of "non-quality" issues, leading to higher quality requirements. In this paper, we propose a deep learning approach based on neural networks for defect detection. In fact, we have developed a new methodology for detecting DENIM weaving defects to facilitate fast, efficient, and accurate decision making. Our neural network model is based on the VGG16 architecture, which generates convolutional feature maps. A classifier was used for our specific classification case. The proposed methodology was tested on a dataset containing a large number of images with and without defects. This approach showed an excellent overall model accuracy of 98.52%.

**Keywords** Textile defect · Deep learning · Convolutional neural network

## Abbreviations

| | |
|---|---|
| CNN | Convolutional Neural Network |
| DL | Deep Learning |
| ANN | Artificial Neural Network |
| ReLU | Rectified Linear Unit |
| DA | Data Augmentation |

K. Hammami (✉) · I. Feki · S. B. Abdessalem
Materials and Textile Processes Research Unit MPTex, University of Monastir, Monastir, Tunisia
e-mail: kh.hammami@gmail.com

I. Feki
e-mail: fekiimed@yahoo.fr

S. B. Abdessalem
e-mail: saber_ba@yahoo.fr

 
S. B. Abdessalem et al. (eds.), *Proceedings of the Second International Conference of Innovative Textiles and Developed Materials-ITDM'2; 05-06 May 2023; Tunisia*,
https://doi.org/10.1007/978-981-99-7950-9_29

| | |
|---|---|
| CV | Computer Vision |
| SGD | Stochastic Gradient Descent |
| ACC | Accuracy |
| CPU | Control Processing Units |
| GPU | Graphics Processing Units |

## 1 Introduction

Fabric defects can arise from various sources, including raw materials and the production process. These defects can occur at any stage of fabric manufacturing and have a significant impact on the fabric's quality and usability. Faulty yarns or incorrect machine settings can lead to different defects, affecting the fabric's properties and influencing its sale. Detecting these defects is crucial for DENIM fabric production and quality management. However, manual inspection is subjective and often lacks precision, with quality control personnel only able to detect around 70% of fabric defects [1]. The presence of fabric defects not only compromises quality but also diminishes profits. Moreover, prolonged exposure to intense light during inspection poses risks to workers' eyesight. Consequently, there is a growing demand for automated inspection systems to address these challenges.

Fabric defects are typically classified into minor, major, and critical categories based on their impact on product purchase and fabric ranking. Minor defects have minimal influence on the purchase decision, while major defects affect the purchase process, and critical defects significantly lower the fabric's ranking. Surface defects are further classified into two quality standards: surface color change and texture irregularity. These defects, such as yarn defects, weaving defects, isolated defects, pattern defects, and wet processing defects, stem from raw materials or the production process. A collection of common DENIM fabric faults is illustrated in Fig. 1.

The objective of this research is to develop an efficient and fast approach for detecting DENIM fabric defects using deep learning and neural network techniques. To address the variety of production defects, a dataset of high-definition images of

**Fig. 1** Examples of DENIM faults

DENIM fabric was collected from a prominent DENIM fabric company in Tunisia. The dataset encompasses major defects such as knots, holes, and oil stains, which were identified through visual inspection and confirmed by experienced fabric inspectors. The defect details are clearly outlined, covering a comprehensive range of major defects in DENIM fabric.

CNNs, or convolutional neural networks, are a category of deep neural networks used in the field of deep learning. They are specifically designed to process grid-structured data, such as images, using convolutional and pooling layers. CNNs are highly effective in extracting important features from images and performing tasks such as recognition and classification. Compared to linear models like SVM and logistic regression, CNNs have the advantage of learning more complex and nonlinear data representations, making them particularly suitable for solving challenging imaging problems and achieving better performance. It is worth noting that the use of techniques such as transfer learning (VGG16, ResNet, Inception, etc.) can also help overcome the lack of data by leveraging previously acquired knowledge from pre-trained models on large datasets.

Indeed, deep learning (DL) is a powerful method of machine learning that allows algorithms to learn autonomously from datasets. Convolutional neural networks (CNNs or ConvNets) have emerged as the most effective and widely used algorithms in computer vision (CV) since the early 2000s. CNNs consist of layers with diverse functionalities, including trainable parameter layers and layers dedicated to function implementation.

Machine learning (ML) is the field that enables computers to learn and improve from experience without explicit programming. During the training process, the ML model receives input data and extracts natural patterns and salient features from them. Based on these features, the model learns to associate attributes with each sample or assign samples to identified clusters. This enables the model to make predictions on new, unseen data. CNN architectures such as FCN [2], U-Net [3], SegNet [4], and their successors share common components like convolution, pooling, and activation functions. The pooling layer plays a crucial role in preventing overfitting and reducing spatial dimensions. Deep networks can consist of more than 100 convolutional layers, for instance, VGGNet [5] with 16 layers and ResNet [6] with 152 layers. To enhance the segmentation of fine structures, additional processing is incorporated into CNNs to refine the coarse CNN outputs.

## 2 Materials and Methods

We have designed and produced a device that simulates the fabric displacement in an inspection machine and a loom to scroll the DENIM fabric (see Fig. 2). A fixed camera was installed to capture faults. We used a 13 MP camera and D65 10W/IP 65 lighting. Choosing the lighting is essential for the quality of a shot. It facilitates subsequent evaluation of the image.

**Fig. 2** Fabric scrolling device

A resizing of the image base is necessary. In deep learning, resizing images is a critical preprocessing step in CV. Principally, deep learning models train faster on small images.

GPUs are processors suitable for training neural networks. The GPUs accelerate the parallelizable computation-intensive portions of code, while the remaining sequential portions are processed by the CPUs. Neural networks require many matrix operations, which are accelerated when distributed across the thousands of GPU cores. The training of our model was carried out on a workstation with Intel(R) Xeon(R) W-1250P CPU @ 4.10 GHz, 16 GB installed RAM (15.7 GB usable), 64-bit operating system, × 64 processor and Intel UHD graphics P630 with 6 cores and 12 threads, and a maximum turbo frequency of 4.7 Ghz. The increase in calculations offered by our workstation, allowed us to speed up our processing of standard routines such as activation functions.

Our model uses transfer learning with the VGG16 model. We start by loading the pre-trained VGG16 model and then add fully connected layers on top of it. The model is created using the sequential class from Keras, where the first layer is the VGG16 model added using the add() method. Next, a flatten layer is added to convert the outputs of the VGG16 model into a one-dimensional vector. To introduce non-linearity and learn more abstract representations of the data, two fully connected layers are added. The first layer has 256 neurons with ReLU activation, and the second layer has 128 neurons also with ReLU activation. Finally, a last fully connected layer is added, corresponding to the number of classes to predict, using the softmax activation function to obtain a probability distribution over the classes.

The learning rate used in this model is set to 0.001. The learning rate controls the magnitude of weight updates in the model during training, which affects the

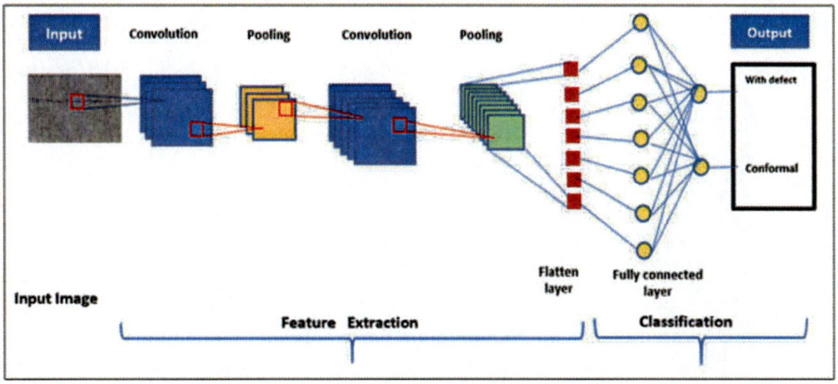


**Fig. 3** CNN model flowchart

convergence speed. A higher learning rate allows for faster convergence but may lead to oscillations or instability. On the other hand, a lower learning rate allows for slower but more precise convergence. In this model, a learning rate of 0.001 is chosen for the SGD optimizer to strike a balance between speed and accuracy.

The neural model (see Fig. 3) is trained using the stochastic gradient descent (SGD) optimization algorithm, and the weights are updated using the error back-propagation algorithm. The goal of the gradient descent algorithm is to adjust the weights in such a way that the error is reduced at each iteration, effectively following the slope of the error surface. The training process of our model is facilitated by the KERAS fit generator libraries, which handle the gradient descent and optimize the network's cost function.

To train our network, we divide the images into an 80–20% ratio for training and validation, respectively. This means that the model learns from the training set and then evaluates its predictions on the validation set. Both datasets are labeled, indicating whether or not the images belong to the defect class. The dataset comprises 331 images of size 4160 × 3120 pixels, which are resized to 224 × 224 pixels. The dataset is further divided into 230 training images, 71 test images, and 25 prediction images. The distribution between the training and validation sets is illustrated in the following diagram (see Fig. 4).

## 3 Results and Discussion

By applying transformations to the input images, we were able to improve the performance of the network. We specifically utilized image augmentation techniques with the KERAS image data generator to increase the variability in our dataset. This approach is commonly used in deep learning to generate additional samples and enhance the efficiency of the network, particularly when working with a limited

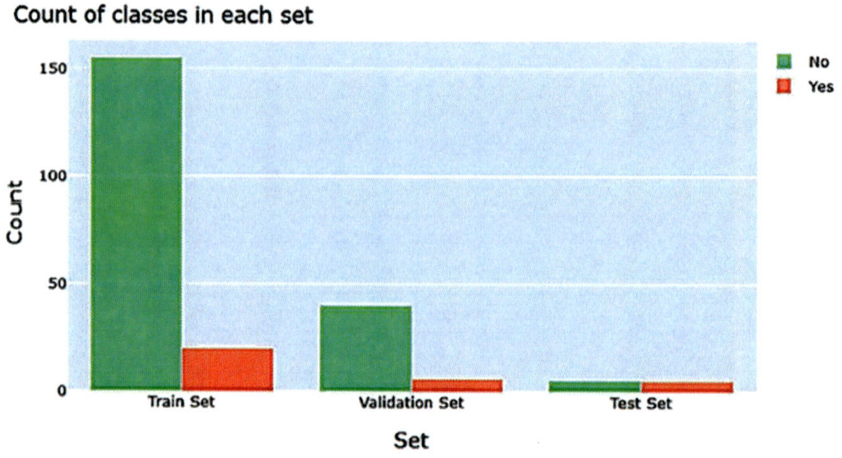


**Fig. 4** Distribution of dataset 80% train set and 20% validation set

dataset. As a result, our approach achieved a high level of accuracy in accurately identifying DENIM defects. The outcomes of our model are illustrated in Fig. 5.

The accuracy of the model, represented by the curve (see Fig. 6), demonstrates the progress over the course of multiple epochs. An epoch refers to a complete iteration of the training data through the algorithm, evaluating the model's prediction accuracy against the actual data. In this study, we had 331 samples for training, out of which the model correctly classified 326 samples, resulting in an accuracy of 98.52%.

The curve (refer to Fig. 7) demonstrates a good fit, which is characterized by a gradual decrease in both training and validation losses, reaching a point of stability with minimal deviation between the two final loss values.

Comparing the accuracy of our model with some previous works, as shown in Table 1, we notice that the effectiveness of our method is better than the others, which is based on the combination of a CNN model classification to predict whether the image has a defect or not with the automation of the image extraction process, is clear.

The equipment used, which works with the developed method, can be easily installed on an inspection machine or a loom to respond to a real industrial need and to allow the online detection of DENIM defects on these machines. On a loom, the developed device allows not only the detection of defects but also the readjustment of the machine settings in order to obtain the desired fabric quality.

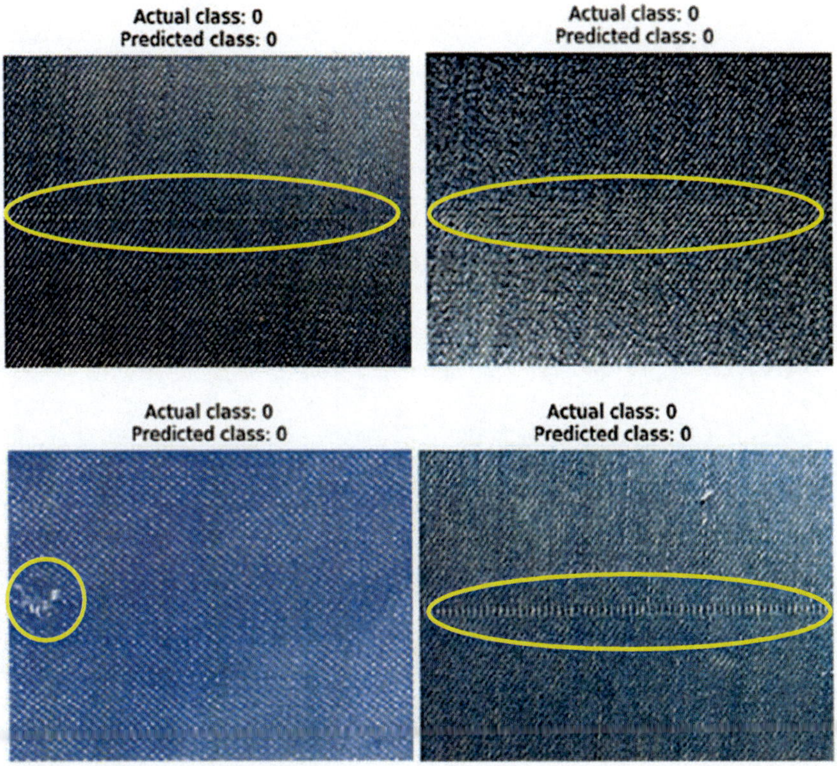


**Fig. 5** Examples of prediction results

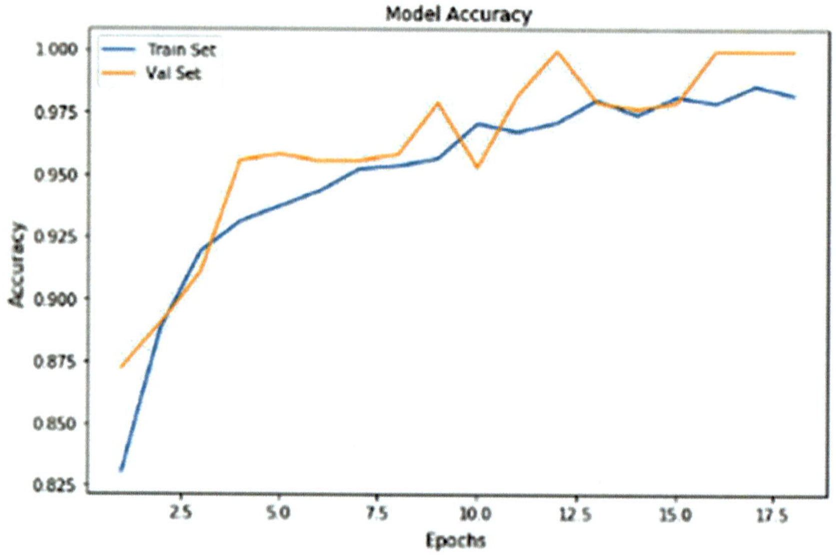


**Fig. 6** Accuracy of CNN model learning

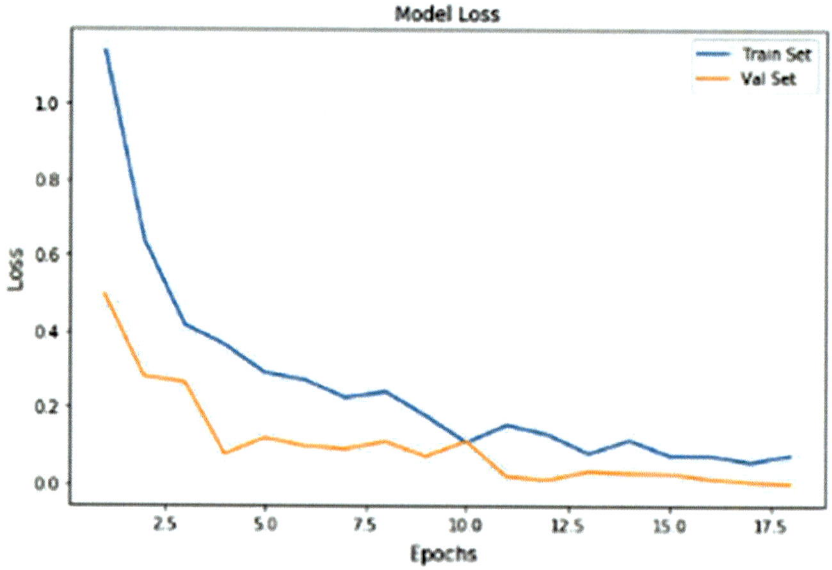


**Fig. 7** Training loss

**Table 1** Accuracy reached by some previous works

| Paper | Reference | Year | Accuracy |
|---|---|---|---|
| Computer vision for automatic detection and classification of fabric defects employing deep learning algorithm | [7] | 2019 | 96.55% |
| Fabric defect detection using computer vision techniques | [8] | 2020 | 97.8% |
| Fabric defect detection based on a deep convolutional neural network using a two-stage strategy | [9] | 2021 | 96% |
| Method of wavelet transform for fabric defect classification | [10] | 2022 | 92.98% |
| Our method | | 2023 | 98.52% |

## 4 Conclusion

Fabric defects are the main problem in DENIM production. Today, human inspection is still insufficient, and fabric inspection by a CV-based fabric defect detection system would allow better quality assurance. In this work, a deep learning algorithm has been developed for an on-loom fabric defect inspection system based on the VGG16 method, chosen according to the analysis of feature extractors. The choice of the extractor was crucial because the number of parameters and the types of layers directly affect the speed and performance of the detector. Our approach was able to identify DENIM defects with an excellent model accuracy of 98.52%.

In order to obtain satisfactory results, the use of a high-resolution image acquisition tool is recommended for the detection of certain delicate DENIM defects, such as the boot bar defect or the dropped weft. The developed methodology can also be successfully applied to knitting images.

## References


1. Dorrity, L., Vachtsevanos, G., Jasper, W., Real-Time Fabric Defect Detection and Control in Weaving Processes, In: National Textile Center, 1995, G94–2, 143–152 (70% detect by worker revistaindustriatextila.ro/)
2. J. Long, E. Shelhamer, and T. Darrell, "Fully convolutional networks for semantic segmentation," in Proc. IEEE Conf. Computer. Screw. Pattern Recognition, Jun. 2015, p. 3431–3440.
3. O. Ronneberger, P. Fischer, and T. Brox, "U-net: Convolutional networks for biomedical image segmentation," in Proc. Int. Conf. Med. Image Computer. Comput.-Assist. Intervent., Cham, Switzerland, 2015, p. 234–241.
4. V. Badrinarayanan, A. Kendall, and R. Cipolla. (2015). "SegNet: A deep convolutional encoder-decoder architecture for image segmentation." [Online]. Available: https://ieeexplore.ieee.org/document/7803544.
5. L. Wang, S. Guo, W. Huang, and Y. Qiao. (2015). "Places205- VGGNet models for scene recognition." [Online]. Available: https://arxiv.org/abs/1508.01667.
6. K. He, X. Zhang, S. Ren, and J. Sun, "Deep residual learning for image recognition," in Proc. IEEE Conf. Computer. Screw. Pattern Recognition, Jun. 2016, p. 770–778.

7. Jeyaraj, PandiaRajan, and Edward Rajan Samuel Nadar. “Computer vision for automatic detection and classification of fabric defect employing deep learning algorithm.“ International Journal of Clothing Science and Technology 31.4 (2019): 510–521.
8. Rasheed, Aqsa, et al. “Fabric defect detection using computer vision techniques: a comprehensive review.“ Mathematical Problems in Engineering 2020 (2020): 1–24.
9. Jun, Xiang, et al. “Fabric defect detection based on a deep convolutional neural network using a two-stagestrategy.“ Textile Research Journal 91.1–2 (2021): 130–142.
10. Kuo C-FJ, Lee CJ, Tsai CC. Using a neural network to identify fabric defects during dynamic fabric inspection. Text Res J 2003; 73: 238–244.

# Modeling of Textile Materials and Structures: Some Numerical and Experimental Aspects

**Hassen Hedfi and Hédi BelHadjSalah**

**Abstract** In this paper, we present, in a synthetic way, the practice of modeling textile materials and structures. To this end, we consider two aspects in this investigation: first, the numerical aspect of the modeling and then the experimental aspect as a tool for the identification of the constitutive equations invested in the modeling. It is a question, initially, of bringing a methodological approach for the modeling of the textile structures, in particular the woven structures, the formulation of laws of behavior making it possible to describe the response of these structures to mechanical solicitations. We present and discuss a review of the main models used. Secondly, we introduce a methodological approach for the experimental identification of the parameters of these constitutive laws.

**Keywords** Numerical modeling · Textile structures · Identification of constitutive equations

## 1 Introduction

Research in fabric modeling began within the textile engineering community since the 1930s. Research in this field underwent a significant evolution in the mid-1980s with the contribution of the computer graphics community. One of the main concerns of this community is to reproduce the appearance of the fabric effectively. Indeed, for the graphic designer, efficiency, stability, and visual realism are more important than the accuracy and physical reality of the material. On the other hand, researchers in textile engineering are interested in modeling fabrics in order to predict, in a sufficiently precise manner, the highly nonlinear behavior of these structures.

We can thus situate the practice of modeling woven structures, for the textile community, on two levels. The first, being the need to find a model that is both

H. Hedfi (✉) · H. BelHadjSalah
Mechanical Engineering Laboratory LGM, National School of Engineers of Monastir, University of Monastir, 5019 Monastir, Tunisia
e-mail: hassen.hedfi@enim.u-monastir.tn

 
S. B. Abdessalem et al. (eds.), *Proceedings of the Second International Conference of Innovative Textiles and Developed Materials-ITDM'2; 05-06 May 2023; Tunisia*,
https://doi.org/10.1007/978-981-99-7950-9_30

representative of the textile structure and simplifying. Indeed, we recognize that textile fabrics are very complex and dependent on several factors (construction, material, and mode of deformation) so trying to study them directly is a difficult challenge to take up without going through simpler models.

The second plan relates to the objectives of the study of textile structures. Indeed, textile fabrics we do the modeling of this structures to predict and study their behavior, to simulate and animate them. We do this practice for applications starting from pure theoretical investigations, to pure animations of virtual objects in the field of computer graphics, passing through the development of computer-aided design (CAD) and/or computer-aided manufacturing (CAM) tools for industrial applications. This makes it possible to reduce the reaction time and provide great flexibility for the entire textile chain. The interest of this practice extends to the performance of virtual characterization tests.

In the field of modeling and simulation of textile fabrics, the first models were static models, and more precisely, they are geometric models allowing the study of the response of the textile structure to particular stresses such as tensile forces or shear forces or bending forces, without any consideration of the effect of the mass of the structure considered immobile.

Static and/or kinematic modeling of textile fabrics is useful in animation applications or simulation of the behavior of these materials under conditions where the inertia effect can really be neglected (simulation of a traction, a shear test, and, at the limit, a bending test). However, in applications where one wants to model more complex modes of deformations (such as fabric drape and fabric formability) or to make the virtual animation of a dress worn by a dancing woman or of a curtain subjected to the action of an air current, the static and kinematic models cease to be applicable.

In this context, we can define the dynamic modeling of textile structures as being a modeling that takes into account the effect of inertia on the deformation and behavior of these materials. A dynamic model will also make it possible to consider the movement of the structure and the effect of this movement on the stresses undergone by the textile. Such modeling will therefore make it possible to predict the shape taken by the textile material following the application of a force (including inertial, collision, and friction forces) and/or when it is in motion.

In this paper, we present a sort of mini-review on the modeling and simulation of the mechanical behavior of structures and textile materials. This mini-review will focus on some characteristic and distinctive aspects of the behavior of this type of materials and structures.

Namely, we can cite their great deformability, under low stresses, their ability to develop curvature and bending deformations that are quite complex and difficult to predict, as well as shear deformations that are so complex and significant that their predictions also become quite difficult. Added to this is the low thickness of these structures and their low surface mass density, which cause and amplify the complexity of the deformations undergone by the textile structures and make their modeling and simulations difficult and complex.

The second point highlights certain aspects relating to the experimental investigation that one can be carried out for the characterization and determination of the physical and mechanical properties necessary and possibly sufficient, at the writing and parametrization of mathematical models of textile structures.

## 2 Modeling of Textile Materials and Structures

### *2.1 Textile Materials and Structures*

Textile materials and structures have several shapes and dimensions in space (i.e., one-dimensional 1D, two-dimensional 2D, or three-dimensional 3D).

We start from one-dimensional discontinuous structures, which are textile fibers. These fibers are of nature, origin, and structure: natural fibers (e.g., wool, cotton, silk, and linen), artificial fibers (e.g., viscose), and especially synthetic fibers (e.g., polyamide, polyester, and acrylic). Therefore, there will arise fibers with different and varying properties.

From these fibers, the transformation into yarns begins, which are continuous one-dimensional structures with different and variable structures and therefore properties.

Subsequently, we can transform threads into two-dimensional structures or textile surfaces. One can do this by knitting, giving rise to knitted structures, called knits, with variable properties depending on the way in which the knit is constructed and according to the finishing treatments that it subsequently undergoes.

A second transformation making it possible to obtain textile surfaces is weaving, giving rise to woven structures, having properties, which depend on the construction parameters, and the properties of the yarns used. However, we can modify and modulate according fabrics properties using appropriate finishing treatments.

Once we have a 2D structures (e.g., woven or knitted structures), we can make it to have a garment occupying a 3D domain in space. We mention the work of Long et al. [1] in which the authors elucidate the hierarchical relationships between the properties of fibers, yarns, and fabrics and their influences on the end-use properties and performance of garments. We also note that Fig. 1 is inspired, in part, from this work.

For further information on the attributes of textile fibers, mainly natural cellulosic fibers (e.g., cotton, hemp, and flax), and their distinctive properties, as well as their impacts on textile properties, we refer to the review article by Shuvo [2].

In Fig. 1, we summarize the main textile structures, their modes of transformation: from fibers to garments, the hierarchical relationships between the properties of fibers, yarns, and fabrics and their impact on the end-use properties and performance of garments, and the different scales of these structures from micro to macroscopic scale. We have also reported the same pattern for nonwovens as textile structures and products obtained from the fibers directly.

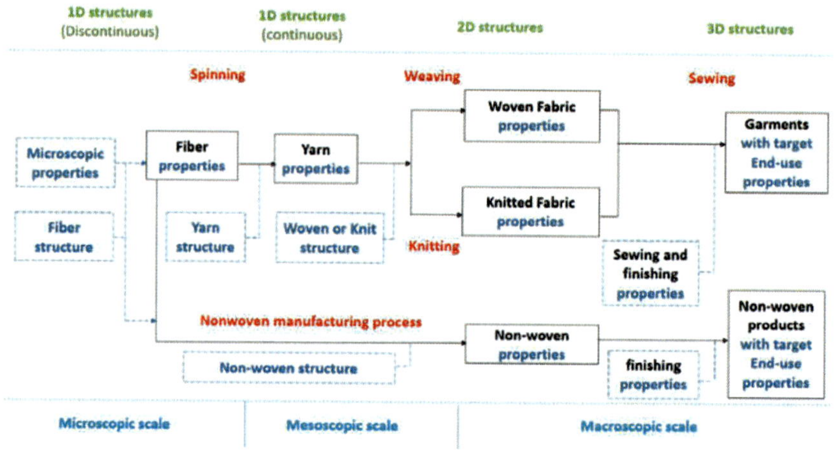


**Fig. 1** Main textile structures, their modes of transformation: from fibers to garments

## 2.2 *Fabric Modeling: Main Historical Milestones, Typology, and Finalities*

In the field of modeling and simulation of textile fabrics, the first models were static models and more precisely, and they are geometric models allowing the study of the response of the textile structure to particular stresses such as tensile, shear, or bending forces without any consideration of the effect of the mass of the structure considered immobile.

Static and/or kinematic modeling of textile fabrics is useful in animation applications or simulation of the behavior of these materials under conditions where the inertia effect can really be neglected (simulation of a traction, a shear test, and, at the limit, a bending test). However, in applications where one wants to model more complex deformation modes such as drape and formability, or to animate a virtual dress worn by a dancing woman, or a curtain subjected to action of an air current, the static and kinematic models cease to be applicable.

However, we can define the dynamic modeling of textile structures as being a modeling that takes into account the effect of inertia on the deformation and behavior of these materials. A dynamic model will also make possible to consider the movement of the structure and the effect of this movement on the forces undergone by the textile. Such modeling will therefore make possible to predict the shape taken by the textile material following the application of a force (including inertial, collision, and friction forces) and/or when a movement animates it.

In Table 1, we have reported, in a summary and general way, the main areas where we can make use, more and more important, frequent and systematic of the modeling and simulation of structures and textile materials, in first approach. However, this extends, in a more general way, to surface structures and deformable objects, as a

**Table 1** Some applications of modeling and simulation of textile materials and structures

| Field | Applications |
|---|---|
| Textile engineering and industry: (clothing or/ and furnishings) | -Clothing computer-aided design CAD<br>-Virtual prototyping (CLO, Maya Clothing)<br>-E-commerce of clothing and deformable objects |
| Composites and technical textiles industry | -Modeling and optimization of composite manufacturing and forming<br>-Prediction of the aging of composite materials |
| Computer graphics and virtual animation | -Cinematography: animated films, video games<br>-Dressed Avatars |
| The medical field | -Simulation of medical devices (such as sutures)<br>-Simulation of interaction with living tissues<br>-Simulation of shape-memory textile used for functional rehabilitation |
| The field of learning and training | -Textile engineering/materials engineering |

second approach. We have also listed, for each area, a certain number of applications, without claiming to be exhaustive.

In Table 2, we have reported a comparative study between the different models used textile fabrics modelling and simulation. We have briefly indicated its principle, its advantages, and its limitations of each model. The last column is devoted to indicating the pioneers and the main contributions. We refer the reader to the articles of Ng et al. [3] and Jevsnik et al. [4] for more details.

# 3 Continuous Models Formulation and Numerical Solving

## 3.1 General Formulation

The examination of the different models and works dedicated to the modeling of textile structures, in particular, woven ones, we propose the following general formulation:

$$\rho \overrightarrow{a} + \overrightarrow{f^{\text{int}}} = \overrightarrow{f^{ext}} \tag{1}$$

In this equation, we note:
-The mass density of the object: $\rho$.
- The acceleration vector: $\overrightarrow{a} = \frac{d^2 \overrightarrow{x}}{dt^2}$.
-The density of forces per unit volume: $\overrightarrow{f^{\text{int}}}$.

**Table 2** Models typology: Principe, advantages, limitation, and pioneers/main contributors

| Type of modelling | Models | Principe | Advantages | Limitations | Pioneers/main contributors |
|---|---|---|---|---|---|
| Geometric models | Models developed by textile researchers: GM-T | **Based** on a representation of the elementary cell of a fabric (usually plain weave)<br>**Used** to determine the response of this cell to a tensile stress, then other mechanical stress: bending, shear | **Better** understanding of the deformations undergone by fabric<br>**Calculation** of the forces undergone during fabrics deformation | **Limited** to the studied cases<br>Static models,<br>**Difficult** and even impossible generalization to other types of fabric | Peirce [5], Olofsson, [6] |
| | Models developed by computer graphics researchers: GM-T | **Purely** geometric representation of textile surfaces: curve fitting, interpolation, and mapping<br>**Aiming** to reproduce the distinctive forms of textiles: curvature, double curvature, crumpling, and draping | **Simple** to implement<br>**Applicable** to a wide variety of deformable surfaces | **Does** not take into account the physics of the material<br>**Rendering** of fabrics not always realistic | Weil [7], Agui et al. [8] |
| Discrete models | Mass-spring models | **Based** on a discrete representation of fabric. This discretization is made by considering mass points distributed uniformly and interconnected by deferent types of springs<br>**The behavior** of the materials is introduced through the springs:<br>- ***Springs*** describing the behavior in elongation–compression<br>- ***Springs*** describing the shear behavior<br>- ***Springs*** describing the bending behavior | **Interactive** animation of deformable objects<br>**Description** representative of the actual behavior of the fabric<br>**More** realism in the animation and simulation of textile fabrics<br>**Modeling** and simulation of very wide variety of deformable objects | **More** complex numerical implementation<br>**Difficulties** in handling large structures<br>**Difficulties** in treatment of contact between the simulated textile fabric and surrounding objects<br>**Difficulties** in controlling self-collisions | Haumann et Parent [9], Provot [10], |

(continued)

**Table 2** (continued)

| Type of modelling | Models | Principe | Advantages | Limitations | Pioneers/main contributors |
|---|---|---|---|---|---|
| Discrete models | Particles models | **Based** on a discrete representation of object by considering it as a distribution of mass points (or particles)<br>**Deformations** are obtained by controlling the particles interactions<br>**These interactions** represented by energy functions | **More** realism obtained<br>**Allowing** the modeling and animation of objects with complex geometries<br>**More** facilities in the processing and control of collisions and self-collisions | **Requiring** an elaborate identification of the physical and mechanical parameters of the material<br>**Requiring** significant machine computing resources | Breen et al. [11], Fontana et al. [12] |
| Continuous models | Deformable models | **Physical-**based model (an energy based formulation)<br>**Based** on differential geometry concepts<br>**Using** tensors to describe the behavior in elongation–compression, shear and bending | **More** realism obtained<br>**More** facilities in the modeling and animation of objects with complex geometries<br>**Models** easily configurable and controllable | **Needing** an elaborate identification of the physical and mechanical material parameters<br>**Necessitating** significant machine computing resources | Terzopoulos et al. [13], Magnenat-Thalmann et al. [14], Hedfi et al. [15] |
| | Continuous mechanical models | **Based** on the mechanics of continuums, considering that the fabric is a continuous medium<br>**Based** on the theory of elasticity to describe fabric dynamic behavior<br>**Using** on finite element method for numerical implementation | **More** realism<br>**More** controllability and easy to parametrize<br>**Useful** to simulate textile reinforced composites | **Needing** an elaborate identification of the physical and mechanical parameters<br>**Necessitating** important machine computing resources | Ascough et al. [16], Boisse et al. [17], Chen et al. [18] |

-The external forces: $\overrightarrow{f^{ext}}$.

We denote by t the time and $\vec{x} = \vec{x}(x_1, x_2, x_3, t)$ is the position vector. The coordinates $x_i$,, (*i* = *1, 2, 3*), are written as a function of material coordinates $a_j$ (*j* = *1*, if the object is 1D structure like a fiber or yarn, *j* = *1, 2*, if the object is a 2D, like a woven or knitted fabric, and *j* = *1, 2, 3*, if the object is a 3D structure, like a cloth). In this work, we are exclusively interested in woven surface structures.

In the expression of external forces applied to textile fabrics, we can consider several types of forces with which the object can interact.

In the expression of internal forces, we must consider two aspects:

The first amounts to describing the various deformations that the fabric can undergo during its movement and manipulation.

The second is to describing the way in which we can take into account the mechanical properties of the fabric.

This will make it possible to have the most faithful and realistic modeling of the textile structure.

In Table 3, we present the expressions of the internal forces in the two types of continuous models: the deformable models and the continuous mechanical models. We refer the reader for further details and calculations to the following papers of Yin et al. [19], Liu et al. [20], and Au et al. [21].

We indicate that, globally, we can easily find an analogy between the two formulations, even a certain equivalence. However, we believe that the formulation of deformable models is advantageous for modeling and simulation aimed at animating

**Table 3** Internal forces in deformable models and continuous mechanical models

| Models | Internal forces |
|---|---|
| Deformable models | $\overrightarrow{f^{\text{int}}} = -\sum_{i,j=1}^{2} \frac{\partial}{\partial a_i}\left(S_{ij} \cdot \frac{\partial \vec{x}}{\partial a_i}\right) + \sum_{i,j=1}^{2} \frac{\partial^2}{\partial a_i \partial a_j}\left(C_{ij} \cdot \frac{\partial^2 \vec{x}}{\partial a_i \partial a_j}\right)$<br>To describe fabric deformations, we use two tensors of surface Euclidian tensor $G = \left(g_{ij}\right)_{1 \le i,j \le 2}$ and curvature tensor $B = \left(b_{ij}\right)_{1 \le i,j \le 2}$<br>$S_{ij} = w_{ij}^s\left(g_{ij}^t - g_{ij}^0\right), C_{ij} = w_{ij}^b\left(b_{ij}^t - b_{ij}^0\right)$<br>To describe material properties, we use $w_{ij}^s$, for stretch and shear deformations, and $w_{ij}^b$, for bending and twisting |
| Continuous mechanical models | $\overrightarrow{f^{\text{int}}} = \overrightarrow{div}(\sigma) = \nabla\sigma$, $\sigma$ denotes a stress tensor<br>The choice of the stress tensor $\sigma$ and the strain tensor $\varepsilon$ that can be associated with it gives a vast field of application for this type of formulation, for the modeling and simulation of a large number of textile materials and deformable structures. The relationship between the two tensors defines the constitutive law of the material. We quote the elastic behavior and the hyper-elastic behavior, largely invested in the modeling and the simulation of the textile structures, in particular the composite materials with textile reinforcement |

**Table 1** Some applications of modeling and simulation of textile materials and structures

| Field | Applications |
|---|---|
| Textile engineering and industry: (clothing or/ and furnishings) | -Clothing computer-aided design CAD<br>-Virtual prototyping (CLO, Maya Clothing)<br>-E-commerce of clothing and deformable objects |
| Composites and technical textiles industry | -Modeling and optimization of composite manufacturing and forming<br>-Prediction of the aging of composite materials |
| Computer graphics and virtual animation | -Cinematography: animated films, video games<br>-Dressed Avatars |
| The medical field | -Simulation of medical devices (such as sutures)<br>-Simulation of interaction with living tissues<br>-Simulation of shape-memory textile used for functional rehabilitation |
| The field of learning and training | -Textile engineering/materials engineering |

second approach. We have also listed, for each area, a certain number of applications, without claiming to be exhaustive.

In Table 2, we have reported a comparative study between the different models used textile fabrics modelling and simulation. We have briefly indicated its principle, its advantages, and its limitations of each model. The last column is devoted to indicating the pioneers and the main contributions. We refer the reader to the articles of Ng et al. [3] and Jevsnik et al. [4] for more details.

## 3 Continuous Models Formulation and Numerical Solving

### 3.1 General Formulation

The examination of the different models and works dedicated to the modeling of textile structures, in particular, woven ones, we propose the following general formulation:

$$\rho \overrightarrow{a} + \overrightarrow{f^{\text{int}}} = \overrightarrow{f^{ext}} \tag{1}$$

In this equation, we note:
-The mass density of the object: $\rho$.
- The acceleration vector: $\overrightarrow{a} = \frac{d^2\overrightarrow{x}}{dt^2}$.
-The density of forces per unit volume: $\overrightarrow{f^{\text{int}}}$.

**Table 2** Models typology: Principe, advantages, limitation, and pioneers/main contributors

| Type of modelling | Models | Principe | Advantages | Limitations | Pioneers/main contributors |
|---|---|---|---|---|---|
| Geometric models | Models developed by textile researchers: GM-T | **Based** on a representation of the elementary cell of a fabric (usually plain weave)<br>**Used** to determine the response of this cell to a tensile stress, then other mechanical stress: bending, shear | **Better** understanding of the deformations undergone by fabric<br>**Calculation** of the forces undergone during fabrics deformation | **Limited** to the studied cases Static models,<br>**Difficult** and even impossible generalization to other types of fabric | Peirce [5], Olofsson, [6] |
| | Models developed by computer graphics researchers: GM-T | **Purely** geometric representation of textile surfaces: curve fitting, interpolation, and mapping<br>**Aiming** to reproduce the distinctive forms of textiles: curvature, double curvature, crumpling, and draping | **Simple** to implement<br>**Applicable** to a wide variety of deformable surfaces | **Does** not take into account the physics of the material<br>**Rendering** of fabrics not always realistic | Weil [7], Agui et al. [8] |
| Discrete models | Mass-spring models | **Based** on a discrete representation of fabric. This discretization is made by considering mass points distributed uniformly and interconnected by deferent types of springs<br>**The behavior** of the materials is introduced through the springs:<br>- ***Springs*** describing the behavior in elongation–compression<br>- ***Springs*** describing the shear behavior<br>- ***Springs*** describing the bending behavior | **Interactive** animation of deformable objects<br>**Description** representative of the actual behavior of the fabric<br>**More** realism in the animation and simulation of textile fabrics<br>**Modeling** and simulation of very wide variety of deformable objects | **More** complex numerical implementation<br>**Difficulties** in handling large structures<br>**Difficulties** in treatment of contact between the simulated textile fabric and surrounding objects<br>**Difficulties** in controlling self-collisions | Haumann et Parent [9], Provot [10], |

(continued)

virtual objects, including textile fabrics and clothing. It is a formulation, rather adapted to researchers in computer graphics and in the development of CAD platform.

As for the continuous mechanical formulation, it has the advantage of being adapted to the simulation and modeling of textile structures, in particular composites with textile reinforcement and which one can implement numerically in the structural calculation software commonly used by mechanical engineers.

### *3.2 Experimental Identification of Constitutive Equations*

The experimental characterization of textile fabrics aims to determine the physical and mechanical properties of these structures. We use these characteristics for the parametrization of numerical models. Subsequently we can to identify the behavior laws of textile structures. The main mechanical characteristics that we introduce in this paper are as follows:

- *Stretch deformation.* We perform this characterization using a tensile test. We can thus obtain the moduli of elasticity in the warp directions, weft direction, and direction at an angle $\theta$. We denote them as $\mathbf{E_0}$, $\mathbf{E_{90,}}$ and $\mathbf{E_\theta}$, respectively. For an angle of 45°, we speak of the bias direction. The same tensile test, but under different conditions (low strain tensile test), gives the values of the Poisson's ratios in the warpdirection$\boldsymbol{\upsilon_0}$and weft direction $\boldsymbol{\upsilon_{90}}$.
- *Shear deformation.* The shear behavior of textile fabrics is very characteristic and decisive for this type of thin structure. We characterize this behavior by determining the shear modulus **H**. We can determine this shear modulus experimentally by one of the following methods: picture frame test or a tensile bias test. We note that we can calculate an estimation of shear modulus using the values of warp, weft, and bias elastic moduli.
- *Bending deformation.* We perform this characterization using a cantilever test. We can thus obtain the bending moduli in the warp directions, weft direction, and direction at an angle $\theta$. We denote them as $\mathbf{Rf_0}$, $\mathbf{Rf_{90,}}$ and $\mathbf{Rf_\theta}$, respectively.

The main physical characteristics that we introduce in this paper are mass density $\boldsymbol{\rho}$ and fabric thickness $\boldsymbol{\delta}$.

### *3.3 Numerical Solving*

The modeling of textile structures leads to a formulation of partial differential equations, in general, nonlinear and involving terms depending on time, terms depending on geometry, and terms depending on spatial coordinates. The resolution of this type of equation is generally done in an approximate way in the numerical sense (we are not looking for analytical solutions; it is difficult and even impossible, moreover useless).

We often use the finite element method for the resolution of this type of equations. The method rests, in a simple way, on the passage from a continuous formulation to a discrete formulation: It is the phase of discretization. At the level of the structure, we also pass from a continuous (or supposedly continuous) structure to a discrete representation, and this is called the meshing phase.

When the model, as is the case with the model of Eq. (1), we must also operate a discretization of the time interval during which we want to make the simulation.

In order to be able to solve Eq. (1) correctly, we must define the boundary conditions of the structure or the object that we are trying to simulate. These boundary conditions of the domain of the object define the forces undergone at the level of each boundary or its state (embedding conditions for example). We also define the initial conditions (i.e. at $\boldsymbol{t} = \mathbf{0}$).

We can summarize the steps of the numerical resolution of the equation by the finite element method:

1. *Preprocessing.* This is a preparatory phase for the calculations. It defines:

   - The geometry of the simulated object
     Initial conditions
     The boundary conditions
     The meshing of the object (type and mesh density)
     The mechanical properties of the object and its physical properties.

2. *Processing.* This is the actual resolution phase. One defines there the algorithm of resolution of the equation. The criteria relating to the conditions for incrementing the iterations and stopping the increments.
3. *3.Post-processing.* This involves the exploitation and processing of the solution obtained by the calculations. This is also the phase where we visualize the results.

We mention that there is a wide variety of numerical calculation software implementing the finite element method. We cite, as examples, the Abaqus software, the Ansys software, and the Comsol multi-physics modeling and simulation software.

## 4 Conclusions

In this work, we have tried to develop a sort of mini-review focusing on the modeling of textile structures, particularly woven structures. This attempt aimed to highlight the particularities of textile structures from the point of view of its complexity as a surface structure, highly deformable, as well as from the point of view of the importance of its modeling and simulation in different fields of research and development.

We have also reviewed the different types of models that researchers have used in the practice of modeling and simulating deformable structures and particularly textile fabrics. We have taken a particular interest in continuous formulations.

In addition, we have introduced, even in a very brief way, certain aspects relating to the resolution and the numerical implementation of these models.

We plan to develop aspects relating to continuous mechanical formulations, in particular the formulation of behavior laws dedicated to textile structures, their identifications, and their numerical implementations.

## References


1. Long J., Burns K. et Yang J., Cloth Modeling and Simulation: A Literature Survey, chez *Digital Human Modeling—Third International Conference, ICDHM 2011,* Held as Part of HCI International 2011, Orlando, FL, USA July 9–14, 2011, 2011.
2. Shuvo I., Fibre attributes and mapping the cultivar influence of different industrial cellulosic crops (cotton, hemp, flax, and canola) on textile properties, Bioresources and Bioprocessing, (17),1-28, 2020.
3. Ng N. H. et Grimsdale R. L.. Computer graphics techniques for modeling cloth. IEEE computer graphics and applications, vol. (16), 28–41, 1996.
4. Jevsnik S, Kalaoglu F, Terliksiz S, Purgaj J. Review of computer models for fabric simulation. Tekstilec, 57(4), 2014, 300–314.
5. Peirce F. T., The handle of cloth as a measurable quantity. Journal of the Textile Institute, (21), 377–416, 1930.
6. Olofsson B., A General Model of a Fabric as a Geometric-Mechanical Structure. Journal of the Textile Institute, (55), 541–557, 1964.
7. Weil J., The Synthesis of Cloth Objects. ACM SIGGRAPH Computer Graphics, (20), 49–54, 1986.
8. Agui T., Nagano Y. et Nakayama M., An expression method of cylindrical cloth objects an expression of folds of a sleeve using computer graphics. Transactions of the Society of Electronics, Information and Communications, vol. J73-D-II, 1095–1097, 1990.
9. Haumann D. R. et Parent R. E., The Behavioral Test-bed : Obtaining Complex Behavior from Simple Rules. The visual computer, vol. 4, 332–347, 1988.
10. Provot X., Deformation constraints in a mass-spring model to describe rigid cloth behavior. In Graphics Interface 95, 147–154, 1995.
11. Breen D. E., House D. H. et Philip H. G., A physically-based particle model of woven cloth. The visual Computer, (8), 264–277, 1992.
12. Fontana M., Rizzi C. et Cugini U., 3D virtual apparel design for industrial applications. Computer-Aided Design, (37), 609–622, 2005.
13. Terzopoulos D., Platt J., Barr A. et Fleischer K., Elastically deformable models ACM SIGGRAPH Computer Graphics, (21), 205–214, 1987.
14. Magnenat-Thalmann N. et Yang Y., New trends in animation and visualization, chapitre Techniques for cloth animation, 242–256. Wiley-Interscience, New York, 1991.
15. Hedfi H., Ghith A. et BelHadjSalah H., Dynamic fabric modelling and simulation using deformable models. Journal of the Textile Institute,102 (8), 647–667, 2011.
16. Ascough J., Bez H. et Bricis A., A Simple beam element, large displacement Model for the finite element simulation of cloth crape. Journal of the Textile Institute, (87), 152–165, 1996.
17. Boisse, P., Hamila, N., Badel, P., et Vidal-Salle, E., Simulations éléments- inis de la déformation de textiles aux échelles macro et mésoscopique. Mécanique & Industries, 10(1), 15–19, 2009.
18. Chen B. et Govindaraj M.. A physically based model of fabric drape using flexible shell theory. Textile Research Journal, (65), 324–330, 1995.
19. Yin H, Varava A, Kragic D. Modeling, learning, perception, and control methods for deformable object manipulation. Sci Robot, 6(54), 2021.
20. Liu, Z., Jagota, A., & Hui, C.-Y., Modeling of surface mechanical behaviors of soft elastic solids: theory and examples. Soft Matter, (16), 6875-6889, 2020.
21. Au, C. K., Wu, Z., & Yuen, M. M. F., Effect of fabric properties on cloth draping modeling. Proceedings Geometric Modeling and Processing 2000. Theory and Applications, 2000.